STATA USER'S GUIDE
RELEASE 10

A Stata Press Publication
StataCorp LP
College Station, Texas

Stata Press, 4905 Lakeway Drive, College Station, Texas 77845

The suggested citation for this software is

StataCorp. 2007. *Stata Statistical Software: Release 10*. College Station, TX: StataCorp LP.

Table of Contents

Stata Basics

Elements of Stata

Advice

Stata Basics

1 Read this—it will help

Contents

(Continued on next page)

3

A complete Documentation Set contains more than 6,000 pages of information in the following manuals:

[GS]   *Getting Started with Stata*

[G]   *User's Guide*

[R]   *Stata Base Reference Manual*
        Volume 1, A–H
        Volume 2, I–P
        Volume 3, Q–Z

[D]   *Stata Data Management Reference Manual*

[G]   *Stata Graphics Reference Manual*

[P]   *Stata Programming Reference Manual*

[XT]   *Stata Longitudinal/Panel-Data Reference Manual*

[MV]   *Stata Multivariate Statistics Reference Manual*

[SVY]   *Stata Survey Data Reference Manual*

[ST]   *Stata Survival Analysis and Epidemiological Tables Reference Manual*

[TS]   *Stata Time-Series Reference Manual*

[I]   *Stata Quick Reference and Index*

[M]   *Mata Reference Manual*
        Volume 1, [M-1]–[M-3]
        Volume 2, [M-4]–[M-6]

Detailed information about each of these manuals and the other Stata manuals may be found online at

http://www.stata-press.com/manuals/

1.1   Getting Started with Stata

There are three *Getting Started* manuals:

Getting Started with Stata for Windows
Getting Started with Stata for Macintosh
Getting Started with Stata for Unix

1. Locate your *Getting Started* manual.

2. Install Stata. The instructions are found in the *Getting Started* manual.

3. Learn how to invoke Stata and use it—read the *Getting Started* manual.

4. Now turn to the other manuals; see [U] **1.2 The User's Guide and the Reference manuals**.

1.2 The User's Guide and the Reference manuals

The *User's Guide* is divided into three sections: *Basics*, *Elements*, and *Advice*. At the beginning of each section is a list of the chapters found in that section. In addition to helping you explore the fundamentals of Stata—information that all users should know—this manual will guide you to other sources for Stata information.

The other manuals are the *Reference* manuals. The Stata *Reference* manuals are each arranged like an encyclopedia—alphabetically. Look at the *Base Reference Manual*. Look under the name of a command. If you do not find the command, look in the index. A few commands are so closely related that they are documented together, such as `ranksum` and `median`, which are both documented in [R] **ranksum**.

Not all the entries in the *Base Reference Manual* are Stata commands; some contain technical information, such as [R] **maximize**, which details Stata's iterative maximization process, or [R] **error messages**, which provides information on error messages and return codes.

Like an encyclopedia, the *Reference* manuals are not designed to be read from cover to cover. When you want to know what a command does, complete with all the details, qualifications, and pitfalls, or when a command produces an unexpected result, read its description. Each entry is written at the level of the command. The descriptions assume that you have little knowledge of Stata's features when they are explaining simple commands, such as those for using and saving data. For more complicated commands, they assume that you have a firm grasp of Stata's other features.

If a Stata command is not in the *Base Reference Manual*, you can find it in one of the other *Reference* manuals. The titles of the manuals indicate the types of commands that they contain. The *Programming Reference Manual*, however, contains commands not only for programming Stata but also for manipulating matrices (not to be confused with the matrix programming language described in the *Mata Reference Manual*).

1.2.1 Sample datasets

Various examples in this manual use what is referred to as the automobile dataset, `auto.dta`. We previously created a dataset on the prices, mileages, weights, and other characteristics of 74 automobiles and saved it in a file called `auto.dta`. (These data originally came from the April 1979 issue of *Consumer Reports* and from the United States Government EPA statistics on fuel consumption; they were compiled and published by Chambers et al., 1983.)

In our examples, you will often see us type

```
. use http://www.stata-press.com/data/r10/auto
```

We include the `auto.dta` file with Stata. If you want to use it from your own computer rather than via the Internet, you can type

```
. sysuse auto
```

See [D] **sysuse**.

You can also access the `auto` dataset by selecting **File > Example Datasets...**, clicking on *Example datasets installed with Stata*, and clicking on `use` beside the `auto.dta` filename.

(Continued on next page)

Here is a list of the example datasets included with Stata:

auto.dta	1978 automobile data
autornd.dta	Subset of 1978 automobile data
bplong.dta	Blood pressure data, fictional
bpwide.dta	Blood pressure data, fictional
cancer.dta	Patient survival in drug trial
census.dta	1980 census data by state
citytemp.dta	U.S. city temperature data
citytemp4.dta	U.S. city temperature data
educ99gdp.dta	Education and GDP
gnp96.dta	U.S. GNP, 1967–2002
lifeexp.dta	Life expectancy
nlsw88.dta	U.S. National Longitudinal Study of Young Women
pop2000.dta	U.S. Census, 2000, extract
sp500.dta	S&P 500 historical data
uslifeexp.dta	U.S. life expectancy, 1900–1999
uslifeexp2.dta	U.S. life expectancy, 1900–1940
voter.dta	1992 presidential voter data

All these datasets may be used or described from the **Example Datasets...** help file.

Most of the datasets that are used in the *Reference* manuals can be found at the Stata Press web site (http://www.stata-press.com/data/). You can download the datasets with your browser, or you can use them directly from the Stata command line; that is,

. use http://www.stata-press.com/data/r10/nlswork

An alternative to the use command is webuse.

. webuse nlswork

For more information, see [D] **webuse**.

1.2.2   Cross-referencing

The *Getting Started* manual, the *User's Guide*, and the *Reference* manuals cross-reference each other.

[R]   **regress**
[D]   **reshape**
[XT]   **xtmixed**

The first is a reference to the regress entry in the *Base Reference Manual*, the second is a reference to the reshape entry in the *Data Management Reference Manual*, and the third is a reference to the xtmixed entry in the *Longitudinal/Panel-Data Reference Manual*.

[GSW] **C. Advanced Stata usage**
[GSM] **C. Advanced Stata usage**
[GSU] **C. Advanced Stata usage**

are instructions to see the appropriate section of the *Getting Started with Stata for Windows*, *Getting Started with Stata for Macintosh*, or *Getting Started with Stata for Unix* manual.

1.2.3 The index

At the end of each manual is an index for that manual. The index for the *Stata Base Reference Manual* is found at the end of the third volume. The index for the *Mata Reference Manual* is found at the end of the second volume.

The *Stata Quick Reference and Index* contains a combined index for the *Getting Started* manuals, the *User's Guide*, and all the *Reference* manuals except the *Mata Reference Manual*. It also contains quick reference information on subjects such as the estimation commands.

To find information and commands quickly, you can use Stata's `search` or `hsearch` commands; see [R] **search** and [R] **hsearch**. At the Stata command prompt, try typing `search geometric mean`. `search` searches Stata's keyword database and, optionally, will search the Internet, too. Try typing `search geometric mean, all`. Searching the Internet lets you find more commands and extensions for Stata written by Stata users. `hsearch`, on the other hand, searches the text of all the Stata help files on your computer, including help files that you have written and help files from user-written extensions that you have installed, and orders the outcome by relevancy. Try typing `hsearch geometric mean`. Between these two commands, if a capability exists, you should be able to find it.

1.2.4 The subject table of contents

A subject table of contents for the *User's Guide* and all the *Reference* manuals except the *Mata Reference Manual* is located in the *Stata Quick Reference and Index*.

If you look under "Functions and expressions", you will see

[U] User's Guide, Chapter 13 .... Functions and expressions
[D] egen .... Extensions to generate
[D] functions .... Functions

1.2.5 Typography

We mix the ordinary printing that you are reading now with a typewriter-style typeface `that looks like this`. When something is printed in the typewriter-style typeface, it means that something is a command or an option—it is something that Stata understands and something that you might actually type into your computer. Differences in typeface are important. If a sentence reads, "You could list the result ...", it is just an English sentence—you *could* list the result, but the sentence provides no clue as to how you might actually do that. On the other hand, if the sentence reads, "You could `list` the result ...", it is telling you much more—you could list the result, and you could do that by using the `list` command.

We will occasionally lapse into periods of inordinate cuteness and write, "We `described` the data and then `listed` the data." You get the idea. `describe` and `list` are Stata commands. We purposely began the previous sentence with a lowercase letter. Because `describe` is a Stata command, it must be typed in lowercase letters. The ordinary rules of capitalization are temporarily suspended in favor of preciseness.

We also mix in words printed in italic type, such as "To perform the rank-sum test, type `ranksum` *varname* , by(*groupvar*)". Italicized words are not supposed to be typed; instead, you are to substitute another word for them.

We would also like users to note our rule for punctuation of quotes. We follow a rule that is often used in mathematics books and British literature. The punctuation mark at the end of the quote is included in the quote only if it is a part of the quote. For instance, the pleased Stata user said she thought that Stata was a "very powerful program". Another user simply said, "I love Stata."

In this manual, however, there is little dialog, and we follow this rule to precisely clarify what you are to type, as in, type "cd c:". The period is outside the quotation mark since you should not type the period. If we had wanted you to type the period, we would have included two periods at the end of the sentence: one inside the quotation and one outside, as in, type "use myfile.".

We have tried not to violate the other rules of English. If you find such violations, they were unintentional and resulted from our own ignorance or carelessness. We would appreciate hearing about them.

We have heard from Nicholas J. Cox of the Department of Geography at Durham University in the United Kingdom and express our appreciation. His efforts have gone far beyond dropping us a note and there is no way with words that we can fully express our gratitude.

1.2.6 Vignettes

If you look, for example, at the entry [R] **brier**, you will see a brief biographical vignette of Glenn Wilson Brier (1913–1998), who did pioneering work on the measures described in that entry. A few such vignettes were added without fanfare in the Stata 8 manuals, just for interest, and many more were added in Stata 9, and even more have been added in this release. A vignette could often appropriately go in several entries, so that G.E.P. Box deserves to be mentioned in entries other than [TS] **arima**, such as [R] **boxcox**. To save space, each is given once only, and an index of all vignettes is given in the *Quick Reference Manual and Index*.

All the vignettes were written by Nicholas J. Cox, Durham University, and were compiled using a wide range of reference books, articles in the literature, Internet sources, and information from individuals. Especially useful were the dictionaries of Upton and Cook (2006) and Everitt (2006) and the compilations of statistical biographies edited by Heyde and Seneta (2001) and Johnson and Kotz (1997). Of these, only the first provides information on people living at the time of publication.

1.3 What's new

This section is intended for previous Stata users. If you are new to Stata, you may as well skip to *What's more*, below.

As always, Stata is 100% compatible with the previous releases, but we remind programmers that it is important that you put version 9, or version 8, etc., at the top of your old do- and ado-files so that they continue to work as you expect. You were supposed to do that when you wrote them, but if you did not, go back and do it now.

We will list all the changes, item by item, but first, here are the highlights:

1. Stata 10 has an interactive, point-and-click editor for your graphs. You do not need to type anything; you just right-click within the Graph window and select **Start Graph Editor**. You can do that any time, either when the graph is drawn or when you have graph used it from disk. You can add text, lines, markers, titles, and annotations, outside the plot region or inside; you can move axes, titles, legends, etc.; you can change colors and sizes, number of tick marks, etc.; and you can even change scatters to lines or bars, or vice versa. See [G] **graph editor**.

2. You can now save estimation results to disk. After fitting a model, whether with regress, logistic, ..., or even a user-written command, you type estimates save *filename* to save it. You type estimates use *filename* to reload it later. See [R] **estimates**.

3. Stata now fits nested, hierarchical, and mixed models with binary and count responses; i.e., you can fit logistic and Poisson models with complex, nested error components. See [XT] **xtmelogit** and [XT] **xtmepoisson**.

4. Stata now has exact logistic and exact Poisson regression. In small samples, exact methods have better coverage than asymptotic methods, and exact methods are the only way to obtain point estimates, tests, and confidence intervals from covariates that perfectly predict the observed outcome. See [R] **exlogistic** and [R] **expoisson**.

5. Stata now supports LIML and GMM estimation in addition to 2SLS. Tests of instrumental relevance and tests of overidentifying restrictions are available. See [R] **ivregress** and [R] **ivregress postestimation**.

6. Stata now has more estimators for dynamic panel-data models, including the Arellano–Bover/Blundell–Bond system estimator. This estimator is an extension of the Arellano–Bond GMM estimator for dynamic panel models. It is more efficient and has smaller bias when the AR process is too persistent. These new estimators can also be used to fit models with serially correlated idiosyncratic errors and where the structure of the predetermined variables is complicated. These new estimators can compute the Windmeijer biased-corrected two-step robust VCE in addition to the standard one-step i.i.d., one-step robust, and two-step i.i.d. VCEs. See [XT] **xtabond**, [XT] **xtdpdsys**, and [XT] **xtdpd**.

7. New estimation command **nlsur** fits a system of nonlinear equations by feasible generalized least squares, allowing for covariances among the equations. See [R] **nlsur**.

8. Stata has new estimation commands for fitting categorical and ranked outcomes, often used for choice models. The new commands allow separate equations for each outcome and support the easy-to-use alternative-specific notation. Joining Stata's alternative-specific multinomial probit are alternative-specific conditional logit (McFadden's choice model) and alternative-specific rank-ordered probit (for modeling ordered outcomes). See [R] **asclogit** and [R] **asroprobit**.

9. Stata's **svy:** prefix now works with 48 estimators, 27 more than previously. Most importantly, **svy:** now works with Cox proportional hazards regression models (**stcox**) and parametric survival models (**streg**). See [SVY] **svy estimation**.

10. The new **stpower** command provides sample-size and power calculations for survival studies that use Cox proportional hazards regressions, log-rank tests for two groups, or differences in exponentially distributed hazards or log hazards. Available are (1) required sample size (given power), (2) power (given sample size), and (3) the minimal detectable effect (given power and sample size). **stpower** allows automated production of customizable tables and has options to assist with creating graphs and power curves. See [ST] **stpower**.

11. Stata 10 provides several discriminant analysis techniques, including linear discriminant analysis (LDA), quadratic discriminant analysis (QDA), logistic discriminant analysis, and kth-nearest-neighbor (KNN) discrimination. See [MV] **discrim**.

12. Stata now provides multiple correspondence analysis (MCA) and joint correspondence analysis (JCA). See [MV] **mca**.

13. Stata now provides modern as well as classical multidimensional scaling (MDS), including metric and nonmetric MDS. Available loss functions include stress, normalized stress, squared stress, normalized squared stress, and Sammon. Available transformations include identity, power, and monotonic. See [MV] **mds**.

14. Stata 10 has time/date variables in addition to date variables, so now you can work with data that say an event happened on 12may2007 14:03:22.234 or events that happen every day at 14:03:22.234. Note the millisecond resolution. Time variables are available in two forms: adjusted for leap seconds and unadjusted. See [D] **dates and times**.

15. New Mata function `optimize()` performs minimization and maximization. You can code just the function, the function and its first derivatives, or the function and its first and second derivatives. Optimization techniques include Newton–Raphson, Davidon–Fletcher–Powell, Broyden–Fletcher–Goldfarb–Shanno, Berndt–Hall–Hall–Hausman, and the simplex method Nelder–Mead. See [M-5] **optimize( )**.

16. Stata's online help now provides saved results and examples that you can run.

17. Stata for Windows now supports Automation, formerly known as OLE Automation, which means that programmers can control Stata from other applications and retrieve results. See [P] **automation**.

18. Stata for Unix now supports unixODBC [*sic*], making it easier to connect to databases such as Oracle, MySQL, and PostgreSQL; see [D] **odbc**.

Another change is the introduction of Stata/MP, but that really happened during Stata 9. Stata/MP is the parallel version of Stata for multiple-CPU and multicore computers; see [U] **5 Flavors of Stata**. Stata/MP runs faster. In Stata 10, many more commands now exploit the multiple processors, which means that they run faster, too. This includes both survey and nonsurvey mean, total, ratio, and proportion estimators and the survey linearized variance estimator, which is available with many Stata estimation commands.

There is much more, and the important changes for you may not be what we list as highlights. Below are the details.

1.3.1   What's new in the GUI and command interface

1. The Review window has been redesigned. It now shows the return code of each previous command and highlights errors. From the window, you can now select multiple commands—not just single ones—to save or execute.

2. The Variables window has been redesigned. It nows shows the storage type and display format for each variable in addition to the variable's name and label. You can change any of these from the window, including the name; just right-click.

3. Stata's Viewer has been redesigned. In addition to an all-new look, it has a **Forward** button and the search capability is now built in rather than provided by a dialog box.

4. The Graph window has been redesigned. In addition to providing an interactive editor, under Windows, it now allows tabs. You can have either one window containing multiple graphs each on its own tab, or each graph in a separate window.

5. You can copy and paste from Stata's Results and Viewer windows in AS-IS mode, meaning that you can include Stata output in documents and slides looking exactly as it looked on your screen.

6. Multiple Do-file Editors can now be opened simultaneously; just click on the files in the Open dialog. (Unix users: you too can now open multiple Do-file Editors.)

7. Concerning dialogs,

    a. Stata now uses child dialogs, making dialogs for Stata commands easier to use.

    b. Programmers can program child dialogs, too; see `help dialog programming`.

    c. Graph dialogs have an all-new look that makes specifying the most important items and options easier, yet provides access to even more of `graph`'s capabilities.

    d. Dialogs that need matrices now allow the user to create the matrix via a matrix-input child dialog and to show that new matrix in the original dialog box after it has been created.

    e. Dialogs that need formats now allow the user to specify the format via a format builder.

    f. Data from ODBC sources can now be accessed via a dialog.

    g. Dialogs now scale with Microsoft Windows' DPI settings.

    h. Dialogs now load faster.

8. Stata for Unix has an all new, more modern GUI.

9. Stata now executes `sysprofile.do`, if it exists, in addition to `profile.do` when Stata is launched. This allows system administrators to provide global customization. See [GS] **Appendix A: More on starting and exiting Stata**.

10. New command `adoupdate` automates the process of updating user-written ado-files; see [R] **adoupdate**.

11. New command `hsearch` searches the help files for specified words and presents a ranked, clickable list in the Viewer; see [R] **hsearch**.

12. Stata's help files are now named `*.sthlp` rather than `*.hlp`, meaning that user-written help files can be sent via email more easily. Many email filters flag `.hlp` files as potential virus carriers because Stata was not the only one to use the `.hlp` suffix. You need not rename your old help files. See [R] **help**.

13. There are now console versions of Stata/SE and Stata/MP for Macintosh, just as there are for Unix. They are included on the installation CD and installed automatically.

14. Stata's `in` command modifier now accepts F and L as synonyms for f and l, meaning first and last observations.

15. Multiple log files may be opened simultaneously; see [R] **log**.

16. Intercooled Stata has been renamed to Stata/IC.

1.3.2   What's new in data management

1. Stata 10 has new date/time variables, so you can now record values like 14jun2007 09:42:41.106 in one variable. They are called %tc and %tC variables. The first is unadjusted for leap seconds; the second is adjusted.

   What used to be called "daily variables" are now called %td variables. This is just a jargon change; daily (%td) variables continue to work as they did before—0 means 01jan1960, 1 means 02jan1960, and so on.

   %tc and %tC variables work similarly: 0 means 01jan1960 00:00:00. Here, however, 1 means 01jan1960 00:00:00.001, 1000 means 01jan1960 00:00:01.000, and 02jan1960 08:00:00 is 115,200,000. The underlying values are big—so it is important you store them as `doubles`—but the %tc and %tC formats make the values readable, just as the %td format makes daily (%td) values readable.

   There are many new functions to go along with this new value type. `clock()`, for instance, converts strings such as "02jan1960 08:00:00" (or even "8:00 a.m., 1/2/1960") to their numeric equivalents. `dofc()` converts a %tc value (such as 115,200,000, meaning 02jan1960 08:00:00) to its %td equivalent (namely, 1, meaning 02jan1960). `cofd()` does the reverse (the result would be 86,400,000, meaning 02jan1960 00:00:00).

   See [D] **dates and times**.

2. The previously existing `date()` function, which converts strings to %td values, is now smarter. In addition to being able to convert strings such as "21aug2005", "August 21, 2005", it can convert "082105", "08212005", "210805", and "21082005". See [D] **dates and times**.

3. New command `datasignature` allows you to sign datasets and later use that signature to determine whether the data have changed. An early version of the command was made available during the Stata 9 release. That command is now called `_datasignature` and was used as the building block for the new, improved `datasignature`. See [D] **datasignature** and [P] **_datasignature**.

4. Existing command `clear` now clears data and value labels only. Type `clear all` to clear everything. This change will bite you the first few times you type `clear` expecting it to `clear all`. The problem was that new users were surprised when `clear` by itself cleared everything, whereas use *filename*, `clear` loaded new data and value labels but left everything else in place. The new users were right.

   `clear` now has the following subcommands:

   a. `clear all` clears everything from memory.

   b. `clear ado` clears automatically loaded ado-file programs.

   d. `clear programs` clears all programs, automatically loaded or not.

   c. `clear results` clears saved results.

   d. `clear mata` clears Mata functions and objects from memory.

   See [D] **clear**.

5. Stata for Unix now supports unixODBC [*sic*], making it easier to connect to databases such as Oracle, MySQL, and PostgreSQL; see [D] **odbc**.

6. Existing command `describe` now allows option `varlist` that was previously allowed only by `describe using`. Existing command `describe using` *filename* now allows option `simple` that was previously allowed only by `describe`. Option `varlist` saves the variable names in `r(varlist)`, and option `simple` displays the variable names in a compact format. See [D] **describe**.

7. Existing command `collapse` now supports four additional *stat*s: `first`, the first value; `last`, the last value; `firstnm`, the first nonmissing value; and `lastnm`, the last nonmissing value. See [D] **collapse**.

8. Existing command `cf` (compare files) now provides a detailed listing of observations that differ when the `verbose` option is specified. Setting version to less than 10.0 restores the earlier behavior. See [D] **cf**.

9. Existing command `codebook` has new option `compact` that produces more compact output. See [D] **codebook**.

10. Existing command `insheet` has new option `case` that preserves the case of variable names when importing data; see [D] **insheet**.

11. Existing command `outsheet` has new option `delimiter()` that specifies an alternative delimiter; see [D] **outsheet**.

12. Existing commands `infile` and `infix` can now read up to 524,275 characters per line; the previous limit was 32,765. See [D] **infile** and [D] **infix (fixed format)**.

13. Existing commands `icd9` and `icd9p` have now been updated to use the V24 codes; see [D] **icd9**.

14. New function `itrim()` returns the string with consecutive, internal spaces collapsed to one space; see *String functions* in [D] **functions**.

15. New functions `lnnormal()` and `lnnormalden()` provide the natural logarithm of the cumulative standard normal distribution and of the standard normal density; see *Probability distributions and density functions* in [D] **functions**.

16. New functions for calculating cumulative densities are now available:

binomial(n, k, p)	lower tail of the binomial distribution
ibetatail(a, b, x)	reverse (upper tail) of the cumulative beta distribution
gammaptail(a, x)	reverse (upper tail) of the cumulative gamma distribution
invgammaptail(a, p)	inverse reverse of the cumulative gamma distribution
invibetatail(a, b, p)	inverse reverse of the cumulative beta distribution
invbinomialtail(n, k, p)	inverse of right cumulative binomial

See *Probability distributions and density functions* in [D] **functions**.

17. Existing function Binomial(n, k, p) has been renamed binomialtail(n, k, p), thus making its name consistent with the naming convention for probability functions. The accuracy of the function has also been improved for very large values of n. At the other end of the number line, the function now returns the appropriate 0 or 1 value when $n = 0$, rather than returning missing. Binomial() continues to work as a synonym for binomialtail().

18. The behavior and accuracy of the following probability functions have been improved:

    a. F(n_1, n_2, f) and Ftail(n_1, n_2, f) are more accurate for small values of n_1 and large values of n_2. Also, F() is more accurate for large f where n_1 and n_2 are less than 1.

    b. gammap(a, x) is more accurate when a is large and x is near a.

    c. ibeta(a, b, x) now is more accurate when x is near $a/(a + b)$ and a or b is large.

    d. invbinomial(n, k, p), invchi2(n, p), invchi2tail(n, p), invF(n_1, n_2, p), and invgammap(a, p) are more accurate for small values of p or for returned values close to zero.

    e. invFtail(n_1, n_2, p) and invibeta(a, b, p) are more accurate for small values of p or for returned values close to zero.

    f. invttail(n, p) is more accurate for small values of p or for returned values close to zero.

    g. ttail(n, t) is more accurate for exceedingly large values of n.

19. Existing function invbinomial(n, k, p) now returns the probability of a success on one trial such that the probability of observing k or *fewer* successes in n trials is p. The previous behavior of invbinomial() is restored under version control.

20. New function fmtwidth() returns the display width of a %*fmt* string; see *Programming functions* in [D] **functions**.

21. The maximum length of a %*fmt* has increased from 12 to 48 characters; see [D] **format**. (This change was necessitated by the new date/time variables.)

22. Existing commands corr2data and drawnorm now allow singular correlation (or covariance) structures. New option forcepsd modifies a matrix to be positive semidefinite and thus to be a proper covariance matrix. See [D] **corr2data** and [D] **drawnorm**.

23. Existing command hexdump, analyze now saves the number of \r\n characters in r(Windows) rather than in r(DOS). r(DOS) is still set when version is less than 10. See [D] **hexdump**.

(*Continued on next page*)

1.3.3   What's new in statistics (general)

1. As mentioned above, you can now save estimation results to disk. You type `estimates save`
   filename to save results and `estimates use` *filename* to reload them. In fact, the entire `estimates`
   command has been reworked. The new command `estimates notes` allows you to add notes
   to estimation results just as you add them to datasets. The new command `estimates esample`
   allows you to restore `e(sample)` after reloading estimates, should that be necessary (usually it
   is not). The maximum number of estimation results that can be held in memory (as opposed to
   saved on disk) is increased to 300 from 20. See [R] **estimates**.

2. Stata now has exact logistic and exact Poisson regression. Rather than having their inference based
   on asymptotic normality, exact estimators enumerate the conditional distribution of the sufficient
   statistics and then base inference upon that distribution. In small samples, exact methods have better
   coverage than asymptotic methods, and exact methods are the only way to obtain point estimates,
   tests, and confidence intervals from covariates that perfectly predict the observed outcome.

   Postestimation command `estat se` reports odds ratios and their asymptotic standard errors. `estat`
   `predict`, available only after `exlogistic`, computes predicted probabilities, asymptotic standard
   errors, and exact confidence intervals for single cases.

   See [R] **exlogistic** and [R] **expoisson**.

3. New estimation command `asclogit` performs alternative-specific conditional logistic regression,
   which includes McFadden's choice model. Postestimation command `estat alternatives` reports alternative-specific summary statistics. `estat mfx` reports marginal effects of regressors on
   probabilities of each alternative. See [R] **asclogit** and [R] **asclogit postestimation**.

4. New estimation command `asroprobit` performs alternative-specific rank-ordered probit regression. `asroprobit` is related to rank-ordered logistic regression (`rologit`) but allows modeling
   alternative-specific effects and modeling the covariance structure of the alternatives. Postestimation
   command `estat alternatives` provides summary statistics about the alternatives in the estimation sample. `estat covariance` displays the variance–covariance matrix of the alternatives.
   `estat correlation` displays the correlation matrix of the alternatives. `estat mfx` computes
   the marginal effects of regressors on the probability of the alternatives. See [R] **asroprobit** and
   [R] **asroprobit postestimation**.

5. New estimation command `ivregress` performs single-equation instrumental-variables regression
   by two-stage least squares, limited-information maximum likelihood, or generalized method of
   moments. Robust and HAC covariance matrices may be requested. Postestimation command `es-`
   `tat firststage` provides various descriptive statistics and tests of instrument relevance. `estat`
   `overid` tests overidentifying restrictions. `ivregress` replaces the previous `ivreg` command. See
   [R] **ivregress** and [R] **ivregress postestimation**.

6. New estimation command `nlsur` fits a system of nonlinear equations by feasible generalized least
   squares, allowing for covariances among the equations; see [R] **nlsur**.

7. Existing estimation command `nlogit` was rewritten and has new, better syntax and runs faster
   when there are more than two levels. Old syntax is available under version control. `nlogit` now
   fits the random utilities maximization (RUM) model by default as well as the nonnormalized
   model that was available previously. The new `nlogit` now allows unbalanced groups and allows
   groups to have different sets of alternatives. `nlogit` now excludes entire choice sets (cases) if
   any alternative (observation) has a missing value; use new option `altwise` to exclude just the
   alternatives (observations) with missing values. Finally, `vce(robust)` is allowed regardless of the
   number of nesting levels. See [R] **nlogit**.

8. Existing estimation command `asmprobit` has the following enhancements:

   a. The new default parameterization estimates the covariance of the alternatives differenced from the base alternative, making the estimates invariant to the choice of base. New option `structural` specifies that the previously structural (nondifferenced) covariance parameterization be used.

   b. `asmprobit` now permits estimation of the constant-only model.

   c. `asmprobit` now excludes entire choice sets (cases) if any alternative (observation) has a missing value; use new option `altwise` to exclude just the alternatives (observations) with missing values.

   d. New postestimation command `estat mfx` computes marginal effects after `asmprobit`.

   See [R] **asmprobit** and [R] **asmprobit postestimation**.

9. Existing estimation command `clogit` now accepts `pweights` and may be used with the `svy:` prefix.

   Also, `clogit` used to be willing to produce cluster–robust VCEs when the groups were not nested within the clusters. Sometimes, this VCE was consistent, and other times it was not. You must now specify the new `nonest` option to obtain a cluster–robust VCE when the groups are not nested within panels.

   `predict` after `clogit` now accepts options that calculate the $\Delta\beta$ influence statistic, the $\Delta\chi^2$ lack-of-fit statistic, the Hosmer and Lemeshow leverage, the Pearson residuals, and the standardized Pearson residuals.

   See [R] **clogit** and [R] **clogit postestimation**.

10. Existing estimation command `cloglog` now accepts `pweights`, may now be used with the `svy:` prefix, and has new option `eform` that requests that exponentiated coefficients be reported; see [R] **cloglog**.

11. Existing estimation command `cnreg` now accepts `pweights`, may be used with the `svy:` prefix, and is now noticeably faster (up to five times faster) when used within loops, such as by `statsby`. See [R] **cnreg**.

12. Existing estimation commands `cnsreg` and `tobit` now accept `pweights`, may be used with the `svy:` prefix, and are now noticeably faster (up to five times faster) when used within loops, such as by `statsby`. Also, `cnsreg` now has new advanced option `mse1` that sets the mean squared error to 1. See [R] **cnsreg** and [R] **tobit**.

13. Existing estimation command `regress` is now noticeably faster (up to five times faster) when used with loops, such as by `statsby`. Also,

    a. Postestimation command `estat hettest` has new option `iid` that specifies that an alternative version of the score test be performed that does not require the normality assumption. New option `fstat` specifies that an alternative F test be performed that also does not require the normality assumption.

    b. Existing postestimation command `estat vif` has new option `uncentered` that specifies that uncentered variance inflation factors be computed.

   See [R] **regress postestimation**.

14. Existing estimation commands `logit`, `mlogit`, `ologit`, `oprobit`, and `probit` are now noticeably faster (up to five times faster) when used within loops, such as by `statsby`.

15. For existing estimation command `probit`, `predict` now allows the `deviance` option; see [R] **probit postestimation**.

16. Existing estimation command `nl` has the following enhancements:

   a. Option `vce`(*vcetype*) is now allowed, with supported *vcetype*s that include types derived from asymptotic theory, that are robust to some kinds of misspecification, that allow for intragroup correlation, and that use bootstrap or jackknife methods. Also, three heteroskedastic- and autocorrelation-consistent variance estimators are available.

   b. `nl` no longer reports an overall model F test because the test that all parameters other than the constant are jointly zero may not be appropriate in arbitrary nonlinear models.

   c. The coefficient table now reports each parameter as its own equation, analogous to how `ml` reports single-parameter equations.

   d. `predict` after `nl` has new options that allow you to obtain the probability that the dependent variable lies within a given interval, the expected value of the dependent variable conditional on its being censored, and the expected value of the dependent variable conditional on its being truncated. These predictions assume that the error term is normally distributed.

   e. `mfx` can be used after `nl` to obtain marginal effects.

   f. `lrtest` can be used after `nl` to perform likelihood-ratio tests.

   See [R] **nl** and [R] **nl postestimation**.

17. Existing estimation command `mprobit` now allows `pweights`, may now be used with the `svy:` pre-fix, and has new option `probitparam` that specifies that the probit variance parameterization, which fixes the variance of the differenced latent errors between the scale and the base alternatives to one, be used. See [R] **mprobit**.

18. Existing estimation command `rologit` now allows `vce(bootstrap)` and `vce(jackknife)`. See [R] **rologit**.

19. Existing estimation command `truncreg` now allows `pweights` and now works with the `svy:` prefix. See [SVY] **svy estimation**.

20. After existing estimation command `ivprobit`, postestimation commands `estat classification`, `lroc`, and `lsens` are now available. Also, in `ivprobit`, the order of the ancillary parameters in the output has been changed to reflect the order in `e(b)`. See [R] **ivprobit** and [R] **ivprobit postestimation**.

21. All estimation commands that allowed options `robust` and `cluster()` now allow option `vce`(*vcetype*). `vce()` specifies how the variance–covariance matrix of the estimators (and hence standard errors) are to be calculated. This syntax was introduced in Stata 9, with options such as `vce(bootstrap)`, `vce(jackknife)`, and `vce(oim)`.

   In Stata 10, option `vce()` is extended to encompass the robust (and optionally clustered) variance calculation. Where you previously typed

   . *estimation-command* ..., `robust`

   you are now to type

   . *estimation-command* ..., `vce(robust)`

   and where you previously typed

   . *estimation-command* ..., `robust cluster`(*clustervar*)

   with or without the `robust`, you are now to type

   . *estimation-command* ..., `vce(cluster` *clustervar*`)`

You can still type the old syntax, but it is undocumented. The new syntax emphasizes that the robust and cluster calculation affects standard errors, not coefficients. See [R] *vce_option*.

Going along with this change, estimation commands now have a term for their default variance calculation. Thus, you will see things like vce(ols), and vce(gnr). Here is what they all mean:

    a. vce(ols). The variance estimator for ordinary least squares; an $s^2(X'X)^{-1}$-type calculation.

    b. vce(oim). The observed information matrix based on the likelihood function; a $(-H)^{-1}$-type calculation, where H is the Hessian matrix.

    c. vce(conventional). A generic term to identify the conventional variance estimator associated with the model. For instance, in the Heckman two-step estimator, vce(conventional) means the Heckman-derived variance matrix from an augmented regression. In two different contexts, vce(conventional) does not necessarily mean the same calculation.

    d. vce(analytic). The variance estimator derived from first principles of statistics for means, proportions, and totals.

    e. vce(gnr). The variance matrix based on an auxiliary regression, which is analogous to $s^2(X'X)^{-1}$ generalized to nonlinear regression. gnr stands for Gauss–Newton regression.

    f. vce(linearized). The variance matrix calculated by a first-order Taylor approximation of the statistic, otherwise known as the Taylor linearized variance estimator, the sandwich estimator, and the White estimator. This is identical to vce(robust) in other contexts.

The above are used for defaults. vce() may also be

    g. vce(robust). The variance matrix calculated by the sandwich estimator of variance, VDV-type calculation, where V is the conventional variance matrix and D is the outer product of the gradients, $\sum_i g_i g_i'$.

    h. vce(cluster *varname*). The cluster-based version of vce(robust) where sums are performed within the groups formed by *varname*, which is equivalent to assuming that the independence is between groups only, not between observations.

    i. vce(hc2) and vce(hc3). Calculated similarly as vce(robust) except that different scores are used in place of the gradient vectors g_i.

    j. vce(opg). The variance matrix calculated by the outer product of the gradients; a $(\sum_i g_i g_i')^{-1}$-type calculation.

    k. vce(jackknife). The variance matrix calculated by the jackknife, including delete one, delete n, and the cluster-based jackknife.

    l. vce(bootstrap). The variance matrix calculated by bootstrap resampling.

You do not need to memorize the above; the documentation for the individual commands, and their corresponding dialog boxes, make clear what is the default and what is available.

22. Estimation commands specified with option vce(bootstrap) or vce(jackknife) now report a note when a variable is dropped because of collinearity.

23. The new option collinear, which has been added to many estimation commands, specifies that the estimation command not remove collinear variables. Typically, you do not want to specify this option. It is for use when you specify constraints on the coefficients such that, even though the variables are collinear, the model is fully identified. See [R] **estimation options**.

24. Estimation commands having a model Wald test composed of more than just the first equation now save the number of equations in the model Wald test in e(k_eq_model).

25. All estimation commands now save macro e(cmdline) containing the command line as originally typed.

26. Concerning existing estimation command ml;

    a. ml now saves the number of equations used to compute the model Wald test in e(k_eq_model), even when option lf0() is specified.

    b. ml score has new option missing that specifies that observations containing variables with missing values not be eliminated from the estimation sample.

    c. ml display has new option showeqns that requests that equation names be displayed in the coefficient table.

    See [R] **ml**.

27. New command lpoly performs a kernel-weighted local polynomial regression and displays a graph of the smoothed values with optional confidence bands; see [R] **lpoly**.

28. New prefix command nestreg: reports comparison tests of nested models; see [R] **nestreg**.

29. Existing commands fracpoly, fracgen, and mfp have new features:

    a. fracpoly and mfp now support cnreg, mlogit, nbreg, ologit, and oprobit.

    b. fracpoly and mfp have new option all that specifies that out-of-sample observations be included in the generated variables.

    c. fracpoly, compare now reports a closed-test comparison between fractional polynomial models by using deviance differences rather than reporting the gain; see [R] **fracpoly**.

    d. fracgen has new option restrict() that computes adjustments and scaling on a specified subsample.

    See [R] **fracpoly** and [R] **mfp**.

30. For existing postestimation command hausman, options sigmaless and sigmamore may now be used after xtreg. These options improve results when comparing fixed- and random-effects regressions based on small to moderate samples because they ensure that the differenced covariance matrix will be positive definite. See [R] **hausman**.

31. Existing postestimation command testnl now allows expressions that are bound in parentheses or brackets to have commas. For example, testnl _b[x] = M[1,3] is now allowed. See [R] **testnl**.

32. Existing postestimation command nlcom has a new option noheader that suppresses the output header; see [R] **nlcom**.

33. Existing command statsby now works with more commands, including postestimation commands. statsby also has new option forcedrop for use with commands that do not allow if or in. forcedrop specifies that observations outside the by() group be temporarily dropped before the command is called. See [D] **statsby**.

34. Existing command mkspline will now create restricted cubic splines as well as linear splines. New option displayknots will display the location of the knots. See [R] **mkspline**.

35. In existing command kdensity, kernel(*kernelname*) is now the preferred way to specify the kernel, but the previous method of simply specifying *kernelname* still works. See [R] **kdensity**.

36. Existing command ktau's computations are now faster; see [R] **spearman**.

37. In existing command ladder, the names of the transformations in the output have been renamed to match those used by gladder and qladder. Also, the returned results r(raw) and r(P_raw) have been renamed to r(ident) and r(P_ident), respectively. See [R] **ladder**.

38. Existing command `ranksum` now allows the *groupvar* in option by(*groupvar*) to be a string; see [R] **ranksum**.

39. Existing command `tabulate, exact` now allows exact computations on larger tables. Also, new option `nolog` suppresses the enumeration log. See [R] **tabulate twoway**.

40. Existing command `tetrachoric`'s default algorithm for computing tetrachoric correlations has been changed from the Edwards and Edwards estimator to a maximum likelihood estimator. Also, standard errors and two-sided significance tests are produced. The Edwards and Edwards estimator is still available by specifying the new `edwards` option. A new `zeroadjust` option requests that frequencies be adjusted when one cell has a zero count. See [R] **tetrachoric**.

1.3.4   What's new in statistics (longitudinal/panel data)

1. New command `xtset` declares a dataset to be panel data and designates the variable that identifies the panels. In previous versions of Stata, you specified options i(*groupvar*) and sometimes t(*timevar*) to identify the panels. You specified the i() and t() options on the xt command you wanted to use. Now you `xtset` *groupvar* or `xtset` *groupvar timevar* first. The values you set will be remembered from one session to the next if you save your dataset.

   `xtset` also provides a new feature. `xtset` allows option `delta()` to specify the frequency of the time-series data, something you will need to do if you are using Stata's new date/time variables.

   Finally, you can still specify old options i() and t(), but they are no longer documented. Similarly, old commands `iis` and `tis` continue to work but are no longer documented. See [XT] **xtset**.

2. New estimation commands `xtmelogit` and `xtmepoisson` fit nested, hierarchical, and mixed models with binary and count responses; i.e., you can fit logistic and Poisson models with complex, nested error components. Syntax is the same as for Stata's linear mixed model estimator, `xtmixed`. To fit a model of graduation with a fixed coefficient on x1 and random coefficient on x2 at the school level, and with random intercepts at both the school and class-within-school level, you type

       . xtmelogit graduate x1 x2 || school: x2 || class:

   `predict` after `xtmelogit` and `xtmepoisson` will calculate predicted random effects. See [XT] **xtmelogit**, [XT] **xtmelogit postestimation**, [XT] **xtmepoisson**, and [XT] **xtmepoisson postestimation**.

3. New estimation commands are available for fitting dynamic panel-data models:

   a. Existing estimation command `xtabond` fits dynamic panel-data models by using the Arellano–Bond estimator but now reports results in levels rather than differences. Also, `xtabond` will now compute the Windmeijer biased-corrected two-step robust VCE. See [XT] **xtabond**.

   b. New estimation command `xtdpdsys` fits dynamic panel-data models by using the Arellano–Bover/Blundell–Bond system estimator. `xtdpdsys` is an extension of `xtabond` and produces estimates with smaller bias when the AR process is too persistent. `xtdpdsys` is also more efficient than `xtabond`. Whereas `xtabond` uses moment conditions based on the differenced errors in producing results, `xtdpdsys` uses moment conditions based on differences and levels. See [XT] **xtdpdsys**.

   c. New estimation command `xtdpd` fits dynamic panel-data models extending the Arellano–Bond or the Arellano–Bover/Blundell–Bond system estimator and allows a richer syntax for specifying models and so will fit a broader class of models then either `xtabond` or `xtdpdsys`. `xtdpd` can be used to fit models with serially correlated idiosyncratic errors, whereas `xtdpdsys` and `xtabond` assume no serial correlation. `xtdpd` can be used with models where the structure of the predetermined variables is more complicated than that assumed by xtdpdsys or xtabond. See [XT] **xtdpd**.

   d. New postestimation command estat abond tests for serial correlation in the first-differenced errors. See [XT] **xtabond postestimation**, [XT] **xtdpdsys postestimation**, and [XT] **xtdpd postestimation**.

   e. New postestimation command estat sargan performs the Sargan test of overidentifying restrictions. See [XT] **xtabond postestimation**, [XT] **xtdpdsys postestimation**, and [XT] **xtdpd postestimation**.

4. Existing estimation command xtreg, fe now accepts aweights, fweights, and pweights. Also, new option dfadj specifies that the cluster–robust VCE be adjusted for the within transform. This was previously the default behavior. See [XT] **xtreg**.

5. Existing estimation commands xtreg, fe and xtreg, re used to be willing to produce cluster–robust VCEs when the panels were not nested within the clusters. Sometimes this VCE is consistent and other times it is not. You must now specify the new nonest option to obtain a cluster–robust VCE when the panels are not nested within the clusters.

6. The numerical method used to evaluate distributions, known as quadrature, has been improved. This method is used by the xt random-effects estimation commands xtlogit, xtprobit, xtcloglog, xtintreg, xttobit, and xtpoisson, re normal.

   a. For the estimation commands, the default method is now intmethod(mvaghermite). The old default was intmethod(aghermite).

   b. Option intpoints(#) for the commands now allows up to 195 quadrature points. The default is 12, and the old upper limit was 30. (Models with large random effects often require more quadrature points.)

   c. The estimation commands may now be used with constraints regardless of the quadrature method chosen.

   d. Command quadchk, for use after estimation to verify that the quadrature approximation was sufficiently accurate, now produces a more informative comparison table. Before, four fewer and four more quadrature points were used, and that was reasonable if the number of quadrature points was, say, $n_q = 12$. Now you may specify significantly larger n_q and the ± 4 is not useful. Now quadchk uses $n_q - \text{int}(n_q/3)$ and $n_q + \text{int}(n_q/3)$.

   e. quadchk has new option nofrom that forces refitted models to start from scratch rather than starting from the previous estimation results. This is important if you use the old intmethod(aghermite), which is sensitive to starting values, but not important if you are using the new default intmethod(mvaghermite).

   See [XT] **quadchk**.

7. All xt estimation commands now accept option vce(*vcetype*). As mentioned in the *What's new in statistics (general)*, vce(robust) and vce(cluster *varname*) are the right ways to specify the old robust and cluster() options, and new option vce() allows other VCE calculations as well.

8. Existing estimation command xtcloglog has new option eform that requests exponentiated coefficients be reported; see [XT] **xtcloglog**.

9. Existing estimation command xthtaylor now allows users to specify only endogenous time-invariant variables, only endogenous time-varying variables, or both. Previously, both were required. See [XT] **xthtaylor**.

10. All xt estimation commands have new option collinear, which specifies that collinear variables are not to be removed. Typically, you do not want to specify this option. It is for use when you specify constraints on the coefficients such that, even though the variables are collinear, the model is fully identified. See [XT] **estimation options**.

11. Existing command xtdes has been renamed to xtdescribe. xtdes continues to work as a synonym for xtdescribe. See [XT] **xtdescribe**.

12. The [XT] manual has an expanded glossary.

1.3.5 What's new in statistics (time series)

1. All time-series analysis commands now support data with frequencies as high as 1 millisecond (ms), corresponding to Stata's new date/time variables. Since your data are probably not recorded at the millisecond level, existing command tsset has new option delta() that allows you to specify the frequency of your data. Previously, time was recorded as t_0, $t_0 + 1$, $t_0 + 2$, ..., and if time $= t$ in some observation, then the corresponding lagged observation was the observation for which time $= t - 1$. That is still the default. When you specify delta(), time is assumed to be recorded as t_0, $t_0 + \delta$, $t_0 + 2\delta$, and if time $= t$ in some observation, then the corresponding lagged observation is the observation for which time $= t - \delta$. Say that you are analyzing hourly data and time is recorded using Stata's new %tc values. One hour corresponds to 3,600,000 ms, and you would want to specify tsset t, delta(3600000). Option delta() is smart; you can specify tsset t, delta(1 hour). See [TS] **tsset**.

2. tsset now reports whether panels are balanced when an optional panel variable is specified.

3. Many ts estimation commands now accept option vce(*vcetype*). As mentioned in *What's new in statistics (general)*, vce(robust) and vce(cluster *varname*) are the right ways to specify the old robust and cluster() options, and option vce() allows other VCE calculations as well.

4. Options vce(hc2) and vce(hc3) are now the preferred way to request alternative bias corrections for the robust variance calculation for existing estimation command prais. See [TS] **prais**.

5. Existing estimation commands arch and arima have new option collinear that specifies that the estimation command not remove collinear variables. Typically, you do not want to specify this option. It is for use when you specify constraints on the coefficients such that, even though the variables are collinear, the model is fully identified. See [TS] **estimation options**.

6. Existing command irf now estimates and reports dynamic–multiplier functions and cumulative dynamic–multiplier functions, as well as their standard errors. See [TS] **irf**.

7. The [TS] manual has an expanded glossary.

(Continued on next page)

1.3.6 What's new in statistics (survey)

1. Stata's `svy:` prefix now works with 48 estimators, 28 more than previously. Most importantly, `svy:` now works with Cox regression (`stcox`) and parametric survival models (`streg`). Other commands with which `svy:` now works include

`biprobit`	bivariate probit regression
`clogit`	conditional (fixed effects) logistic regression
`cloglog`	complementary log-log regression
`cnreg`	censored-normal regression
`cnsreg`	constrained linear regression
`glm`	generalized linear models
`hetprob`	heteroskedastic probit model
`ivregress`	instrumental-variables regression
`ivprobit`	probit model with endogenous regressors
`ivtobit`	tobit model with endogenous regressors
`mprobit`	multinomial probit regression
`nl`	nonlinear least-squares estimation
`scobit`	skewed logistic regression
`slogit`	stereotype logistic regression
`stcox`	Cox proportional hazards regression
`streg`	parametric survival models (5 estimators)
`tobit`	tobit regression
`treatreg`	treatment-effects model
`truncreg`	truncated regression
`zinb`	zero-inflated negative binomial regression
`zip`	zero-inflated Poisson regression
`ztnb`	zero-truncated negative binomial regression
`ztp`	zero-truncated Poisson regression

   See [SVY] **svy estimation**.

2. `svy:` prefix now calculates the linearized variance estimator 2 to 100 times faster, the larger multiplier applying to large datasets with many sampling units; see [SVY] **svy**.

3. `svy: mean`, `svy: proportion`, `svy: ratio`, and `svy: total` are considerably faster when the `over()` option identifies many subpopulations.

4. `svy:`, `svy: mean`, `svy: proportion`, `svy: ratio`, and `svy: total` now take advantage of multiple processors in Stata/MP, making them even faster in that case.

5. Concerning `svyset`,

   a. New option `singleunit(method)` provides three methods for handling strata with one sampling unit. If not specified, the default in such cases is to report standard errors as missing value.

   b. New option `fay(#)` specifies that Fay's adjustment be made to the BRR weights.

   See [SVY] **svyset**.

6. `estat` has two new subcommands for use with svy estimation results:

   a. `estat sd`, used after `svy: mean`, reports subpopulation standard deviations.

   b. `estat strata` reports the number of singleton and certainty strata within each sampling stage.

   See [SVY] **estat**.

7. `svy: tabulate` now allows string variables. See [SVY] **svy: tabulate oneway** and [SVY] **svy: tabulate twoway**.

8. Existing command `svydes` has been renamed `svydescribe`; `svydes` continues to work. `svydescribe` now puts missing values in the `generate(newvar)` variable for observations outside the specified estimation sample. Previously, the variable would contain a zero for observations outside the estimation sample. See [SVY] **svydescribe**.

9. The [SVY] manual has been reorganized. Stata's survey estimation commands are now documented in [SVY] **svy estimation**. All model-specific information is now documented in the manual entry for the corresponding estimation command.

1.3.7 What's new in statistics (survival analysis)

1. Existing estimation commands `stcox` and `streg` may now be used with the `svy:` prefix and so can fit models for complex survey data; see [ST] **stcox** and [ST] **streg**.

2. New command `stpower` provides sample-size and power calculations for survival studies that use Cox proportional hazards regressions, log-rank tests for two groups, or differences in exponentially distributed hazards or log hazards.

   a. `stpower cox` estimates required sample size (given power) or power (given sample size) or the minimal detectable coefficient (given power and sample size) for models with multiple covariates. The command provides options to account for possible correlation between the covariate of interest and other predictors and for withdrawal of subjects from the study. See [ST] **stpower cox**.

   b. `stpower logrank` estimates required sample size (given power) or power (given sample size) or the minimal detectable hazard ratio (given power and sample size) for studies comparing survivor functions of two groups by using the log-rank test. Both the Freedman (1982) and the Schoenfeld (1981) methods are provided. The command allows for unequal allocation of subjects between the groups and possible withdrawal of subjects. Estimates can be adjusted for uniform accrual. See [ST] **stpower logrank**.

   c. `stpower exponential` estimates sample size (given power) or power (given sample size) of tests of the difference between hazards or log hazards of two groups under the assumption of exponential survivor functions (also known as the exponential test). Both the Lachin–Foulkes (1986) and Rubinstein–Gail–Santner(1981) methods are provided. Unequal group allocation, uniform or truncated exponential accrual, and different exponential losses due to follow-up in each group are allowed. See [ST] **stpower exponential**.

   The `stpower` commands allow automated production of customizable tables and have options to assist with creating graphs of power curves. See [ST] **stpower**.

3. Concerning existing command `sts graph`,

   a. New option `risktable()` places a subjects-at-risk table underneath and aligned with the survivor or hazard plot.

   b. New option `ci` replaces old options `gwood`, `cna`, and `cihazard`. `sts graph` will choose the appropriate confidence interval on the basis of the function being graphed.

   c. Confidence intervals are now graphed using shaded areas and new options `plotopts()` and `ciopts()` allow you to control how plots and confidence intervals look.

   d. Overlaid confidence intervals are now allowed and are produced when new option `ci` is combined with existing option `by(varlist)`.

e. New option censopts() controls the appearance of ticks and markers produced by existing option censored().

f. Boundary computations for smoothing hazards have been improved. New option noboundary specifies that no boundary correction be done.

g. The lower bound of the range to plot the hazard function now extends to zero.

h. Option na has been renamed cumhaz. na may still be used.

See [ST] **sts graph**. Setting version to less than 10 restores previous behavior.

4. For sts list, option na has been renamed cumhaz. na may be used as a synonym for cumhaz. See [ST] **sts list**.

5. Improvements to stcurve analogous to those of sts graph have been made.

a. Boundary computations for smoothing hazards have been improved. New option noboundary specifies that no boundary correction be done.

b. The lower bound of the range to plot the hazard function now extends to zero.

See [ST] **stcurve**.

6. All st estimation commands accept option vce(*vcetype*). As mentioned in the *What's new in statistics (general)*, vce(robust) and vce(cluster *varname*) are the right ways to specify the old robust and cluster() options, and option vce() now allows other VCE calculations as well.

7. Existing command predict after stcox has a new option, scores, that allows generating variables with the partial efficient score residuals; see [ST] **stcox postestimation**.

8. Existing command ltable has new options byopts(), plotopts(), plot#opts(), and ci#opts() that allow for more customization of the graph. New option ci adds confidence intervals to the graph. See [ST] **ltable**.

9. Existing command stphplot has a new option plot#opts() that allows for more customization of the graph. See [ST] **stcox diagnostics**.

10. Existing command stcoxkm has new options byopts(), obsopts(), obs#opts(), predopts(), and pred#opts() that allow for more customization of the graph. See [ST] **stcox diagnostics**.

11. Existing command cc has new option tarone that produces Tarone's (1985) adjustment of the Breslow–Day test for homogeneity of odds ratios. See [ST] **epitab**.

12. Existing command stdes has been renamed to stdescribe. stdes continues to work. See [ST] **stdescribe**.

13. The [ST] manual has an expanded glossary.

1.3.8   What's new in statistics (multivariate)

1. New estimation commands discrim and candisc provide several discriminant analysis techniques, including linear discriminant analysis (LDA), quadratic discriminant analysis (QDA), logistic discriminant analysis, and kth nearest neighbor discrimination. See [MV] **discrim**, [MV] **discrim estat**, and [MV] **candisc**.

2. Existing estimation commands mds, mdslong, and mdsmat now provide modern as well as classical multidimensional scaling (MDS), including metric and nonmetric MDS. Available loss functions include stress, normalized stress, squared stress, normalized squared stress, and Sammon. Available transformations include identity, power, and monotonic. mdslong also now allows aweights and fweights, and mdsmat has a weight() option. See [MV] **mds**, [MV] **mdslong**, and [MV] **mdsmat**.

3. New estimation command mca provides multiple correspondence analysis (MCA) and joint correspondence analysis (JCA); see [MV] **mca** and [MV] **mca postestimation**. You can use existing command screeplot afterward to graph principal inertias; see [MV] **screeplot**.

4. Concerning existing estimation command ca (correspondence analysis),

   a. ca now allows crossed (stacked) variables. This provides a way to automatically combine two or more categorical variables into one crossed variable and perform correspondence analysis with it.

   b. ca's existing option normalize() now allows normalize(standard) to provide normalization of the coordinates by singular vectors divided by the square root of the mass.

   c. ca's new option length() allows you to customize the length of labels with crossed variables in output.

   d. New postestimation command estat loadings, used after ca and camat, displays correlations of profiles and axes.

   e. Existing postestimation command cabiplot has new option origin that displays the origin within the plot. cabiplot also now accepts originlopts(*line_options*) to customize the appearance of the origin on the graph.

   f. Existing postestimation commands cabiplot and caprojection now allow row and column marker labels to be specified using the mlabel() suboption of rowopts() and colopts().

   See [MV] **ca** and [MV] **ca postestimation**.

5. Existing commands cluster, matrix dissimilarity, and mds now allow the Gower measure for a mix of binary and continuous data; see [MV] *measure_option*.

6. Existing command biplot has new options. dim() specifies the dimensions to be displayed. negcol specifies that negative column (variable) arrows be plotted. negcolopts(*col_options*) provides graph options for the negative column arrows. norow and nocolumn suppress the row points or column arrows. See [MV] **biplot**.

7. New postestimation command estat rotate after canon performs orthogonal varimax rotation of the raw coefficients, standard coefficients, or canonical loadings. After estat rotate, new postestimation command estat rotatecompare displays the rotated and unrotated coefficients or loadings and the most recently rotated coefficients or loadings. See [MV] **canon postestimation**.

8. Existing commands pcamat and factormat now allow singular correlation or covariance structures. New option forcepsd modifies a matrix to be positive semidefinite and thus to be a proper covariance matrix. See [MV] **pca** and [MV] **factor**.

9. Existing commands rotate and rotatemat now refer to "Kaiser normalization" rather than "Horst normalization". A search of the literature indicates that Kaiser normalization is the preferred terminology. Previously option horst was a synonym for normalize. Now option horst is not documented. See [MV] **rotate** and [MV] **rotatemat**.

10. Existing command procrustes now saves the number of y variables in scalar e(ny); see [MV] **procrustes**.

1.3.9   What's new in graphics

1. Stata 10 has an interactive, point-and-click editor for your graphs. You do not need to type anything; you just right-click within the Graph window and select **Start Graph Editor**. You can do that any time, either when the graph is drawn or when you have graph used it from disk. You can add text, lines, markers, titles, and annotations, outside the plot region or inside; you can move

axes, titles, legends, etc.; you can change colors and sizes, number of tick marks, etc.; and you can even change scatters to lines or bars, or vice versa. See [G] **graph editor**.

2. New command `graph twoway lpoly` plots a local polynomial smooth; see [G] **graph twoway lpoly**. New command `graph twoway lpolyci` plots a local polynomial smooth along with a confidence interval; see [G] **graph twoway lpolyci**.

3. Concerning command `graph twoway`,

   a. `graph twoway` now allows more than 100 variables to be plotted.

   b. New suboption `custom` of *axis_label_options* allows you to create custom axis ticks and labels that have a different color, size, tick length, etc., from the standard ticks and labels on the axis. Such custom ticks can be used to emphasize points in the scale, such as important dates, physical constants, or other special values. See the `custom` suboption in [G] *axis_label_options*.

   c. New suboption `norescale` of *axis_label_options* specifies that added ticks and labels be placed directly on the graph without rescaling the axis or associated plot region for the new values; see [G] *axis_label_options*.

   d. New advanced options `yoverhangs` and `xoverhangs` adjust the graph region margins to prevent long labels on the y or x axis from extending off the edges of the graphs; see [G] *advanced_options*.

4. `graph twoway pcarrow` and `graph twoway pcbarrow` may now be drawn on plot regions with log scales or reversed scales; see [G] **graph twoway pcarrow**.

5. `graph bar` and `graph dot` no longer require user-provided names when a variable is repeated with more than one statistic; see [G] **graph bar** and [G] **graph dot**.

6. `graph twoway lfit` and `graph twoway qfit` now use value labels to annotate the x axis when using the existing suboption `valuelabel` of `xmlabel()`; see [G] *axis_label_options*.

7. `graph export` has new options `width(#)` and `height(#)` that specify the width and height of the graph when exporting to PNG or TIFF, thus allowing the resolution to be greater than screen resolution; see [G] *png_options* and [G] *tif_options*.

8. `graph twoway area` and `graph twoway rarea` now allow option `cmissing()` to control whether missing values produce breaks in the areas or are ignored; see [G] **graph twoway area** and [G] **graph twoway rarea**.

9. Typing `help graph` *option* now displays the help file for the specified option of the `graph` command. See [R] **help**.

1.3.10 What's new in programming

1. First, a warning for time-series programmers: Stata's new date/time values, which contain the number of milliseconds since 01jan1960 00:00:00, result in large numbers. 21apr2007 12:14:07.123 corresponds to 1,492,776,847,123. Date/time values must be stored as `doubles`. Programmers should use scalars to store these values whenever possible. If you must use a macro, exercise caution that the value is not rounded. It would not do at all for 1,492,776,847,123 to be recorded as "1.493e+12" (which would be 24apr2007 02:13:20). If you must use macros, our recommendations are

   a. If a date/time value is stored in one macro and you need it in another, code

   ```
   local new 'old'
   ```

b. If a date/time value is the result of an expression, and you must store it as a macro, code

```
local new = string(exp, "%21x")
```

or

```
local new : display %21x (exp)
```

Now we will continue with What's new.

2. Stata for Windows now supports Automation, formerly known as OLE Automation, which means that programmers can control Stata from other applications and retrieve results. See [P] **automation**.

3. New command confirm {numeric|string|date} format verifies that the format is of the specified type; see [P] **confirm**.

4. New function fmtwidth(s) returns the display width of a *%fmt* string, including date formats; see *Programming functions* in [D] **functions**.

5. Expression limits have been increased in Stata/MP, Stata/SE, and Stata/IC. The limit on the number of dyadic operators has increased from 500 to 800, and the limit on the number of numeric literals has increased from 150 to 300. See help limits.

6. Intercooled Stata has been renamed to Stata/IC. c(flavor) now contains "IC" rather than "Intercooled" if version ≥ 10. Backward-compatibility old global macro $S_FLAVOR continues to contain "Intercooled". See [P] **creturn** and [P] **macro**.

7. c() now contains values associated with Stata/MP: c(MP) (1 or 0 depending on whether you are running Stata/MP), c(processors) (the number of processors Stata will use), c(processors_mach) (the number of processors on the computer), c(processors_lic) (the maximum number of processors the license will allow you to use), and c(processors_max) (the maximum number of processors that could be used on this computer with this license).

8. New command include is a variation on do and run. include executes the commands stored in a file just as if they were entered from the keyboard or the current do-file. It is most often used in advanced Mata situations to create the equivalent of #include files. See [P] **include**.

9. New commands signestimationsample and checkestimationsample are useful in writing estimation/postestimation commands that need to identify the estimation sample; see [P] **signestimationsample**.

10. New command _datasignature is the building block for Stata's datasignature command and the programming commands signestimationsample and checkestimationsample. In advanced situations, you may wish to call it directly. See [P] **_datasignature**.

11. New extended macro function :copy copies one macro to another and is faster when the macro being copied is long. That is, coding

```
local new : copy local old
```

is faster than

```
local new `old'
```

See [P] **macro**.

12. New command timer times sections of code; see [P] **timer**.

13. Existing command matrix accum is now faster when some observations contain zeros; see [P] **matrix accum**.

14. Existing command ml display has new option showeqns that requests that equation names be displayed in the coefficient table; see [R] **ml**.

15. Existing command `mkmat` has new options `rownames()`, `roweq()`, `rowprefix()`, `obs`, and `nchar()` that specify the row names of the created matrix; see [P] **matrix mkmat**.

16. Existing command `_rmdcoll`'s `nocollin` option has been renamed to `normcoll`. `nocollin` will continue to work as a synonym for `normcoll`. See [P] **_rmcoll**.

17. Existing command `describe`'s option `simple` no longer saves the names of the variables in `r(varlist)`; you must specify option `varlist` if you want that. Also, existing command `describe using` *filename* now allows options `simple` and `varlist`. See [D] **describe**.

18. New extended macro function `adosubdir` returns the subdirectory in which Stata would search for a file along the ado-path. Determining the subdirectory in which Stata stores files is now a function of the file's extension. `adosubdir` returns the subdirectory in which to look. See [P] **macro**.

19. Existing command `syntax { [` *optionname*`(real` ...`)] }` now returns the number specified in `%18.0g` format if `version` is set to 10.0 or higher. For `version` less than 10, the number is returned in `%9.0g` format. See [P] **syntax**.

20. New functions `_byn1()` and `_byn2()`, available within a `byable(recall)` program, return the beginning and ending observation numbers of the by-group currently being executed; see [P] **byable**.

21. Existing command `program drop` may now specify `program drop _allado` to drop programs that were automatically loaded from ado-files; see [P] **program**.

22. Concerning SMCL,

    a. Existing directive `{synoptset}` has new optional argument `notes` that specifies that some of the table entries will have footnotes and results in a larger indentation of the first column.

    b. Existing directive `{p}` now has an optional fourth argument specifying the paragraph's width.

    See [P] **smcl**.

23. Concerning classes, you can now define an `oncopy` member program to perform operations when a copy of an object is being created. See [P] **class**.

24. Concerning programmable menus, the maximum number of menu items that can be added to Stata has increased to 1,250 from 1,000; see `help window programming`.

25. Concerning programmable dialogs,

    a. Child dialogs can now be created.

    b. New control `TEXTBOX` allows displaying multiline text.

    c. In the dialog programming language, (1) `if` now allows `else` and (2) new command `close` closes the dialog programmatically.

    d. Messages can be passed to dialogs when they are launched; see `help db`.

    e. Dialogs can now be designated as modal, meaning that this dialog must be dealt with by the user before new dialogs (other than children) can be launched.

    f. Several controls have new options and new member programs. For instance, `FILE` and `LISTBOX` now have option `multiselect`, which lets the user pick more than one item.

    See `help dialog programming`.

26. Stata's help files are now named `*.sthlp` rather than `*.hlp`, meaning that user-written help files can be sent via email more easily. Many email filters flag `.hlp` files as potential virus carriers because Stata was not the only one to use the `.hlp` suffix. You need not rename your old help files. See [R] **help**.

27. Two new C functions have been exposed from Stata for use by plugins: sstore() and sdata(). sstore() stores string data in the Stata dataset and sdata() reads them. See http://www.stata.com/plugins/.

1.3.11 What's new in Mata

1. New Stata command include is a variation on do and run and is useful for implementing #include for header files in advanced programming situations. See [P] **include** and type viewsource optimize.mata for an example of use.

2. Mata now has structures, which will be of special interest to those writing large systems. See [M-2] **struct** and [M-5] **liststruct( )**.

3. Mata now engages in more thorough type checking, and produces better code, for those who explicitly declare arguments and variables.

4. Mata inherits all of Stata's formats and functions for dealing with the new date/time variables and values; see [M-5] **date( )** and [M-5] **fmtwidth( )**.

5. New functions inbase() and frombase() perform base conversions; see [M-5] **inbase( )**.

6. New function floatround() returns values rounded to float precision. This is Mata's equivalent of Stata's float() function. See [M-5] **floatround( )**.

7. New function nameexternal() returns the name of an external; see [M-5] **findexternal( )**.

8. Concerning matrix manipulation,

   a. Matrix multiplication is now faster when one of the matrices contains many zeros, as is function cross().

   b. Appending rows or columns to a matrix using , and \ is now faster.

   c. New function _diag() replaces the principal diagonal of a matrix with a specified vector or scalar; see [M-5] **_diag( )**.

   d. New functions select() and st_select() select rows or columns of a matrix on the basis of a criterion; see [M-5] **select( )**.

   e. Existing functions rowsum(), colsum(), sum(), quadrowsum(), quadcolsum(), and quadsum() now allow an optional second argument that determines how missing values are handled; see [M-5] **sum( )**.

   f. New functions runningsum(), quadrunningsum(), _runningsum(), and _quadrunningsum() return the running sum of a vector; see [M-5] **runningsum( )**.

   g. New functions minindex() and maxindex() return the indices of minimums and maximums (including tied values) of a vector; see [M-5] **minindex( )**.

9. Concerning statistics,

   a. New Mata function optimize() performs minimization and maximization. You can code just the function, the function and its first derivatives, or the function and its first and second derivatives. Optimization techniques include Newton–Raphson, Davidon–Fletcher–Powell, Broyden–Fletcher–Goldfarb–Shanno, Berndt–Hall–Hall–Hausman, and the simplex method Nelder–Mead. See [M-5] **optimize( )**.

   b. New function cvpermute() forms permutations; see [M-5] **cvpermute( )**.

   c. New function ghk() provides the Geweke–Hajivassiliou–Keane multivariate normal simulator; see [M-5] **ghk( )**. New function ghkfast() is faster but a little more difficult to use; see [M-5] **ghkfast( )**.

d. New functions `halton()`, `_halton()`, and `ghalton()` compute Halton and Hammersley sets; see [M-5] **halton( )**.

e. The new density and probability functions available in Stata are also available in Mata, including `binomial()`, `binomialtail()`, `gammaptail()`, `invgammaptail()`, `invbinomialtail()`, `ibetatail()`, `invibetatail()`, `lnnormal()`, and `lnnormalden()`; see [M-5] **normal( )**. Also, as in Stata, convergence and accuracy of many of the cumulatives, reverse cumulatives, and density functions have been improved.

f. Existing Mata functions `mean()`, `variance()`, `quadvariance()`, `meanvariance()`, `quadmeanvariance()`, `correlation()`, and `quadcorrelation()` now make the weight argument optional. If not specified, unweighted estimates are returned. See [M-5] **mean( )**.

10. Concerning string processing,

a. New function `stritrim()` replaces multiple, consecutive internal spaces with one space; see [M-5] **stritrim( )**.

b. New functions `strtoreal()` and `_strtoreal()` convert strings to numeric values; see [M-5] **strtoreal( )**.

c. New function `_substr()` substitutes a substring into an existing string; see [M-5] **_substr( )**.

d. New function `invtokens()` is the inverse of the existing function `tokens()`; see [M-5] **invtokens( )**.

e. New function `tokenget()` performs advanced parsing; see [M-5] **tokenget( )**.

11. Concerning I/O,

a. New function `adosubdir()` returns the subdirectory in which Stata would search for a file; see [M-5] **adosubdir( )**. New function `pathsearchlist(fn)` returns a row vector of full paths/filenames specifying all the locations, in order, where Stata would look for the specified *fn* along the official Stata ado-path; see [M-5] **pathjoin( )**.

b. New function `byteorder()` returns 1 if the computer is HILO and returns 2 if the computer is LOHI; see [M-5] **byteorder( )**.

c. New functions `bufput()` and `bufget()` copy elements into and out of buffers; see [M-5] **bufio( )**.

d. Existing function `cat()` now allows optional second and third arguments that specify the beginning and ending lines of the file to read; see [M-5] **cat( )**.

12. New functions `ferrortext()` and `freturncode()` obtain error messages and return codes following an I/O error; see [M-5] **ferrortext( )**.

13. Concerning the Stata interface,

a. New function `stataversion()` returns the version of Stata that you are running, and new function `statasetversion()` allows setting it. See [M-5] **stataversion( )**.

b. New function `setmore()` allows turning `more` on and off. New function `setmoreonexit()` allows restoring `more` to its original setting when execution ends. See [M-5] **more( )**.

1.3.12   What's more

We have not listed all the changes, but we have listed the important ones.

Stata is continually being updated, and those updates are available for free over the Internet. All you have to do is type

```
. update query
```

and follow the instructions.

To learn what has been added since this manual was printed, select **Help > What's new?** or type

```
. help whatsnew
```

We hope that you enjoy Stata 10.

1.4 References

Chambers, J. M., W. S. Cleveland, B. Kleiner, and P. A. Tukey. 1983. *Graphical Methods for Data Analysis.* Belmont, CA: Wadsworth.

Everitt, B. S. 2006. *The Cambridge Dictionary of Statistics.* 3rd ed. Cambridge: Cambridge University Press.

Freedman, L. S. 1982. Tables of the number of patients required in clinical trials using the logrank test. *Statistics in Medicine* 1: 121–129.

Heyde, C. C., and E. Seneta, ed. 2001. *Statisticians of the Centuries.* New York: Springer.

Johnson, N. L., and S. Kotz, ed. 1997. *Leading Personalities in Statistical Sciences.* New York: Wiley.

Lachin, J. M., and M. A. Foulkes. 1986. Evaluation of sample size and power for analysis of survival with allowance for nonuniform patient entry, losses to follow-up, noncompliance, and stratification. *Biometrics* 42: 507–519.

Rubinstein, L. V., M. H. Gail, and T. J. Santner. 1981. Planning the duration of a comparative clinical trial with loss to follow-up and a period of continued observation. *Journal of Chronic Diseases* 34: 469–479.

Schoenfeld, D. 1981. The asymptotic properties of nonparametric tests for comparing survival distributions. *Biometrika* 68: 316–319.

Tarone, R. E. 1985. On heterogeneity tests based on efficient scores. *Biometrika* 72: 91–95.

Upton, G., and I. Cook. 2006. *A Dictionary of Statistics.* 2nd ed. Oxford: Oxford University Press.

2 A brief description of Stata

Stata is a statistical package for managing, analyzing, and graphing data.

Stata is available for a variety of platforms. Stata may be used either as a point-and-click application or as a command-driven package.

Stata's GUI provides an easy interface for those new to Stata and for experienced Stata users who wish to execute a command that they seldom use.

The command language provides a fast way to communicate with Stata and to communicate more complex ideas.

Here is an extract of a Stata session using the GUI:

(Throughout the Stata manuals, we will refer to various datasets. These datasets are all available from http://www.stata-press.com/data/r10/. For easy access to them within Stata, you may type `webuse` *dataset_name*, or select **File > Example Datasets...** and click on Stata 10 manual datasets.)

```
. webuse lbw2
(Hosmer & Lemeshow data)
```

We select **Data > Describe data > Summary statistics** and choose to summarize variables `low`, `age`, and `smoke`, whose names we obtained from the Variables window. We press **OK**.

```
. summarize low age smoke
```

Variable	Obs	Mean	Std. Dev.	Min	Max
low	189	.3121693	.4646093	0	1
age	189	23.2381	5.298678	14	45
smoke	189	.3915344	.4893898	0	1

Stata shows us the command that we could have typed in command mode—summarize low age smoke—before displaying the results of our request.

Next, we fit a logistic regression model of low on age and smoke. We select **Statistics > Binary outcomes > Logistic regression (reporting odds ratios)**, fill in the fields, and press **OK**.

```
. logistic low age smoke
```

Logistic regression				Number of obs	=	189
				LR chi2(2)	=	7.40
				Prob > chi2	=	0.0248
Log likelihood = -113.63815				Pseudo R2	=	0.0315

low	Odds Ratio	Std. Err.	z	P>|z|	[95% Conf. Interval]	
age	.9514394	.0304194	-1.56	0.119	.8936481	1.012968
smoke	1.997405	.642777	2.15	0.032	1.063027	3.753081

Here's an extract of a Stata session using the command language:

```
. use http://www.stata-press.com/data/r10/auto
(1978 Automobile Data)

. summarize mpg weight
    Variable |       Obs        Mean    Std. Dev.       Min        Max
-------------+--------------------------------------------------------
         mpg |        74     21.2973    5.785503         12         41
      weight |        74    3019.459    777.1936       1760       4840
```

The user typed `summarize mpg weight` and Stata responded with a table of summary statistics. Other commands would produce different results:

```
. correlate mpg weight
(obs=74)
             |      mpg    weight
-------------+------------------
         mpg |   1.0000
      weight |  -0.8072    1.0000

. gen w_sq = weight^2

. regress mpg weight w_sq
      Source |       SS       df       MS              Number of obs =      74
-------------+------------------------------           F(  2,    71) =   72.80
       Model |  1642.52197     2  821.260986           Prob > F      =  0.0000
    Residual |  800.937487    71  11.2808097           R-squared     =  0.6722
-------------+------------------------------           Adj R-squared =  0.6630
       Total |  2443.45946    73  33.4720474           Root MSE      =  3.3587

------------------------------------------------------------------------------
         mpg |      Coef.   Std. Err.      t    P>|t|     [95% Conf. Interval]
-------------+----------------------------------------------------------------
      weight |  -.0141581   .0038835    -3.65   0.001    -.0219016   -.0064145
        w_sq |   1.32e-06   6.26e-07     2.12   0.038     7.67e-08    2.57e-06
       _cons |   51.18308   5.767884     8.87   0.000     39.68225    62.68392
------------------------------------------------------------------------------

. scatter mpg weight, by(foreign, total row(1))
```

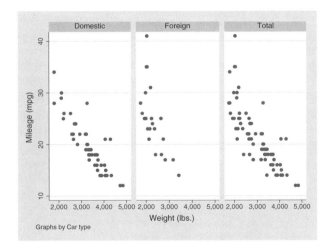

The user-interface model is type a little, get a little, etc., so that the user is always in control.

Stata's model for a dataset is that of a table—the rows are the observations and the columns are the variables:

```
. list mpg weight in 1/10
```

	mpg	weight
1.	22	2,930
2.	17	3,350
3.	22	2,640
4.	20	3,250
5.	15	4,080
6.	18	3,670
7.	26	2,230
8.	20	3,280
9.	16	3,880
10.	19	3,400

Observations are numbered; variables are named.

Stata is fast. That speed is due partly to clever programming, and partly it is because Stata keeps the data in memory. Stata's data model is that of a word processor: a dataset may exist on disk, but that is just a copy. The dataset is loaded into memory, where it is worked on, analyzed, changed, and then perhaps stored back on disk.

Working on a copy of the data in memory makes Stata safe for interactive use. The only way to harm the permanent copy of your data on disk is if you explicitly save over it.

Having the data in memory means that the dataset size is limited by the amount of memory. Stata stores the data in memory in a compressed format—you will be surprised how much data can fit into a given region of memory. Nevertheless, if you work with large datasets, you may run into memory constraints. There are two solutions to this problem:

1. By default, Stata/MP and Stata/SE allocates 10 MB to Stata's data areas, and you can change it; see [U] **6 Setting the size of memory**.

   By default, Stata/IC allocates 1 MB to Stata's data areas, and you can change it; see [U] **6 Setting the size of memory**.

   By default, Small Stata allocates about 300 K to Stata's data areas, and you cannot change it.

2. You will want to learn how to compress your data as much as possible; see [D] **compress**.

3 Resources for learning and using Stata

3.1 Overview

The *Getting Started* manual, *User's Guide*, and *Reference* manuals are the primary tools for learning about Stata; however, there are many other sources of information. A few are

1. Stata itself. Stata has a subject table of contents online with links to the help system and dialog boxes that make it easy to find and to execute a Stata command. See [U] **4 Stata's online help and search facilities**.

2. The Stata web site. Visit http://www.stata.com. Much of the site is dedicated to user support; see [U] **3.2 The http://www.stata.com web site**.

3. The Stata Press web site. Visit http://www.stata-press.com. This site contains the datasets used throughout the Stata manuals; see [U] **3.3 The http://www.stata-press.com web site**.

4. The Stata listserver. An active group of Stata users communicate over an Internet listserver, which you can join for free; see [U] **3.4 The Stata listserver**.

5. The Stata software distribution site and other user-provided software distribution sites. Stata itself can download and install updates and additions. We provide official updates to Stata—type update or select **Help > Official Updates**. We also provide user-written additions to Stata and links to other user-provided sites—type net or select **Help > SJ and User-written Programs**; see [U] **3.6 Updating and adding features from the web**.

6. The *Stata Journal* and the *Stata Technical Bulletin*. The *Stata Journal* contains reviewed papers, regular columns, book reviews, and other material of interest to researchers applying statistics in a variety of disciplines. The *Stata Technical Bulletin*, the predecessor to the *Stata Journal*, contains articles and user-written commands. See [U] **3.5 The Stata Journal and the Stata Technical Bulletin**.

7. NetCourses. We offer training via the Internet. Details are in [U] **3.7 NetCourses** below.

8. Books and support materials. Supplementary Stata materials are available; see [U] **3.8 Books and other support materials**.

9. Technical support. We provide technical support by email, telephone, and fax; see [U] **3.9 Technical support**.

3.2   The http://www.stata.com web site

Point your browser to http://www.stata.com and click on **Support**. More than half our web site is dedicated to providing support to users.

1. The web site provides FAQs (frequently asked questions) on Windows, Macintosh, Unix, statistics, programming, Internet capabilities, graphics, and data management. These FAQs run the gamut from "I cannot save/open files" to "What does 'completely determined' mean in my logistic regression output?" Most users will find something of interest.

2. Visiting the web site is one way that you can subscribe to the Stata listserver; see [U] **3.4 The Stata listserver**.

3. The web site provides detailed information about NetCourses, along with the current schedule; see [U] **3.7 NetCourses**.

4. The web site provides information about Stata courses and meetings, both in the United States and elsewhere.

5. The web site provides an online bookstore for Stata-related books and other supplementary materials; see [U] **3.8 Books and other support materials**.

6. The web site provides links to information about statistics: other statistical software providers, book publishers, statistical journals, statistical organizations, and statistical listservers.

7. The web site provides links to Stata resources for learning Stata at http://www.stata.com/links/resources.html. Be sure to look at these materials, as many valuable resources about Stata are listed here, including the UCLA Stata portal, which includes a set of links about Stata, and the Statistical Software Components (SSC) archive, which has become the premier Stata download site for user-written software on the web.

In short, the web site provides up-to-date information on all support materials and, where possible, provides the materials themselves. Visit http://www.stata.com if you can.

3.3   The http://www.stata-press.com web site

Point your browser to http://www.stata-press.com. This site is devoted to the publications and activities of Stata Press.

1. Datasets that are used in the Stata *Reference* manuals and other books published by Stata Press may be downloaded. Visit http://www.stata-press.com/data/. These datasets can be used in Stata by simply typing use `http://www.stata-press.com/data/r10/auto`. You could also type `webuse auto`; see [D] **webuse**.

2. An online catalog of all our books and multimedia products is at http://www.stata-press.com/catalog.html. We have tried to include enough information, such as table of contents and preface material, so that you may tell whether the book is appropriate for you.

3. Information about forthcoming publications is posted at http://www.stata-press.com/forthcoming.html.

3.4   The Stata listserver

The Stata listserver (Statalist) is an independently operated, real-time list of Stata users on the Internet. Anyone may join. Instructions for doing so can be found at http://www.stata.com by clicking on Support and then Statalist or by emailing service@stata.com.

Many knowledgeable users are active on the list, as are the StataCorp technical staff. We recommend that new users subscribe, observe the exchanges, and, if it turns out not to be useful, unsubscribe.

3.5   The Stata Journal and the Stata Technical Bulletin

The *Stata Journal* (SJ) is a printed and electronic journal, published quarterly, containing articles about statistics, data analysis, teaching methods, and effective use of Stata's language. The *Journal* publishes reviewed papers together with shorter notes and comments, regular columns, tips, book reviews, and other material of interest to researchers applying statistics in a variety of disciplines. The *Journal* is a publication for all Stata users, both novice and experienced, with different levels of expertise in statistics, research design, data management, graphics, reporting of results, and of Stata, in particular.

Tables of contents for past issues and abstracts of the articles are available at http://www.stata-journal.com/archives.html.

We recommend that all users subscribe to the SJ. Visit http://www.stata-journal.com to learn more about the *Stata Journal* and to order your subscription.

To obtain any programs associated with articles in the SJ, type

```
. net from http://www.stata-journal.com/software
```

or

1. Select **Help > SJ and User-written Programs**
2. Click on **Stata Journal**

The Stata Technical Bulletin

For 10 years, the *Stata Technical Bulletin* (STB) served as the means of distributing new commands and Stata upgrades, both user-written and "official". After 10 years of continual publication, the STB evolved into the *Stata Journal*. The Internet provided an alternative delivery mechanism for user-written programs, so the emphasis shifted from user-written programs to more expository articles. Although the STB is no longer published, many of the programs and articles that appeared in it are still valuable today. Reprints of past issues are available from http://www.stata.com/bookstore/stbr.html. To obtain the programs that were published in the STB, type

```
. net from http://www.stata.com
. net cd stb
```

(Continued on next page)

3.6   Updating and adding features from the web

Stata itself is web aware.

First, try this:

```
. use http://www.stata.com/manual/oddeven, clear
```

That will load an uninteresting dataset into your computer from our web site. If you have a home page, you can use this feature to share datasets with coworkers. Save a dataset on your home page, and researchers worldwide can use it. See [R] **net**.

3.6.1   Official updates

Although we follow no formal schedule for the release of updates, we typically provide updates to Stata approximately every 2 weeks. Installing the updates is easy. Type

```
. update
```

or select **Help > Official Updates**. Do not be concerned; nothing will be installed unless and until you say so. Once you have installed the update, you can type

```
. help whatsnew
```

or select **Help > What's New?** to find out what has changed. We distribute official updates to fix bugs and to add new features.

3.6.2   Unofficial updates

There are also "unofficial" updates—additions to Stata written by Stata users, which includes members of the StataCorp technical staff. Stata is programmable, and even if you never write a Stata program, you may find these additions useful, some of them spectacularly so. You start by typing

```
. net from http://www.stata.com
```

or select **Help > SJ and User-written Programs**.

Be sure to try visit the SSC Archive. The `ssc` command makes it easy for you to install and uninstall packages from SSC. Type

```
. ssc whatsnew
```

to find out what's new at the site. If you find something that interests you, type

```
. ssc describe pkgname
```

for more information.

Periodically, you can type

```
. news
```

or select **Help > News** to display a short message from our web site telling you what is newly available.

See [U] **28 Using the Internet to keep up to date**.

3.7   NetCourses

We offer courses on Stata at the introductory and advanced levels. Courses on software are typically expensive and time consuming. They are expensive because, in addition to the direct costs of the course, participants must travel to the course site. We have found that it is better to organize courses over the Internet—saving everyone time and money.

We offer courses over the Internet and call them Stata NetCourses^{TM}.

1. **What is a NetCourse?**
    A NetCourse is a course offered through the Stata web site that varies in length from 7 to 8 weeks. You must have an email address and web browser to participate.

2. **How does it work?**
    Every Friday a lecture is posted on a password-protected web site. After reading the lecture over the weekend or perhaps on Monday, participants then post questions and comments on a message board. Course leaders typically respond to the questions and comments on the same day they are posted. The other participants are encouraged to amplify or otherwise respond to the questions or comments as well. The next lecture is then posted on Friday, and the process repeats.

3. **How much of my time does it take?**
    It depends on the course, but the introductory courses are designed to take roughly 3 hours per week.

4. **There are three of us here—can just one of us enroll and then redistribute the NetCourse materials ourselves?**
    We ask that you not. NetCourses are priced to cover the substantial time input of the course leaders. Moreover, enrollment is typically limited to prevent the discussion from becoming unmanageable. The value of a NetCourse, just like a real course, is the interaction of the participants, both with each other and with the course leaders.

5. **I've never taken a course by Internet before. I can see that it might work, but then again, it might not. How do I know I'll benefit?**
    All Stata NetCourses come with a 30-day satisfaction guarantee. The 30 days begins after the conclusion of the final lecture.

You can learn more about the current NetCourse offerings by visiting http://www.stata.com. Our offerings include

>   NC-101 Introduction to Stata
>   NC-151 Introduction to Stata Programming
>   NC-152 Advanced Stata Programming
>   NC-461 Introduction to Univariate Time Series with Stata

NetCourseNow

A NetCourseNow offers the same material as NetCourses but it allows you to choose the time and pace of the course, and you have a personal NetCourse instructor.

1. **What is a NetCourseNow?**
    A NetCourseNow offers the same material as a Netcourse, but allows you to move at your own pace and to specify a starting date. With a NetCourseNow, you also have the added benefit of a personal NetCourse instructor whom you can email directly with questions about lectures and exercises. You must have an email address and a web browser to participate.

2. **How does it work?**

   All course lectures and exercises are posted at once, and you are free to study at your own pace. You will be provided with the email address of your personal NetCourse instructor to contact when you have questions.

3. **How much of my time does it take?**

   A NetCourseNow allows you to set your own pace. How long the course takes and how much time you spend per week is up to you.

3.8   Books and other support materials

3.8.1   For readers

There are books published about Stata, both by us and by others. Visit the Stata bookstore at http://www.stata.com/bookstore. For the books that we carry, we include the table of contents and comments written by a member of our technical staff, explaining why we think this book might interest you.

3.8.2   For authors

If you have written a new book related to Stata and would like us to consider carrying it in our bookstore, email bookstore@stata.com.

If you are writing a book, join our free Author Support Program. Stata professionals are available to review your Stata code to ensure that it is efficient and reflects modern usage, production specialists are available to help format Stata output, and editors and statisticians are available to ensure the accuracy of Stata-related content. Visit http://www.stata.com/authorsupport.

If you are thinking about writing a Stata-related book, consider publishing it with Stata Press. Email submissions@statapress.com.

3.9   Technical support

We are committed to providing superior technical support for Stata software. To assist you as efficiently as possible, please follow the procedures listed below.

3.9.1   Register your software

You must register your software to be eligible for technical support, updates, special offers, and other benefits. By registering, you will receive the *Stata News*, and you may access our support staff for free with any question that you encounter. You may register your software either electronically or by mail.

Electronic registration: after installing Stata and successfully entering your License and Authorization Key, your default web browser will open to the online registration form at the Stata web site. You may also manually point your web browser to http://www.stata.com/register/ if you wish to register your copy of Stata at a later time.

Mail-in registration: fill in the registration card that came with Stata and mail it to StataCorp.

3.9.2   Before contacting technical support

Before you spend the time gathering the information our technical support department needs, make sure that the answer does not already exist in the help files. You can use the `help` and `search` commands to find all the entries in Stata that address a given subject. Be sure to try selecting **Help > Contents**. Check the manual for a particular command. There are often examples that address questions and concerns. Another good source of information is our web site. You should keep a bookmark of our frequently asked questions page (http://www.stata.com/support/faqs/) and check it occasionally for new information.

Our technical department will need some information from you to provide detailed assistance. Most important is your serial number, but they will also need the following information:

1. The system information on the computer that you are using is especially important if you are having hardware problems. This includes the make and model of various hardware components such as the computer manufacturer, the video driver, the operating system and its version number, relevant peripherals, and the version number of any other software with which you experience a conflict.

2. The version of Stata that you are running. Type `about` at the Stata prompt, and Stata will display this information.

3. The types of variables in your dataset and the number of observations.

4. The command that is causing the error along with the exact error message and return code (error number).

3.9.3   Technical support by email

This is the preferred method of asking a technical support question. It has the following advantages:

- You will receive a prompt response from us saying that we have received your question and that it has been forwarded to *Technical Services* to answer.

- We can route your question to a specialist for your particular question.

- Questions submitted via email may be answered after normal business hours, or even on weekends or holidays. Although we cannot promise that this will happen, it may, and your email inquiry is bound to receive a faster response than leaving a message on Stata's voicemail.

- If you are receiving an error message or an unexpected result, it is easy to include a log file that demonstrates the problem.

Please see the FAQ at http://www.stata.com/support/faqs/techsup/ for some suggestions to follow that will aid Technical Services in promptly answering your question.

3.9.4   Technical support by phone or fax

Our technical support telephone number is 979-696-4600. Please have your serial number handy. It is also best if you are at your computer when you call. If your question involves an error message from a command, please note the error message and number, as this will greatly help us in assisting you. Telephone support is reserved for nonstatistical questions. If your question requires the attention of a statistician, the question should be submitted via email or fax.

Send fax requests to 979-696-4601. If possible, collect the relevant information in a log file and include the file in your fax.

Please see the FAQ at http://www.stata.com/support/faqs/techsup/ for some suggestions to follow that will aid Technical Services in promptly answering your question.

3.9.5   Comments and suggestions for our technical staff

By all means, send in your comments and suggestions. Your input is what determines the changes that occur in Stata between releases, so if we don't hear from you, we may not include your most desired new estimation command! Email is preferred, as this provides us with a permanent copy of your request. When requesting new commands, please include any references that you would like us to review should we develop those new commands. Email your suggestions to service@stata.com.

4 Stata's online help and search facilities

Contents

4.1 Introduction

To access Stata's online help, you will either

1. select **Help** form the menu bar, or

2. use the `help`, `search`, and `hsearch` commands.

Regardless of the method you use, results will be shown in the Viewer or Results windows. Blue text indicates a hypertext link, so you can click to go to related entries.

4.2 Getting started

The first time you use help, try one of the following:

1. Select **Help > Advice** form the menu bar, or

2. Type `help advice`.

(*Continued on next page*)

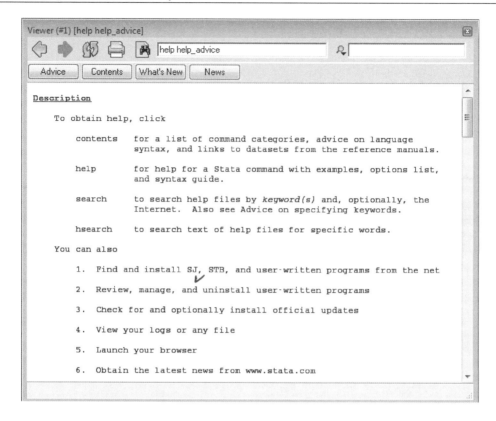

Advice presents you with four alternatives:

1. contents: Click on that and you will be taken to a system organized by topic: Basics, Data management, Statistics, Graphics, and Programming and matrices.

2. help: Click on that and you can type the name of a Stata command and go directly to its online help.

3. search: Click on that and you can perform a keyword search.

4. hsearch: Click on that and you can do a search of the contents of the help files much like search engines on the Internet search the contents of the web.

The above are the four main features of Stata's help system, and you can access each in various ways:

1. Topical help:

    Select **Help > Advice**, click on contents, or

    Select **Help > Contents**, or

    Type help contents

2. Help on a command or topic:

    Select **Help > Advice**, click on help, or

    Select **Help > Stata Command...**, or

    Type help commandname

3. Keyword search:

> Select **Help > Advice**, click on `search`, or
>
> Select **Help > Search...**, or
>
> Type `search` *keyword(s)*

4. Text search:

> Select **Help > Advice**, click on `hsearch`, or
>
> Type `hsearch` *whatever*

4.3   help: Stata's online manual pages

When you

1. Select **Help > Stata command...**
   Type a command name in the Command edit field
   Click OK, or

2. Type `help` followed by a command name

you access Stata's online help files. These files provide shortened versions of what is in the printed manuals. Let's access the online help for Stata's `ttest` command. Do one of the following:

1. Select **Help > Stata command...**
   Type `ttest` in the Command edit field
   Click OK, or

2. Type `help ttest`

Regardless of which you do, the result will be

(Continued on next page)

The trick is in already knowing that Stata's command for testing equality of means is `ttest` and not, say `meanstest`. There are two solutions to that problem, topical help and searching. Topical help you access by selecting **Help > Contents** or by typing `help contents`.

4.4   Searching

If you do not know the name of the Stata command you are looking for, you can search for it by keyword,

　　1. Select **Help > Search...**
　　　Type keywords in the edit field
　　　Click OK

　　2. Type `search` followed by the keywords

or you can search the text of Stata's help files,

　　3. Type `hsearch` followed by text

`search` and `hsearch` each have their own advantages.

`search` is available both by menu and command; `hsearch` is available by command only (and from **Help > Advice**).

`search` matches the keywords you specify to a database and returns matches found in Stata commands, FAQs at www.stata.com, and articles that have appeared in the *Stata Journal*. Optionally, it can also find user-written additions to Stata available over the web. To do that, from the Search dialog box select *Search all* or, on the `search` command, specify the `all` option.

hsearch searches the text of the Stata help files on your computer. That does not include FAQs at www.stata.com and articles that have appeared in the *Stata Journal*, but it does include any help files you have written and the help files associated with any user-written additions you may have installed.

search does a better job when what you want is based on terms commonly used or when what you are looking for might not already be installed on your computer.

hsearch does a better job when the search terms are odder or when what you want might already be installed.

It is sometimes a good idea to use both.

4.5   More on search

However you access search—command or menu—it does the same thing. You tell search what you want information about, and it searches for relevant entries. If you want search to look for the topic across all sources, including the online help, the FAQs at the Stata web site, the *Stata Journal*, and all Stata-related Internet sources including user-written additions, specify "Search all" from the Search dialog or, on the search command, specify the all option. (Command findit is a synonym for search, all, so many users type findit ... rather than search ..., all.)

search can be used broadly or narrowly. For instance, if you want to perform the Kolmogorov–Smirnov test for equality of distributions, you could type

```
. search Kolmogorov-Smirnov test of equality of distributions
[R]      ksmirnov . . . . . . Kolmogorov-Smirnov equality of distributions test
         (help ksmirnov)
```

In fact, we did not have to be nearly so complete—typing search Kolmogorov-Smirnov would have been adequate. Had we specified our request more broadly—looking up equality of distributions—we would have obtained a longer list that included ksmirnov.

Here are guidelines for using search.

1. Capitalization does not matter. Look up Kolmogorov-Smirnov or kolmogorov-smirnov.

2. Punctuation does not matter. Look up kolmogorov smirnov.

3. Order of words does not matter. Look up smirnov kolmogorov.

4. You may abbreviate, but how much depends. Break at syllables. Look up kol smir. search tends to tolerate a lot of abbreviation; it is better to abbreviate than to misspell.

5. The prepositions for, into, of, on, to, and with are ignored. Use them—look up equality of distributions—or omit them—look up equality distributions—it makes no difference.

6. search tolerates plurals, especially when they can be formed by adding an *s*. Even so, it is better to look up the singular. Look up normal distribution, not normal distributions.

7. Specify the search criterion in English, not in computer jargon.

8. Use American spellings. Look up color, not colour.

9. Use nouns. Do not use -ing words or other verbs. Look up median tests, not testing medians.

10. Use few words. Every word specified further restricts the search. Look up distribution, and you get one list; look up normal distribution, and the list is a sublist of that.

11. Sometimes words have more than one context. The following words can be used to restrict the context:

    a. `data`, meaning in the context of data management. Order could refer to the order of data or to order statistics. Look up `order data` to restrict order to its data-management sense.

    b. `statistics` (abbreviation `stat`), meaning in the context of statistics. Look up `order statistics` to restrict order to the statistical sense.

    c. `graph` or `graphs`, meaning in the context of statistical graphics. Look up `median graphs` to restrict the list to commands for graphing medians.

    d. `utility` (abbreviation `util`), meaning in the context of utility commands. The `search` command itself is not data management, not statistics, and not graphics; it is a utility.

    e. `programs` or `programming` (abbreviation `prog`), to mean in the context of programming. Look up `programming scalar` to obtain a sublist of scalars in programming.

`search` has other features, as well; see [U] **4.8 search: All the details**.

4.6   More on help

Both `help` and `search` are understanding of some mistakes. For instance, you may abbreviate a command name. If you type either `help regres` or `help regress`, you will bring up the help file for `regress`.

When `help` cannot find the command you are looking for, try `search`. In this case typing `search regres` would find the command, but that would because 'regres' is an abbreviation of the word regression, which word probably appears in the keyword database.

Stata can run into some problems with abbreviations. For instance, Stata has a command with the inelegant name `ksmirnov`. You forget and think the command is called `ksmir`:

```
. help ksmir
help for ksmir not found
try help contents or search ksmir
```

This is a case where `help` gives bad advice because typing `search ksmir` will do you no good. You should type `search` followed by what you are really looking for: `search kolmogorov smirnov`.

4.7   help contents: Stata's online table of contents

Typing `help contents` or selecting **Help > Contents** provides another way of locating entries in the documentation and online help. Whichever you use, you will be presented with a long table of contents, organized topically.

. help contents

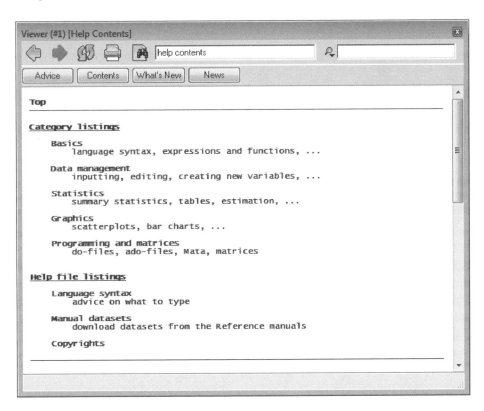

4.8   search: All the details

The search command actually provides a few features that are not available from the **Help** menu.
The full syntax of the search command is

search *word* [*word* ...] [, [local | net | all] author entry exact faq

historical or manual sj]

where underlining indicates the minimum allowable abbreviation and [brackets] indicate optional.

local, the default (unless changed by set searchdefault), specifies that the search be performed
   using only Stata's keyword database.

net specifies that the search be performed across the materials available via Stata's net command.
   Using search word [*word* ...], net is equivalent to typing net search word [*word* ...]
   (without options); see [R] **net**.

all specifies that the search be performed across both the local keyword database and the net materials.

author specifies that the search be performed on the basis of author's name rather than keywords.

entry specifies that the search be performed on the basis of entry IDs rather than keywords.

exact prevents matching on abbreviations.

faq limits the search to entries found in the FAQs at http://www.stata.com.

historical adds to the search entries that are of historical interest only. By default, such entries are not listed. Past entries are classified as historical if they discuss a feature that later became an official part of Stata. Updates to historical entries will always be found, even if historical is not specified.

or specifies that an entry be listed if any of the words typed after search are associated with the entry. The default is to list the entry only if all the words specified are associated with the entry.

manual limits the search to entries in the *User's Guide* and all the *Reference* manuals.

sj limits the search to entries in the *Stata Journal* and the *Stata Technical Bulletin*.

4.8.1 How search works

search has a database—a file—containing the titles, etc., of every entry in the *User's Guide*, *Reference* manuals, articles in the *Stata Journal* and in the *Stata Technical Bulletin*, and FAQs at http://www.stata.com. In this file is a list of words associated with each entry, called keywords.

When you type search *xyz*, search reads this file and compares the list of keywords with *xyz*. If it finds *xyz* in the list or a keyword that allows an abbreviation of *xyz*, it displays the entry.

When you type search *xyz abc*, search does the same thing but displays an entry only if it contains both keywords. The order does not matter, so you can search linear regression or search regression linear.

How many entries search finds depends on how the search database was constructed. We have included a plethora of keywords under the theory that, for a given request, it is better to list too much rather than risk listing nothing at all. Still, you are in the position of guessing the keywords. Do you look up normality test, normality tests, or tests of normality? Normality test would be best, but all would work. In general, use the singular, and strike the unnecessary words. We provide guidelines for specifying keywords in [U] **4.5 More on search** above.

4.8.2 Author searches

search ordinarily compares the words following search with the keywords for the entry. If you specify the author option, however, it compares the words with the author's name. In the search database, we have filled in author names for *Stata Journal* and STB articles and for FAQs.

For instance, in [R] **kdensity**, you will discover that Isaías H. Salgado-Ugarte wrote the first version of Stata's kdensity command and published it in the STB. Assume that you have read his original and find the discussion useful. You might now wonder what else he has written in the STB. To find out, type

```
. search Salgado-Ugarte, author
  (output omitted )
```

Names like Salgado-Ugarte are confusing to some people. search does not require you specify the entire name; what you type is compared with each "word" of the name, and, if any part matches, the entry is listed. The hyphen is a special character, and you can omit it. Thus, you can obtain the same list by looking up Salgado, Ugarte, or Salgado Ugarte without the hyphen.

Actually, to find all entries written by Salgado-Ugarte, you need to type

```
. search Salgado-Ugarte, author historical
  (output omitted )
```

Prior inserts in the STB that provide a feature that later was superseded by a built-in feature of Stata are marked as historical in the search database and, by default, are not listed. The `historical` option ensures that all entries are listed.

4.8.3 Entry ID searches

If you specify the `entry` option, `search` compares what you have typed with the entry ID. The entry ID is not the title—it is the reference listed to the left of the title that tells you where to look. For instance, in

```
[R]    regress . . . . . . . . . . . . . . . . . . . . . . . Linear regression
       (help regress)
```

"[R] regress" is the entry ID. In

```
GS         . . . . . . . . . . . . . . . . . . . . . . . . . Getting Started manual
```

"GS" is the entry ID. In

```
SJ-6-4   st0113  . Testing for cross-sectional dependence in panel-data models
         (help xtcsd if installed) . . . . . . R. E. De Hoyos and V. Sarafidis
         Q4/06   SJ 6(4): 482--496
         tests for the presence of cross-sectional dependence in
         panels with many cross-sectional units and few time-series
         observations
```

"SJ-6-4 st0113" is the entry ID.

`search` with the `entry` option searches these entry IDs.

Thus, you could generate a table of contents for the *Reference* manuals by typing

```
. search [R], entry
(output omitted)
```

You could generate a table of contents for the 16th issue of the STB by typing

```
. search STB-16, entry historical
(output omitted)
```

The `historical` option here is possibly important. STB-16 was published in November 1993, and perhaps some of its inserts have been marked as historical.

You could obtain a list of all inserts associated with *dm36* by typing

```
. search dm36, entry historical
(output omitted)
```

Again, we include the `historical` option in case any of the relevant inserts have been marked historical.

4.8.4 FAQ searches

To search across the FAQs, specify the `faq` option:

```
. search logistic regression, faq
(output omitted)
```

4.8.5 Return codes

In addition to indexing the entries in the *User's Guide* and all the *Stata Reference* manuals, `search` also can be used to look up return codes.

To see information about return code 131, type

```
. search rc 131
```
```
[R]      error messages  . . . . . . . . . . . . . . . . . . . .  Return code 131
         not possible with test;
         You requested a test of a hypothesis that is nonlinear in the
         variables.  test tests only linear hypotheses.  Use testnl.
```

To get a list of all Stata return codes, type

```
. search rc
```
       (*output omitted*)

4.9 net search: Searching net resources

When you select **Help > Search**, there are two types of searches to choose. The first, which has been discussed in the previous sections, is to **Search documentation and FAQs**. The second is to **Search net resources**. This feature of Stata searches resources over the Internet.

When you choose **Search net resources** in the search dialog box and enter *keywords* in the field, Stata searches all user-written programs on the Internet, including user-written additions published in the *Stata Journal* and the STB. The results are displayed in the Viewer, and you can click to go to any of the matches found.

Equivalently, you can type `net search` *keywords* on the Stata command line to display the results in the Results window. For the full syntax for using the `net search` command, see [R] **net search**.

4.10 hsearch: An alternative to search

`hsearch` is available from command mode, and by clicking on `hsearch` from **Help > Advice**. Just like you can type `search` *anything*, you can type `hsearch` *anything*. Whereas `search` bases its results on a file of keywords for official Stata commands, official FAQs, and the *Stata Journal*, `hsearch` searches the text of the help files on your computer. `hsearch` searches all the help files, whether official or not, i.e., whether the help files are part of Stata as supplied by StataCorp or from user-written additions you have installed or from help files you wrote yourself.

`hsearch` shows what is installed on your computer.

`search` shows what is official and includes FAQs and *Stata Journal* contributions.

5 Flavors of Stata

Contents

5.1   Platforms

Stata is available for a variety of computers, including

Stata for Windows, 32-bit x86
Stata for Windows, 64-bit x86-64
Stata for Windows, 64-bit Itanium

Stata for Macintosh, Intel
Stata for Macintosh, PowerPC

Stata for AIX, 64-bit
Stata for HP-UX, 64-bit Itanium
Stata for Linux, 32-bit x86
Stata for Linux, 64-bit x86-64
Stata for Linux, 64-bit Itanium
Stata for Solaris, 64-bit Sparc
Stata for Solaris, 64-bit x86-64

Which version of Stata you run does not matter—Stata is Stata. You instruct Stata in the same way and Stata produces the same results, right down to the random-number generator. Even files can be shared. A dataset created on one computer can be used on any other computer, and the same goes for graphs, programs, or any file Stata uses or produces. Moving files across platforms is simply a matter of copying them; no translation is required.

Some computers, however, are faster than others. Some computers have more memory than others. Computers with more memory, and faster computers, are better.

The list above includes both 32- and 64-bit computers. Sixty-four bit Stata runs faster than 32-bit Stata and 64-bit Stata will allow processing data in excess of 2 gigabytes, assuming you have enough memory. Thirty-two bit Stata will run on 64-bit hardware.

5.2   Stata/MP, Stata/SE, Stata/IC, and Small Stata

Stata is available in four flavors, although perhaps sizes would be a better word. The flavors are, from largest to smallest, Stata/MP, Stata/SE, Stata/IC, and Small Stata. Small Stata is not available under Unix.

Stata/MP is the multiprocessor version of Stata. It runs on multiple CPUs or on multiple cores, from two to 32. Stata/MP uses however many CPUs or cores you tell it to use (even one). Stata/MP is the fastest version of Stata. Even so, all the details of parallelization are handled internally and you use Stata/MP just like you use any of the other flavors of Stata.

Stata/SE is like Stata/MP, but for single CPUs. Stata/SE will run on multiple CPUs or multiple-core computers, but it will use only one of the CPUs or cores. SE stands for special edition.

Both Stata/MP and Stata/SE have the same limits and the same capabilities and are intended for those who work with large datasets. You may have up to 32,767 variables with either. Statistical models may have up to 11,000 variables.

Stata/IC is standard Stata. IC stands for intercooled, a name used before the introduction of Stata/SE and later of Stata/MP to indicate that it was the top version of Stata. Up to 2,047 variables are allowed. Statistical models may have up to 800 variables.

Stata/MP, Stata/SE, and Stata/IC all allow up to 2,147,583,647 observations, assuming you have enough memory.

Small Stata is intended for students and limited to 99 variables and 1,000 observations.

5.2.1   Determining which version you own

Check your license. Included with every copy of Stata is a paper license that contains codes that you will input during installation. This license determines which flavor of Stata you have and for which platform.

Contact us or your distributor if you want to upgrade from one flavor to another or switch from one platform to another. Usually, all you need is an upgraded paper license with the appropriate codes. All flavors of Stata for a platform are on the same CD.

If you purchased one flavor of Stata and want to use a lesser version, you may. You might want to do this if you had a large computer at work and a smaller one at home. Please remember, however, that you have only one license (or however many licenses you purchased). You may, both legally and ethically, install on both and then use one or the other, but you should not use them both simultaneously.

5.2.2   Determining which version is installed

If Stata is already installed, you can find out which Stata you are using by entering Stata as you normally do and typing about:

```
. about

Stata/IC 10.0 for Windows
Born date
Copyright (C) 1985-2007

10-user Windows (network) perpetual license:
      Serial number:  1910040000
        Licensed to:  Marsha Martinez
                      StataCorp
```

5.3   Size limits of Stata/MP, SE, IC, and Small Stata

Here are some of the different size limits for Stata/MP, Stata/SE, Stata/IC, and Small Stata. Type help limits for a longer list.

Maximum size limits for Stata/MP, Stata/SE, Stata/IC, and Small Stata

	Stata/MP and SE	Stata/IC	Small Stata
Number of observations	limited only by memory	limited only by memory	fixed at approx. 1,000
Number of variables	32,767	2,047	fixed at 99
Width of a dataset	393,192	24,564	200
Maximum matrix size (matsize)	11,000	800	fixed at 40
Number of characters in a macro	1,081,511	165,200	8,681
Number of characters in a command	1,081,527	165,216	8,697

Stata/MP and Stata/SE allow more variables, larger matrices, longer macros, and a longer command line than Stata/IC. The longer command line and macro length are required because of the greater number of variables allowed. The larger matrix size means that Stata/MP and Stata/SE can fit statistical models with more independent variables.

Small Stata is limited. It is intended for student use and often used in undergraduate labs.

5.4   Speed comparison of Stata/MP, SE, IC, and Small Stata

We have written a 190-page white paper comparing the performance of Stata/MP with Stata/SE, see http://www.stata.com/statamp/report.pdf. The white paper includes command-by-command performance measurements.

In summary, on a 2-CPU or dual-core computer, Stata/MP will run commands in 71% of the time required by Stata/SE. There is variation; some commands run in half the time and others are not sped up at all. Statistical estimation commands run in 59% of the time. Numbers quoted are medians. Average performance gains are higher because commands that take longer to execute are generally sped up more.

Stata/MP running on four CPUs runs in 50% (all commands) and 36% (estimation commands) of the time required by Stata/SE. Both numbers are median measures.

Stata/MP supports up to 32 CPUs.

Stata/IC is slower than Stata/SE, but those differences emerge only when processing datasets that are pushing the limits of Stata/IC. Stata/SE has a larger memory footprint and uses that extra memory for larger look-aside tables to more efficiently process large datasets. The real benefits of the larger tables become apparent only after exceeding the limits of Stata/IC. Stata/SE was designed for processing large datasets.

Small Stata is, by comparison to all of the above, slow, but given its limits, no one notices. Small Stata was designed to have a minimal memory footprint, and to achieve that, different logic is sometimes used. For instance, in Stata's test command, it must compute the matrix calculation $\mathbf{RZR}'$ (where $\mathbf{Z} = (\mathbf{X}'\mathbf{X})^{-1}$). Stata/MP, Stata/SE, and Stata/IC make the calculation in a straightforward way, which is to form $\mathbf{T} = \mathbf{RZ}$ and then calculate $\mathbf{TR}'$. This requires temporarily storing the matrix $\mathbf{T}$. Small Stata, on the other hand, goes into more complicated code to form the result directly—code that requires temporary storage of only one scalar. This code, in effect, recalculates intermediate results over and over again, and so it is slower. Also, because Small Stata is designed to work with small datasets, it uses different memory-management routines. These memory-management routines use 2-byte rather than 4-byte offsets and therefore require only half the memory to track locations.

In all cases, the differences are all technical and internal. From the user's point of view, Stata/MP, Stata/SE, Stata/IC, and Small Stata work the same way.

5.5 Feature comparison of Stata/MP, SE, and IC

The features of all flavors of Stata on all platforms are the same. The differences are in speed and in limits. The differences in limits are

Parameter	Stata/IC			Stata/SE and /MP			
	Default	min	max	Default	min	max	
maxvar	2,047	2,047	2,047	5,000	2,047	**32,767**	
matsize	200	10	800	400	10	**11,000**	
memory	1M	500K	...	10M	500K	...	
processors	1	1	1	1	1	1	/SE
				2	1	**32**	/MP

NOTE: the default number of processors for Stata/MP is the minimum of processors licensed and processors available.

The limits on Stata/MP and /SE are settable. You reset the limits temporarily by typing

```
. set maxvar #
. set matsize #
. set memory #                    (where # may be #g, #m, or #k)
. set processors #
```

About the last, Stata/MP users sometimes want to use fewer processors to leave some free for other applications.

You reset the limits permanently by typing

```
. set maxvar #, permanently
. set matsize #, permanently
. set memory #, permanently       (where # may be #g, #m, or #k)
```

If you use Stata/SE or Stata/MP, type

```
. help stata/se
```

or

```
. help stata/mp
```

to learn more.

6 Setting the size of memory

6.1   Memory size considerations

Stata works with a copy of the data that it loads into memory.

By default, Stata/MP and Stata/SE allocates 10 megabytes and Stata/IC allocates 1 megabyte to Stata's data areas, and you can change it.

By default, Small Stata allocates about 300 K to Stata's data areas, and you cannot change it.

You can even change the allocation to be larger than the physical amount of memory on your computer because Windows, Macintosh, and Unix systems provide virtual memory.

Virtual memory is slow but adequate in rare cases when you have a dataset that is too large to load into real memory. If you use large datasets often, we recommend that you add more memory to your computer.

One way to change the allocation is when you start Stata. Instructions for doing this are provided in

> Windows   [GSW] **C.4 Specifying the amount of memory allocated**
> Macintosh   [GSM] **C.2 Specifying the amount of memory allocated**
> Unix   [GSU] **C.2 Specifying the amount of memory allocated**

Also, if you use Stata/MP, Stata/SE, or Stata/IC for Windows, Unix, or Macintosh, you can change the total amount of memory allocated while Stata is running. That is the topic of this chapter.

It does not matter which method you use. Being able to change the total on the fly is convenient, but even if you cannot do this, it just means that you specify it ahead of time. If later you need more, simply exit Stata and reinvoke it with the larger total.

6.2   Setting memory size on the fly: Stata/MP and Stata/SE

There are three limits in Stata/MP and Stata/SE that affect memory allocation and usage. The three limits are

1. maxvar, the maximum number of variables allowed in a dataset. This limit is initially set to 5,000; you can increase it up to 32,767.

2. `matsize`, the largest dimension of a matrix: this limit is initially set to 400, and you can increase it up to 11,000. This usually relates to the maximum number of independent variables allowed in the models that you fit, and thus, the dimension of the estimated variance–covariance matrix. However, in some panel-data models, covariance or correlation matrices must be fitted, and their dimensions depend on either the number of panels (groups) in your data or the number of observations in your dataset. This limit does not apply to Mata matrices; see the *Mata Reference Manual*.

3. `memory`, the amount of memory Stata requests from the operating system to store your data. This limit is initially set to 10 megabytes in Stata/MP and Stata/SE. You may set it to as large a number as your operating system will allow.

You set the limits using the

$$\text{set maxvar } \# \qquad \left[, \underline{\text{permanently}}\right]$$

$$\text{set matsize } \# \qquad \left[, \underline{\text{permanently}}\right]$$

$$\text{set memory } \#\left[\text{b}|\text{k}|\text{m}|\text{g}\right] \left[, \underline{\text{permanently}}\right]$$

commands. For instance, you might type

```
. set maxvar  5000
. set matsize 900
. set memory  50m
```

The order in which you set the limits does not matter. If you specify the `permanently` option when you set a limit, in addition to making the change right now, Stata will remember the new limit and use it in the future when you invoke Stata:

```
. set maxvar  5000, permanently
. set matsize 900, permanently
. set memory  50m, permanently
```

You can reset the current or permanent limits whenever and as often as you wish.

6.2.1   Advice on setting maxvar

$$\text{set maxvar } \# \left[, \underline{\text{permanently}}\right]$$

where $2{,}048 \leq \# \leq 32{,}767$

Why is there a limit on `maxvar`? Why not just set `maxvar` to 32,767 and be done with it? Because simply allowing room for variables, even if they do not exist, causes Stata to consume memory and, if you will be using only datasets with a lot fewer variables, you will be wasting memory.

The formula for the amount of memory consumed by `set maxvar` is approximately

$$\text{megs} = .3147 * (\text{maxvar}/1000) + .002$$

For instance, if you set `maxvar` to 20,000, the memory would be approximately

$$\text{megs} = .3147 * 20 + .002 = 6.296 \text{ megs}$$

and if you left it at the default, the memory use would be roughly

$$megs = .3147 * 5 + .002 = 1.575 \text{ megs}$$

Thus, how big you set `maxvar` does not dramatically affect memory usage. Still, at `maxvar`=32,000, memory use is 10.072 MB.

Recommendation: think about datasets with the most variables that you typically use. `set maxvar` to a few hundred or even 1,000 above that. (The memory cost of an extra 1,000 variables is only .315 megs.)

Remember, you can always reset `maxvar` temporarily by typing `set maxvar`.

❏ Technical Note

The formula above is only approximate, and the formula given is the formula appropriate for 32-bit computers. When you `set maxvar`, Stata/MP and Stata/SE will give you a memory report showing the exact amount of memory used:

```
. set maxvar 10000
```

Current memory allocation

settable	current value	description	memory usage (1M = 1024k)
set maxvar	10000	max. variables allowed	3.149M
set memory	10M	max. data space	10.000M
set matsize	400	max. RHS vars in models	1.254M
			14.403M

❏

6.2.2 Advice on setting matsize

`set matsize #` [`, permanently`]

where $10 \leq \# \leq 11{,}000$

Although `matsize` can theoretically be set up to 11,000, on all but 64-bit computers, you will be unable to do that, and, even if you succeeded, Stata/SE would probably run out of memory subsequently. The value of `matsize` has a dramatic effect on memory usage, the formula being

$$megs = (8 * \mathtt{matsize}^2 + 88 * \mathtt{matsize})/(1024^2)$$

This formula is valid across all computers, 32-bit and 64-bit. For instance, the above formula states

matsize	memory use
400	1.254M
800	4.950M
1,600	19.666M
3,200	78.394M
6,400	313.037M
11,000	924.080M

The formula, in fact, is an understatement of the amount of memory certain Stata commands use and is an understatement of what you will certainly use yourself if you use matrices directly. The formula gives the amount of memory required for one matrix and 11 vectors. If two matrices are required, the numbers above are nearly doubled. When you `set matsize`, if you specify too large a value, Stata will refuse, but remember that just because Stata does not complain, you still may run into problems later. What might happen is that Stata could be running a command and then complain, "op. sys. refuses to provide memory"; r(909).

For `matsize` = 11,000, nearly 1 gigabyte of memory is required, and doubling that would require nearly 2 gigabytes of memory. On most 32-bit computers, 2 gigabytes is the maximum amount of memory the operating system will allocate to one task, so nearly nothing would be left for all the rest of Stata.

Why, then, is `matsize` allowed to be set so large? Because on 64-bit computers, such large amounts cause no difficulty.

For "reasonable" values of matsize (say, up to 3,200), memory consumption is not too great. Choose a reasonable value given the kinds of models you fit, and remember that you can always reset the value.

6.2.3   Advice on setting memory

set memory #[b|k|m|g] [, permanently]

where # $\geq$ 500k and (b, k, m, and g may be typed in uppercase.)

The advice for setting memory is the same as for Stata/IC: set enough so that your datasets fit easily, and do not set so much that you exceed physical memory present on your computer, except in emergencies.

You may set memory in bytes (b), kilobytes (k), megabytes (m), or gigabytes (g), but the number specified must be an integer, so if you want to set 1.5g, you set 1500m. Actually, 1.5g is 1536m because the formulas for a kilobyte, megabyte, and gigabyte are

```
1024 bytes     = 1 kilobyte
1024 kilobytes = 1 megabyte
1024 megabytes = 1 gigabyte
```

This detail does not matter, but this is the rule that Stata uses when presenting numbers, so do not be surprised when 2000k is not displayed as 2M, or 2M is displayed as 2048k.

If you have a large 32-bit computer, the maximum amount of memory you can set may surprise you. Many people think that 32-bit computers can allow up to 4 gigabytes of memory and that, in a sense, is true. Some 32-bit computers will even allow you to install 4 gigabytes of physical memory. Nevertheless, most modern operating systems will allocate a maximum of one-half the theoretical maximum to individual tasks; i.e., most operating systems will allow Stata only 2 gigabytes of memory (even if they have 4 gigabytes of memory)!

The same one-half rule applies to 64-bit computers, but half the theoretical limit is still 536,870,912 gigabytes, so no one much cares.

This one-half limit is imposed by the operating system, not by Stata, and the operating system developers have good technical reasons for imposing the rule.

6.3 Setting memory size on the fly: Stata/IC

You can reallocate memory on the fly. Assume you have changed nothing about how Stata starts, so you get the default 1 megabyte of memory allocated to Stata's data areas. You are working with a large dataset and now wish to increase it to 32 megabytes. You can type

```
. set memory 32m
(32768k)
```

and, if your operating system can provide the memory to Stata, Stata will work with the new total. Later in the session, if you want to release that memory and work with only 2 megabytes, you could type

```
. set memory 2m
(2048k)
```

There is only one restriction on the set memory command: whenever you change the total, there cannot be any data already in memory. If you have a dataset in memory, save it, clear memory, reset the total, and then use it again. We are getting ahead of ourselves, but you might type

```
. save mydata, replace
file mydata.dta saved
. clear
. set memory 32m
(32768k)
. use mydata
```

When you request the new allocation, your operating system might refuse to provide it:

```
. set memory 128m
op. sys. refuses to provide memory
r(909);
```

If that happens, you are going to have to take the matter up with your operating system. In the above example, Stata asked for 128 megabytes and the operating system said no.

For most 32-bit computers, the absolute maximum amount of memory that can theoretically be allocated will be approximately $2^{(32-1)}$ bytes (2 gigabytes), regardless of the operating system. In practice, the amount of memory available to any one application on a system is affected by many factors and may be somewhat less than the theoretical maximum.

Sixty-four bit computers can theoretically have up to $2^{(64-1)}$ bytes (more than 2 billion gigabytes). In practice, 64-bit computers will have as much memory as is affordable. We are aware of sites using Stata with datasets consuming more than 10 gigabytes of memory.

6.4 The memory command

memory helps you figure out whether you have sufficient memory to do something.

(Continued on next page)

```
. use http://www.stata-press.com/data/r10/regsmpl
(NLS Women 14-26 in 1968)

. memory
```

	bytes	
Details of set memory usage		
overhead (pointers)	114,136	10.88%
data	913,088	87.08%
data + overhead	1,027,224	97.96%
free	21,344	2.04%
Total allocated	1,048,568	100.00%
Other memory usage		
system overhead	677,289	
set matsize usage	16,320	
programs, saved results, etc.	505	
Total	694,114	
Grand total	1,742,682	

Having 21,344 bytes free is not much. You might increase the amount of memory allocated to Stata's data areas by specifying set memory 2m.

```
. save regsmpl
file regsmpl.dta saved

. clear

. set memory 2m
(2048k)

. use regsmpl
(NLS Women 14-26 in 1968)

. memory
```

	bytes	
Details of set memory usage		
overhead (pointers)	114,136	5.44%
data	913,088	43.54%
data + overhead	1,027,224	48.98%
free	1,069,920	51.02%
Total allocated	2,097,144	100.00%
Other memory usage		
system overhead	677,289	
set matsize usage	16,320	
programs, saved results, etc.	667	
Total	694,276	
Grand total	2,791,420	

More than 1 MB free: that's better. See [D] **memory** for more information.

6.5   Virtual memory and speed considerations

When you use more memory than is physically available on your computer, Stata slows down. If you are only using a little more memory than on your computer, performance is probably not too bad. On the other hand, when you are using a lot more memory than is on your computer, performance will be noticeably affected. In these cases, we recommend that you

```
. set virtual on
```

Virtual memory systems exploit locality of reference, which means that keeping objects closer together allows virtual memory systems to run faster. set virtual controls whether Stata should perform extra work to arrange its memory to keep objects close together. By default, virtual is set off. set virtual can be used with Stata/MP, Stata/SE, and Stata/IC on all supported operating systems.

In general, you want to leave set virtual set to the default of off so that Stata will run faster.

When you set virtual on, you are asking Stata to arrange its memory so that objects are kept closer together. This requires Stata to do a substantial amount of work. We recommend setting virtual on only when the amount of memory in use drastically exceeds what is physically available. In these cases, setting virtual on will help, but performance will still be slow. If you are using virtual memory often, you should consider adding memory to your computer.

6.6   An issue when returning memory to Unix

There is a surprising issue of returning memory that Unix users need to understand. Let's say that you set memory to 128 megabytes, went along for a while, and then, being a good citizen, returned most of it:

```
. set memory 2m
(2048k)
```

Theoretically, 126 megabytes just got returned to the operating system for use by other processes. If you use Windows, that is exactly what happens and, with some Unix systems, that is what happens, too.

Other Unix systems, however, are strange about returned memory in a misguided effort to be efficient: they do not really take the memory back. Instead, they leave it allocated to you in case you ask for it back later. Still other Unix systems sort of take the memory back: they put it in a queue for your use, but, if you do not ask for it back in 5 or 10 minutes, then they return it to the real system pool!

The unfortunate situation is that we at Stata cannot force the operating system to take the memory back. Stata returns the memory to Unix, and then Unix does whatever it wants with it.

So, let's review: you make your Stata smaller in an effort to be a good citizen. You return the memory so that other users can use it, or perhaps so you can use it with some other software.

If you use Windows, the memory really is returned and all works exactly as you anticipated.

If you use Unix, it might go back immediately, it might go back in 5 or 10 minutes, or it might never go back. In the last case, the only way to really return the memory is to exit Stata. All Unix systems agree on that: when a process ends, the memory really does go back into the pool.

To find out how your Unix works, you need to experiment. We would publish a table and just tell you, but we have found that within manufacturer the way their Unix works will vary by subrelease! The experiment is tedious but not difficult:

1. Bring up a Stata and make it really big; use a lot of memory, so much that you are virtually hogging the computer.

2. Go to another window or session and bring up another Stata. Verify that you cannot make it big—that you get the "system limit exceeded" message.

3. Go back to the first Stata, leaving the second running, and make it smaller.

4. Go to the second Stata and try again to make it big. If you succeed, then your Unix returns memory instantly.

5. If you still get the "system limit exceeded" message, wait 5 minutes and try again. If it now works, your system delays accepting returned memory for about 5 minutes.

6. If you still get the "system limit exceeded" message, wait another 5 minutes and try again. If it now works, your system delays accepting returned memory for about 10 minutes.

7. Go to the first Stata and exit from it.

8. Go to the second Stata and try to make it big again. If it now works, your system never really accepts returned memory. If it still does not work, start all over again. Some other process took memory and corrupted your experiment.

If you are one of the unfortunates who have a Unix that never accepts returned memory, you will just have to remember that you must exit and reenter Stata to really give memory back.

7   –more– conditions

Contents

7.1   Description

When you see —more— at the bottom of the screen,

Press ...	and Stata...
letter *l* or *Enter*	displays the next line
letter *q*	acts as if you pressed *Break*
space bar or any other key	displays the next screen

Also, you can press the *clear –more– condition* button, the button labeled **Go** with a circle around it.

—more— is Stata's way of telling you that it has something more to show you, but showing you that something more will cause the information on the screen to scroll off.

7.2   set more off

If you type `set more off`, —more— conditions will never arise and Stata's output will scroll by at full speed.

If you type `set more on`, —more— conditions will be restored at the appropriate places.

Programmers: do-file writers sometimes include `set more off` in their do-files because they do not care to interactively watch the output. They want Stata to proceed at full speed because they plan on making a log of the output that they will review later. Do-filers need not bother to `set more on` at the conclusion of their do-file. Stata automatically restores the previous `set more` when the do-file (or program) concludes.

7.3   The more programming command

Ado-file programmers need take no special action to have —more— conditions arise when the screen is full. Stata handles that automatically.

If, however, you wish to force a —more— condition early, you can include the `more` command in your program. The syntax of `more` is

```
more
```

`more` takes no arguments.

For more information, see [P] **more**.

8 Error messages and return codes

Contents

8.1 Making mistakes

When an error occurs, Stata produces an error message and a *return code*. For instance,

```
. list myvar
no variables defined
r(111);
```

We ask Stata to list the variable named myvar. Since we have no data in memory, Stata responds with the message "no variables defined" and a line that reads "r(111)".

The "no variables defined" is called the error message.

The 111 is called the return code. You can click on blue return codes to get a detailed explanation of the error.

8.1.1 Mistakes are forgiven

After "no variables defined" and r(111), all is forgiven; it is as if the error never occurred.

Typically, the message will be enough to guide you to a solution, but if it is not, the numeric return codes are documented in [P] **error**.

8.1.2 Mistakes stop user-written programs and do-files

Whenever an error occurs in a user-written program or do-file, the program or do-file immediately stops execution and the error message and return code are displayed.

For instance, consider the following do-file:

```
───────────────────────────────────────── top of myfile.do ──────────
    use http://www.stata-press.com/data/r10/auto
    decribe
    list
───────────────────────────────────────── end of myfile.do ──────────
```

Note the second line—you meant to type decribe but typed decribe. Here is what happens when you execute this do-file by typing do myfile:

```
. do myfile
. use http://www.stata-press.com/data/r10/auto
(1978 Automobile Data)
```

```
. decribe
unrecognized command:  decribe
r(199);

end of do-file
r(199);

. _
```

The first error message and return code were caused by the illegal `decribe`. This then caused the do-file itself to be aborted; the valid `list` command was never executed.

8.1.3   Advanced programming to tolerate errors

Errors are not only of the typographical kind; some are substantive. A command that is valid in one dataset might not be valid in another. Moreover, in advanced programming, errors are sometimes anticipated: use one dataset if it is there, but use another if you must.

Programmers can access the return code to determine whether an error occurred, which they can then ignore, or, by examining the return code, code their programs to take the appropriate action. This is discussed in [P] **capture**.

You can also prevent do-files from stopping when errors occur by using the `do` command's `nostop` option.

```
. do myfile, nostop
```

8.2   The return message for obtaining command timings

In addition to error messages and return codes, there is something called a return message, which you normally do not see. Normally, if you typed `summarize tempjan`, you would see

```
. use http://www.stata-press.com/data/r10/citytemp
(City Temperature Data)

. summarize tempjan
```

Variable	Obs	Mean	Std. Dev.	Min	Max
tempjan	954	35.74895	14.18813	2.2	72.6

If you were to type

```
. set rmsg on
r; t=0.00 10:21:22
```

sometime during your session, Stata would display return messages:

```
. summarize tempjan
```

Variable	Obs	Mean	Std. Dev.	Min	Max
tempjan	954	35.74895	14.18813	2.2	72.6
```
r; t=0.01 10:21:26
```

The line that reads `r; t=0.01 10:21:26` is called the return message.

The `r;` indicates that Stata successfully completed the command.

The `t=0.01` shows the amount of time, in seconds, it took Stata to perform the command (timed from the point you pressed *Return* to the time Stata typed the message). This command took a hundredth of a second. Stata also shows the time of day with a 24-hour clock. This command completed at 10:21 a.m.

Stata can run commands stored in files (called do-files) and can log output. Some users find the detailed return message helpful with do-files. They construct a long program and let it run overnight, logging the output. They come back the next morning, look at the output, and discover a mistake in some portion of the job. They can look at the return messages to determine how long it will take to rerun that portion of the program.

You may `set rmsg on` whenever you wish.

When you want Stata to stop displaying the detailed return message, type `set rmsg off`.

9 The Break key

Contents

9.1   Making Stata stop what it is doing

When you want to make Stata stop what it is doing and return to the Stata dot prompt, you press *Break*:

Stata for Windows:	click the **Break** button (it is the button with the big red X), or hold down *Ctrl* and press the *Pause/Break* key
Stata for Macintosh:	click the **Break** button or hold down *Command* and press period
Stata for Unix(GUI):	click the **Break** button or hold down *Ctrl* and press k
Stata for Unix(console):	hold down *Ctrl* and press c or press q

Elsewhere in this manual, we describe this action as simply pressing *Break*. Break tells Stata to cancel what it is doing and return control to you as soon as possible.

If you press *Break* in response to the input prompt or while you are typing a line, Stata ignores it, since you are already in control.

If you press *Break* while Stata is doing something—creating a new variable, sorting a dataset, making a graph, etc.—Stata stops what it is doing, undoes it, and issues an input prompt. The state of the system is the same as if you had never issued the command.

▷ Example 1

You are fitting a logit model, type the command, and, as Stata is working on the problem, realize that you omitted an important variable:

```
. logit foreign mpg weight
Iteration 0:   log likelihood =  -45.03321
Iteration 1:   log likelihood = -29.898968
—Break—
r(1);

. _
```

When you pressed *Break*, Stata responded by typing —Break— and then typing r(1);. Pressing *Break* always results in a return code of 1—that is why return codes are called return codes and not error codes. The 1 does not indicate an error, but it does indicate that the command did not complete its task.

◁

9.2   Side effects of pressing Break

In general, there are none. We said above that Stata undoes what it is doing so that the state of the system is the same as if you had never issued the command. There are two exceptions to this.

If you are reading data from disk by using `insheet`, `infile`, or `infix`, whatever data have already been read will be left behind in memory, the theory being that perhaps you stopped the process so you could verify that you were reading the right data correctly before sitting through the whole process. If not, you can always `drop _all`.

```
. infile v1-v9 using workdata
(eof not at end of obs)
(4 observations read)
—Break—
r(1);
```

The other exception is `sort`. You have a large dataset in memory, decide to sort it, and then change your mind.

```
. sort price
—Break—
r(1);
```

If the dataset was previously sorted by, say, the variable `prodid`, it is no longer. When you press *Break* in the middle of a `sort`, Stata marks the data as unsorted.

9.3   Programming considerations

There are basically no programming considerations for handling Break because Stata handles it all automatically. If you write a program or do-file, execute it, and then press *Break*, Stata stops execution just as it would with an internal command.

Advanced programmers may be concerned about cleaning up after themselves; perhaps they have generated a temporary variable they intended to drop later or a temporary file they intended to erase later. If a Stata user presses *Break*, how can you ensure that these temporary variables and files will be erased?

If you obtain names for such temporary items from Stata's `tempname`, `tempvar`, and `tempfile` commands, Stata will automatically erase the temporary items; see [U] **18.7 Temporary objects**.

There are instances, however, when a program must commit to executing a group of commands without interruption, or the user's data would be left in an intermediate or undefined state. In these instances, Stata provides a

```
        nobreak {
                ...
        }
```

construct; see [P] **break**.

10 Keyboard use

Contents

10.1 Description

The keyboard should operate much the way you would expect, with a few additions:

1. There are some unexpected keys you can press to obtain previous commands you have typed. Also, you can click once on a command in the Review window to reload it, or click it twice to reload and execute; this feature is discussed in the *Getting Started* manuals.

2. There are a host of command-editing features for Stata for Unix(console) users since their user interface does not offer such features.

3. Regardless of operating system or user interface, if there are *F*-keys on your keyboard, they have special meaning and you can change the definitions of the keys.

10.2 F-keys

Windows users: *F10* is reserved internally by Windows; you cannot program this key.

By default, Stata defines the *F*-keys to mean

F-key	Definition
F1	help
F2	#review;
F3	describe;
F7	save
F8	use

The semicolons at the end of some entries indicate an implied *Return*.

`help` shows a Stata help file—you use it by typing `help` followed by the name of a Stata command; see [U] **4 Stata's online help and search facilities**. You can type out `help` or you can press *F1*, type the Stata command, and press *Return*.

`#review` is the command to show the last few commands you issued. It is described in [U] **10.5 Editing previous lines in Stata**. Rather than typing out `#review` and pressing *Return*, you can simply press *F2*. You do not press *Return* following *F2* because the definition of F2 ends in a semicolon—Stata presses the *Return* key for you.

`describe` is the Stata command to report the contents of data loaded into memory. It is explained in [D] **describe**. Normally, you type `describe` and press *Return*. You can also press *F3*.

save is the command to save the data in memory into a file, and use is the command to load data; see [D] **use** and [D] **save**. The syntax of each is the same: save or use followed by a filename. You can type out the commands or you can press *F7* or *F8* followed by the filename.

You can change the definitions of the *F*-keys. For instance, the command to list data is list; you can read about it in [D] **list**. The syntax is list to list all the data, or list followed by the names of some variables to list just those variables (there are other possibilities).

If you wanted *F3* to mean list, you could type

    . global F3 "list "

In the above, F3 refers to the letters *F* followed by *3*, not the *F3* key. Note the capitalization and spacing of the command.

You type global in lowercase, type out the letters F3, and then type "list ". The space at the end of list is important. In the future, rather than typing out list mpg weight, you want to be able to press the *F3* key and then type only mpg weight. You put a space in the definition of F3 so that you would not have to type a space in front of the first variable name after pressing *F3*.

Now say you wanted *F5* to mean list all the data—list followed by *Return*. You could define

    . global F5 "list;"

Now you would have two ways of listing all the data: (1) press *F3*, and then press *Return*, or (2) press *F5*. The semicolon at the end of the definition of *F5* will press *Return* for you.

If you really want to change the definitions of *F3* and *F5*, you will probably want to change the definition every time you invoke Stata. One way would be to type out the two global commands every time you invoked Stata. Another way would be to type the two commands into an ASCII text file called profile.do. Stata executes the commands in profile.do every time it is launched if profile.do is placed in the appropriate directory:

Windows:	put profile.do in the "start-in" directory;
	see [GSW] **C.3 Executing commands every time Stata is started**
Macintosh:	put profile.do in your home directory;
	see [GSM] **C.1 Executing commands every time Stata is started**
Unix:	put profile.do someplace along your shell's PATH;
	see [GSU] **C.1 Executing commands every time Stata is started**

You can use the *F*-keys any way you desire: they contain a string of characters, and pressing the *F*-key is equivalent to typing out those characters.

❑ Technical Note

[*Stata for Unix(console) users.*] Sometimes Unix assigns a special meaning to the *F*-keys, and if it does, those meanings supersede our meanings. Stata provides a second way to get to the *F*-keys. Hold down *Ctrl*, press F, release the keys, and then press a number from 0 through 9. Stata interprets *Ctrl-F* plus 1 as equivalent to the *F1* key, *Ctrl-F* plus 2 as *F2*, and so on. *Ctrl-F* plus 0 means *F10*. These keys will work only if they are properly mapped in your termcap or terminfo entry.

❑

❏ Technical Note

On some international keyboards, the left single quote is used as an accent character. In this case, we recommend mapping this character to one of your function keys. In fact, you might find it convenient to map both the left single quote (`) and right single quote (') characters so that they are next to each other.

Within Stata, open the Do-file Editor. Type the following two lines in the Do-file Editor:

```
global F4 `
global F5 '
```

Save the file as `profile.do` into your Stata directory. If you already have a `profile.do` file, append the two lines to your existing `profile.do` file.

Exit Stata and restart it. You should see the startup message

running C:\Program Files\Stata10\profile.do ...

or some variant of it depending on where your Stata is installed. Press the function keys *F4* and *F5* to verify that they work.

If you did not see the startup message, you did not save the `profile.do` in your home folder.

You can of course map to any other function keys, but *F1*, *F2*, *F3*, *F7*, and *F8* are already used.

❏

10.3   Editing keys in Stata

Users have available to them the standard editing keys for their operating system. So, Stata should just edit what you type in the natural way—the Stata Command window is a standard edit window.

Also, you can fetch commands from the Review window into the Command window. Click on a command in the Review window, and it is loaded into the Command window, where you can edit it. Alternatively, if you double-click a line in the Review window, it is loaded and executed.

Another way to get lines from the Review window into the Command window is with the *PgUp* and *PgDn* keys. Tap *PgUp* and Stata loads the last command you typed into the Command window. Tap it again and Stata loads the line before that, and so on. *PgDn* goes the opposite direction.

Another editing key that may interest users is *Esc*. This key clears the Command window.

In summary,

Press	Result
PgUp	Steps back through commands and moves command from Review window to Command window
PgDn	Steps forward through commands and moves command from Review window to Command window
Esc	Clears Command window

(Continued on next page)

10.4   Editing keys in Stata for Unix(console)

Certain keys allow you to edit the line that you are typing. Since Stata supports a variety of computers and keyboards, the location and the names of the editing keys are not the same for all Stata users.

Every keyboard has the standard alphabet keys (*QWERTY* and so on), and every keyboard has a *Ctrl* key. Some keyboards have extra keys located to the right, above, or left, with names like *PgUp* and *PgDn*.

Throughout this manual we will refer to Stata's editing keys using names that appear on nobody's keyboard. For instance, PrevLine is one of the Stata editing keys—it retrieves a previous line. Hunt all you want, but you will not find it on your keyboard. So, where is PrevLine? We have tried to put it where you would naturally expect it. On keyboards with a key labeled *PgUp*, *PgUp* is the PrevLine key, but on everybody's keyboard, no matter which version of Unix, brand of keyboard, or anything else, *Ctrl-R* also means PrevLine.

When we say press PrevLine, now you know what we mean: press *PgUp* or *Ctrl-R*. The editing keys are the following:

Name for editing key	Editing key	Function
Kill	*Esc* on PCs and *Ctrl-U*	Deletes the line and lets you start over.
Dbs	*Backspace* on PCs and *Backspace*, *Rubout*, or *Delete* on other computers	Backs up and deletes one character.
Lft	←, *4* on the numeric keypad for PCs, and *Ctrl-H*	Moves the cursor left one character without deleting any characters.
Rgt	→, *6* on the numeric keypad for PCs, and *Ctrl-L*	Moves the cursor forward one character.
Up	↑, *8* on the numeric keypad for PCs, and *Ctrl-O*	Moves the cursor up one physical line on a line that takes more than one physical line. Also see PrevLine.
Dn	↓, *2* on the numeric keypad for PCs, and *Ctrl-N*	Moves the cursor down one physical line on a line that takes more than one physical line. Also see NextLine.
PrevLine	*PgUp* and *Ctrl-R*	Retrieves a previously typed line. You may press PrevLine multiple times to step back through previous commands.
NextLine	*PgDn* and *Ctrl-B*	The inverse of PrevLine.
Seek	*Ctrl-Home* on PCs and *Ctrl-W*	Goes to the line number specified. Before pressing Seek, type the line number. For instance, typing *3* and then pressing Seek is the same as pressing PrevLine three times.
Ins	*Ins* and *Ctrl-E*	Toggles insert mode. In insert mode, characters typed are inserted at the position of the cursor.
Del	*Del* and *Ctrl-D*	Deletes the character at the position of the cursor.
Home	*Home* and *Ctrl-K*	Moves the cursor to the start of the line.
End	*End* and *Ctrl-P*	Moves the cursor to the end of the line.
Hack	*Ctrl-End* on PCs, and *Ctrl-X*	Hacks off the line at the cursor.
Tab	⊣ on PCs, *Tab*, and *Ctrl-I*	Moves the cursor forward eight spaces.
Btab	⊢ on PCs, and *Ctrl-G*	The inverse of Tab.

▷ Example 1

It is difficult to demonstrate the use of editing keys on paper. You should try each of them. Nevertheless, here is an example:

. summarize price w̲aht

You typed `summarize price waht` and then tapped the *Lft* key (← key or *Ctrl-H*) three times to maneuver the cursor back to the a of `waht`. If you were to press *Return* right now, Stata would see the command `summarize price waht`, so where the cursor is does not matter when you press *Return*. If you wanted to execute the command `summarize price`, you could back up one more character and then press the Hack key. We will assume, however, that you meant to type `weight`.

If you were now to press the letter *e* on the keyboard, an e would appear on the screen to replace the a, and the cursor would move under the character h. We now have `weht`. You press *Ins*, putting Stata into insert mode, and press *i* and *g*. The line now says `summarize price weight`, which is correct, so you press *Return*. We did not have to press *Ins* before every character we wanted to insert. The *Ins* key is a toggle: If we press it again, Stata turns off insert mode, and what we type replaces what was there. When we press *Return*, Stata forgets all about insert mode, so we do not have to remember from one command to the next whether we are in insert mode.

◁

❑ Technical Note

Stata performs its editing magic from the information about your terminal recorded in `/etc/termcap`(5) or, under System V, `/usr/lib/terminfo`(4). If some feature does not appear to work, the entry for your terminal in the `termcap` file or `terminfo` directory is probably incorrect. Contact your system administrator.

❑

10.5   Editing previous lines in Stata

In addition to what is said below, remember that the Review window also shows the contents of the review buffer.

You may edit previously typed lines, or at least any of the last 25 or so lines. Stata records every line you type in a wraparound buffer. A wraparound buffer is a buffer of finite length in which the most recent thing you type replaces the oldest thing stored in the buffer. Stata's buffer is 8,697 characters long for Small Stata, 165,200 characters long for Stata/IC, and can range from 67,800 to 1,081,527 for Stata/MP and Stata/SE.

One way to retrieve lines is with the PrevLine and NextLine keys. Remember, PrevLine and NextLine are the names we attach to these keys—there are no such keys on your keyboard. You have to look back at the previous section to find out which keys correspond to PrevLine and NextLine on your computer. To save you the effort this time, PrevLine probably corresponds to *PgUp* and NextLine probably corresponds to *PgDn*.

Suppose you wanted to reissue the third line back. You could press PrevLine three times and then press *Return*. If you made a mistake and pressed PrevLine four times, you could press NextLine to go forward in the buffer. You do not have to count lines because, each time you press PrevLine or NextLine, the current line is displayed on your monitor. Simply tap the key until you find the line you want.

Another method for reviewing previous lines, #review, is convenient when you want to see the lines in context.

▷ Example 2

Typing #review by itself causes Stata to list the last five commands you typed. (You need not type out #review—pressing *F2* has the same effect.) For instance,

```
. #review
5 list make mpg weight if abs(res)>6
4 list make mpg weight if abs(res)>5
3 tabulate foreign if abs(res)>5
2 regress mpg weight weight2
1 test weight2=0

. _
```

We can see from the listing that the last command typed by the user was test weight2=0. Or, you may just look at the Review window to see the history of commands you typed.

◁

▷ Example 3

Perhaps the command you are looking for is not among the last five commands you typed. You can tell Stata to go back any number of lines. For instance, typing #review 15 tells Stata to show you the last 15 lines you typed:

```
. #review 15
15 replace resmpg=mpg-pred
14 summarize resmpg, detail
13 drop predmpg
12 describe
11 sort foreign
10 by foreign: summarize mpg weight
9 * lines that start with a * are comments.
8 * they go into the review buffer too.
7 summarize resmpg, detail
6 list make mpg weight
5 list make mpg weight if abs(res)>6
4 list make mpg weight if abs(res)>5
3 tabulate foreign if abs(res)>5
2 regress mpg weight weight2
1 test weight2=0

. _
```

If you wanted to resubmit the 10th previous line, you could type 10 and press Seek, or you could press PrevLine 10 times. No matter which of the above methods you prefer for retrieving lines, you may edit previous lines by using the editing keys.

◁

10.6   Tab expansion of variable names

Another way to quickly enter a variable name is to take advantage of Stata's variable name completion feature. Simply type the first few letters of the variable name in the Command window and press the *Tab* key. Stata will automatically type the rest of the variable name for you. If more than one variable name matches the letters you have typed, Stata will complete as much as it can and beep at you to let you know that you have typed a nonunique variable abbreviation.

Elements of Stata

11 Language syntax

Contents

11.1 Overview

With few exceptions, the basic Stata language syntax is

$$\left[\texttt{by } \textit{varlist:}\right] \quad \textit{command} \quad \left[\textit{varlist}\right] \quad \left[\texttt{=}\textit{exp}\right] \quad \left[\texttt{if } \textit{exp}\right] \quad \left[\texttt{in } \textit{range}\right] \quad \left[\textit{weight}\right] \quad \left[\texttt{, } \textit{options}\right]$$

where square brackets distinguish optional qualifiers and options from required ones. In this diagram, *varlist* denotes a list of variable names, *command* denotes a Stata command, *exp* denotes an algebraic expression, *range* denotes an observation range, *weight* denotes a weighting expression, and *options* denotes a list of options.

11.1.1 varlist

Most commands that take a subsequent *varlist* do not require that you explicitly type one. If no *varlist* appears, these commands assume a *varlist* of _all, the Stata shorthand for indicating all the variables in the dataset. In commands that alter or destroy data, Stata requires that the *varlist* be specified explicitly. See [U] **11.4 varlists** for a complete description.

▷ Example 1

The summarize command lists the mean, standard deviation, and range of the specified variables. In [R] **summarize**, we see that the syntax diagram for summarize is

summarize [*varlist*] [*if*] [*in*] [*weight*] [, *options*]

Farther down on the manual page is a table summarizing options, but let's focus on the syntax diagram itself first. Since everything except the word summarize is enclosed in square brackets, the simplest form of the command is 'summarize'. Typing summarize without arguments is equivalent to typing summarize _all; all the variables in the dataset are summarized. Underlining denotes the shortest allowed abbreviation, so we could have typed just su; see [U] **11.2 Abbreviation rules**.

The table that defines *options* looks like this:

options	description
Main	
detail	display additional statistics
meanonly	suppress the display; calculate only the mean; programmer's option
format	use variable's display format
separator(#)	draw separator line after every # variables; default is separator(5)

Thus we learn we could also type, for instance, summarize, detail or summarize, detail format.

As another example, the drop command eliminates variables or observations from a dataset. When dropping variables, its syntax is

drop *varlist*

drop has no option table because it has no options.

In fact, nothing is optional. Typing drop by itself would result in the error message "varlist or in range required". To drop all the variables in the dataset, we must type drop _all.

Even before looking at the syntax diagram, we could have predicted that the *varlist* would be required—drop is destructive, so Stata requires us to spell out our intent. The syntax diagram informs us that the *varlist* is required since *varlist* is not enclosed in square brackets. Since drop is not underlined, it cannot be abbreviated.

◁

11.1.2   by varlist:

The by *varlist*: prefix causes Stata to repeat a command for each subset of the data for which the values of the variables in the *varlist* are equal. When prefixed with by *varlist*:, the result of the command will be the same as if you had formed separate datasets for each group of observations, saved them, and then given the command on each dataset separately. The data must already be sorted by *varlist*, although by has a sort option; see [U] **11.5 by varlist: construct** for more information.

▷ Example 2

Typing `summarize marriage_rate divorce_rate` produces a table of the mean, standard deviation, and range of `marriage_rate` and `divorce_rate`, using all the observations in the data:

```
. use http://www.stata-press.com/data/r10/census12
(1980 Census data by state)
. summarize marriage_rate divorce_rate
```

Variable	Obs	Mean	Std. Dev.	Min	Max
marriage_r~e	50	.0133221	.0188122	.0074654	.1428282
divorce_rate	50	.0056641	.0022473	.0029436	.0172918

Typing `by region: summarize marriage_rate divorce_rate` produces one table for each region of the country:

```
. sort region
. by region: summarize marriage_rate divorce_rate
```

-> region = N Cntrl

Variable	Obs	Mean	Std. Dev.	Min	Max
marriage_r~e	12	.0099121	.0011326	.0087363	.0127394
divorce_rate	12	.0046974	.0011315	.0032817	.0072868

-> region = NE

Variable	Obs	Mean	Std. Dev.	Min	Max
marriage_r~e	9	.0087811	.001191	.0075757	.0107055
divorce_rate	9	.004207	.0010264	.0029436	.0057071

-> region = South

Variable	Obs	Mean	Std. Dev.	Min	Max
marriage_r~e	16	.0114654	.0025721	.0074654	.0172704
divorce_rate	16	.005633	.0013355	.0038917	.0080078

-> region = West

Variable	Obs	Mean	Std. Dev.	Min	Max
marriage_r~e	13	.0218987	.0363775	.0087365	.1428282
divorce_rate	13	.0076037	.0031486	.0046004	.0172918

The dataset must be sorted on the by variables:

```
. use http://www.stata-press.com/data/r10/census12
(1980 Census data by state)
. by region: summarize marriage_rate divorce_rate
not sorted
r(5);
. sort region
. by region: summarize marriage_rate divorce_rate
(output appears)
```

We could also have asked that by sort the data:

```
. by region, sort: summarize marriage_rate divorce_rate
(output appears)
```

by *varlist*: can be used with most Stata commands; we can tell which ones by looking at their syntax diagrams. For instance, we could obtain the correlations by `region`, between `marriage_rate` and `divorce_rate`, by typing by `region: correlate marriage_rate divorce_rate`.

◁

❏ Technical Note

The *varlist* in by *varlist*: may contain up to 32,767 variables with Stata/MP and Stata/SE or 2,047 variables with Stata/IC; these are the maximum allowed in the dataset. For instance, if we had data on automobiles and wished to obtain means according to market category (`market`) broken down by manufacturer (`origin`), we could type by `market origin: summarize`. That *varlist* contains two variables: `market` and `origin`. If the data were not already sorted on `market` and `origin`, we would first type `sort market origin`.

❏

❏ Technical Note

The *varlist* in by *varlist*: may contain string variables, numeric variables, or both. In the example above, `region` is a string variable, in particular, a `str7`. The example would have worked, however, if `region` were a numeric variable with values 1, 2, 3, and 4, or even 12.2, 16.78, 32.417, and 152.13.

❏

11.1.3   if exp

The `if` *exp* qualifier restricts the scope of a command to those observations for which the value of the expression is *true* (which is equivalent to the expression being nonzero; see [U] **13 Functions and expressions**).

▷ Example 3

Typing `summarize marriage_rate divorce_rate if region=="West"` produces a table for the western region of the country:

```
. summarize marriage_rate divorce_rate if region == "West"
```

Variable	Obs	Mean	Std. Dev.	Min	Max
marriage_r~e	13	.0218987	.0363775	.0087365	.1428282
divorce_rate	13	.0076037	.0031486	.0046004	.0172918

The double equal sign in `region=="West"` is not an error. Stata uses *double* equal signs to denote equality testing and a *single* equal sign to denote assignment; see [U] **13 Functions and expressions**.

A command may have at most one `if` qualifier. If you want the summary for the West restricted to observations with values of `marriage_rate` in excess of 0.015, do *not* type `summarize marriage_rate divorce_rate if region=="West" if marriage_rate>.015`. Type instead

```
. summarize marriage_rate divorce_rate if region == "West" & marriage_rate >.015
```

Variable	Obs	Mean	Std. Dev.	Min	Max
marriage_r~e	1	.1428282	.	.1428282	.1428282
divorce_rate	1	.0172918	.	.0172918	.0172918

You may not use the word *and* in place of the symbol '&' to join conditions. To select observations that meet one condition *or* another, use the '|' symbol. For instance, summarize marriage_rate divorce_rate if region=="West" | marriage_rate>.015 summarizes all observations for which region is West *or* marriage_rate is greater than 0.015.

◁

▷ Example 4

if may be combined with by. Typing by region: summarize marriage_rate divorce_rate if marriage_rate>.015 produces a set of tables, one for each region, reflecting summary statistics on marriage_rate and divorce_rate among observations for which marriage_rate exceeds 0.015:

```
. by region: summarize marriage_rate divorce_rate if marriage_rate >.015
```

-> region = N Cntrl

Variable	Obs	Mean	Std. Dev.	Min	Max
marriage_r~e	0				
divorce_rate	0				

-> region = NE

Variable	Obs	Mean	Std. Dev.	Min	Max
marriage_r~e	0				
divorce_rate	0				

-> region = South

Variable	Obs	Mean	Std. Dev.	Min	Max
marriage_r~e	2	.0163219	.0013414	.0153734	.0172704
divorce_rate	2	.0061813	.0025831	.0043548	.0080078

-> region = West

Variable	Obs	Mean	Std. Dev.	Min	Max
marriage_r~e	1	.1428282	.	.1428282	.1428282
divorce_rate	1	.0172918	.	.0172918	.0172918

The results indicate that there are no states in the Northeast and North Central regions for which marriage_rate exceeds 0.015, whereas there are two such states in the South and one state in the West.

◁

11.1.4   in range

The in *range* qualifier restricts the scope of the command to a specific observation range. A range specification takes the form $\#_1 \left[/\#_2 \right]$, where $\#_1$ and $\#_2$ are positive or negative integers. Negative integers are understood to mean "from the end of the data", with -1 referring to the last observation. The implied first observation must be less than or equal to the implied last observation.

The first and last observations in the dataset may be denoted by f and l (lowercase letter), respectively. F is allowed as a synonym for f, and L is allowed as a synonym for l. A range specifies absolute observation numbers within a dataset. As a result, the in qualifier may not be used when the command is preceded by the by *varlist*: prefix; see [U] **11.5 by varlist: construct**.

▷ Example 5

Typing summarize marriage_rate divorce_rate in 5/25 produces a table based on the values of marriage_rate and divorce_rate in observations 5–25:

```
. summarize marriage_rate divorce_rate in 5/25
    Variable |       Obs        Mean    Std. Dev.         Min         Max
-------------+--------------------------------------------------------------
 marriage_r~e |        21    .0096001    .0013263    .0075757    .0125884
 divorce_rate |        21     .004726    .0012025    .0029436    .0072868
```

This is, admittedly, a rather odd thing to want to do. It would not be odd, however, if we substituted list for summarize. If we wanted to see the states with the 10 lowest values of marriage_rate, we could type sort marriage_rate followed by list marriage_rate in 1/10.

Typing summarize marriage_rate divorce_rate in f/l is equivalent to typing summarize marriage_rate divorce_rate—all observations are summarized.

◁

▷ Example 6

Typing summarize marriage_rate divorce_rate in 5/25 if region=="South" produces a table based on the values of the two variables in observations 5–25 for which the value of region is South:

```
. summarize marriage_rate divorce_rate in 5/25 if region=="South"
    Variable |       Obs        Mean    Std. Dev.         Min         Max
-------------+--------------------------------------------------------------
 marriage_r~e |         4    .0108886    .0015061    .0089201    .0125884
 divorce_rate |         4    .0054448    .0011166    .0041485    .0068685
```

The ordering of the in and if qualifiers is not significant. The command could also have been specified as summarize marriage_rate divorce_rate if region=="South" in 5/25.

◁

▷ Example 7

Negative in ranges can be useful with sort. For instance, we have data on automobiles and wish to list the five with the highest mileage ratings:

```
. use http://www.stata-press.com/data/r10/auto
(1978 Automobile Data)

. sort mpg

. list make mpg in -5/l
```

	make	mpg
70.	Toyota Corolla	31
71.	Plym. Champ	34
72.	Subaru	35
73.	Datsun 210	35
74.	VW Diesel	41

◁

11.1.5   =exp

=exp specifies the value to be assigned to a variable and is most often used with `generate` and `replace`. See [U] **13 Functions and expressions** for details on expressions and [D] **generate** for details on the `generate` and `replace` commands.

▷ Example 8

Expression	Meaning
`generate newvar=oldvar+2`	creates a new variable named `newvar` equal to `oldvar+2`
`replace oldvar=oldvar+2`	changes the contents of the existing variable `oldvar`
`egen newvar=rank(oldvar)`	creates `newvar` containing the ranks of `oldvar` (see [D] **egen**)

◁

11.1.6   weight

weight indicates the weight to be attached to each observation. The syntax of *weight* is

$$[weightword=exp]$$

where you actually type the square brackets and where *weightword* is one of

weightword	Meaning
<u>w</u>eight	default treatment of weights
<u>f</u>weight *or* <u>freq</u>uency	frequency weights
<u>p</u>weight	sampling weights
<u>aw</u>eight *or* <u>cell</u>size	analytic weights
<u>i</u>weight	importance weights

The underlining indicates the minimum acceptable abbreviation. Thus, `weight` may be abbreviated `w` or `we`, etc.

▷ Example 9

Before explaining what the different types of weights mean, let's obtain the population-weighted mean of a variable called median_age from data containing observations on all 50 states of the United States. The dataset also contains a variable named pop, which is the total population of each state.

```
. use http://www.stata-press.com/data/r10/census12
(1980 Census data by state)
. summarize median_age [weight=pop]
(analytic weights assumed)
```

Variable	Obs	Weight	Mean	Std. Dev.	Min	Max
median_age	50	225907472	30.11047	1.66933	24.2	34.7

In addition to telling us that our dataset contains 50 observations, Stata informs us that the sum of the weight is 225,907,472, which was the number of people living in the United States as of the 1980 census. The weighted mean is 30.11. We were also informed that Stata assumed that we wanted "analytic" weights.

◁

weight is each command's idea of what the "natural" weights are and is one of fweight, pweight, aweight, or iweight. When you specify the vague weight, the command informs you which kind it assumes. Not every command supports every kind of weight. A note below the syntax diagram for a command will tell you which weights the command supports.

Stata understands four kinds of weights:

1. fweights, or frequency weights, indicate duplicated observations. fweights are always integers. If the fweight associated with an observation is 5, that means there are really 5 such observations, each identical.

2. pweights, or sampling weights, denote the inverse of the probability that this observation is included in the sample because of the sampling design. A pweight of 100, for instance, indicates that this observation is representative of 100 subjects in the underlying population. The scale of these weights does not matter in terms of estimated parameters and standard errors, except when estimating totals and computing finite-population corrections with the svy commands; see [SVY] **survey**.

3. aweights, or analytic weights, are inversely proportional to the variance of an observation; i.e., the variance of the jth observation is assumed to be σ^2/w_j, where w_j are the weights. Typically, the observations represent averages, and the weights are the number of elements that gave rise to the average. For most Stata commands, the recorded scale of aweights is irrelevant; Stata internally rescales them to sum to N, the number of observations in your data, when it uses them.

4. iweights, or importance weights, indicate the relative "importance" of the observation. They have no formal statistical definition; this is a catch-all category. Any command that supports iweights will define how they are treated. They are usually intended for use by programmers who want to produce a certain computation.

See [U] **20.17 Weighted estimation** for a thorough discussion of weights and their meaning.

❑ Technical Note

When you do not specify a weight, the result is equivalent to specifying [fweight=1].

❑

11.1.7 options

Many commands take command-specific options. These are described along with each command in the *Reference* manuals. Options are indicated by typing a comma at the end of the command, followed by the options you want to use.

▷ Example 10

Typing summarize marriage_rate produces a table of the mean, standard deviation, minimum, and maximum of the variable marriage_rate:

```
. summarize marriage_rate
```

Variable	Obs	Mean	Std. Dev.	Min	Max
marriage_r~e	50	.0133221	.0188122	.0074654	.1428282

The syntax diagram for summarize is

summarize [*varlist*] [*if*] [*in*] [*weight*] [, *options*]

followed by the option table

options	description
Main	
detail	display additional statistics
meanonly	suppress the display; calculate only the mean; programmer's option
format	use variable's display format
separator(#)	draw separator line after every # variables; default is separator(5)

Thus, the options allowed by summarize are detail or meanonly, format, and separator(). The shortest allowed abbreviations for these options are d for detail, mean for meanonly, f for format, and sep() for separator(); see [U] **11.2 Abbreviation rules**.

Typing summarize marriage_rate, detail produces a table that also includes selected percentiles, the four largest and four smallest values, the skewness, and the kurtosis.

```
. summarize marriage_rate, detail
```

marriage_rate

	Percentiles	Smallest		
1%	.0074654	.0074654		
5%	.0078956	.0075757		
10%	.0080043	.0078956	Obs	50
25%	.0089399	.0079079	Sum of Wgt.	50
50%	.0105669		Mean	.0133221
		Largest	Std. Dev.	.0188122
75%	.0122899	.0146266		
90%	.0137832	.0153734	Variance	.0003539
95%	.0153734	.0172704	Skewness	6.718494
99%	.1428282	.1428282	Kurtosis	46.77306

◁

❏ Technical Note

Once you have typed the *varlist* for the command, you can place options anywhere in the command. You can type summarize marriage_rate divorce_rate if region=="West", detail, or you can type summarize marriage_rate divorce_rate, detail, if region=="West". You use a second comma to indicate a return to the command line as opposed to the option list. Leaving out the comma after the word detail would cause an error because Stata would attempt to interpret the phrase if region=="West" as an option rather than as part of the command.

You may not type an option in the middle of a *varlist*. Typing summarize marriage_rate, detail, divorce_rate will result in an error.

Options need not be specified contiguously. You may type summarize marriage_rate divorce_rate, detail, if region=="South", noformat. Both detail and noformat are options.

❏

❏ Technical Note

Most options are toggles—they indicate that something either is or is not to be done. Sometimes it is difficult to remember which is the default. The following rule applies to all options: if *option* is an option, then no*option* is an option as well, and vice versa. Thus, if we could not remember whether detail or nodetail were the default for summarize but we knew that we did not want the detail, we could type summarize, nodetail. Typing the nodetail option is unnecessary, but Stata will not complain.

Some options take *arguments*. The Stata kdensity command has an n(#) option that indicates the number of points at which the density estimate is to be evaluated. When an option takes an argument, the argument is enclosed in parentheses.

Some options take more than one argument. In such cases, arguments should be separated from one another by commas. For instance, you might see in a syntax diagram

saving(*filename* [, replace])

Here replace is the (optional) second argument. *Lists*, such as lists of variables (varlists) and lists of numbers (numlists), are considered to be one argument. If a syntax diagram reported

powers(*numlist*)

the list of numbers would be one argument, so the elements would not be separated by commas. You would type, for instance, powers(1 2 3 4). In fact, Stata will tolerate commas here, so you could type powers(1,2,3,4).

Some options take string arguments. regress has an eform() option that works this way—for instance, eform("Exp Beta"). To play it safe, you should type the quotes surrounding the string, although it is not required. If you do not type the quotes, any sequence of two or more consecutive blanks will be interpreted as one blank. Thus, eform(Exp beta) would be interpreted the same as eform(Exp beta).

❏

11.1.8 numlist

A *numlist* is a list of numbers. Stata allows certain shorthands to indicate ranges:

numlist	meaning
2	just one number
1 2 3	three numbers
3 2 1	three numbers in reversed order
.5 1 1.5	three different numbers
1 3 -2.17 5.12	four numbers in jumbled order
1/3	three numbers: 1, 2, 3
3/1	the same three numbers in reverse order
5/8	four numbers: 5, 6, 7, 8
-8/-5	four numbers: -8, -7, -6, -5
-5/-8	four numbers: -5, -6, -7, -8
-1/2	four numbers: -1, 0, 1, 2
1 2 to 4	four numbers: 1, 2, 3, 4
4 3 to 1	four numbers: 4, 3, 2, 1
10 15 to 30	five numbers: 10, 15, 20, 25, 30
1 2:4	same as 1 2 to 4
4 3:1	same as 4 3 to 1
10 15:30	same as 10 15 to 30
1(1)3	three numbers: 1, 2, 3
1(2)9	five numbers: 1, 3, 5, 7, 9
1(2)10	the same five numbers, 1, 3, 5, 7, 9
9(-2)1	five numbers: 9, 7, 5, 3, and 1
-1(.5)2.5	the numbers -1, $-.5$, 0, .5, 1, 1.5, 2, 2.5
1[1]3	same as 1(1)3
1[2]9	same as 1(2)9
1[2]10	same as 1(2)10
9[-2]1	same as 9(-2)1
-1[.5]2.5	same as -1(.5)2.5
1 2 3/5 8(2)12	eight numbers: 1, 2, 3, 4, 5, 8, 10, 12
1,2,3/5,8(2)12	the same eight numbers
1 2 3/5 8 10 to 12	the same eight numbers
1,2,3/5,8,10 to 12	the same eight numbers
1 2 3/5 8 10:12	the same eight numbers

poisson's constraints() option has syntax constraints(*numlist*). Thus, you could type constraints(2 4 to 8), constraints(2(2)8), etc.

11.1.9 datelist

A *datelist* is a list of dates or times and is often used with graph options when the variable being graphed has a date format. For a description of how dates and times are stored and manipulated in Stata, see [U] **24 Dealing with dates and times**. Calendar dates, also known as %td dates, are recorded in Stata as the number of days since 01jan1960, so 0 means 01jan1960, 1 means 02jan1960, and 16,541 means 15apr2005. Similarly, -1 means 31dec1959, -2 means 30dec1959, and $-16,541$ means 18sep1914. In such a case, a datelist is either a list of dates, as in

15apr1973 17apr1973 20apr1973 23apr1973

or it is a first and last date with an increment between, as in

17apr1973(3)23apr1973

or it is a combination:

15apr1973 17apr1973(3)23apr1973

Dates specified with spaces, slashes, or commas must be bound in parentheses, as in

> (15 apr 1973) (april 17, 1973)(3)(april 23, 1973)

Evenly spaced calendar dates are not especially useful, but with other time units, even spacing can be useful, such as

> 1999q1(1)2005q1

when %tq dates are being used. 1999q1(1)2005q1 means every quarter between 1991q1 and 2005q1. 1999q1(4)2005q1 would mean every first quarter.

To interpret a datelist, Stata first looks at the format of the related variable and then uses the corresponding date-to-numeric translation function. For instance, if the variable has a %td format, the td() function is used to translate the date; if the variable has a %tq format, the tq() function is used; and so on.

11.1.10   Prefix commands

Stata has a handful of commands that are used to prefix other Stata commands. by varlist:, discussed in section [U] **11.1.2 by varlist:**, is in fact an example of a prefix command. In that section, we demonstrated by using

> by region: summarize marriage_rate divorce_rate

and later,

> by region, sort: summarize marriage_rate divorce_rate

and although we did not, we could also have demonstrated

> by region, sort: summarize marriage_rate divorce_rate, detail

Each of the above runs the summarize command separately on the data for each region.

by itself follows standard Stata syntax:

$$\text{by } varlist[, \ options]: \ \ldots$$

In 'by region, sort: summarize marriage_rate divorce_rate, detail', region is by's varlist and sort is by's option, just as marriage_rate and divorce_rate are summarize's varlist and detail is summarize's option.

by is not the only prefix command, and the full list of such commands is

Prefix command	description
by	run command on subsets of data
statsby	same as by, but collect statistics from each run
rolling	run command on moving subsets and collect statistics
bootstrap	run command on bootstrap samples
jackknife	run command on jackknife subsets of data
permute	run command on random permutations
simulate	run command on manufactured data
svy	run command and adjust results for survey sampling
nestreg	run command with accumulated blocks of regressors, and report nested model comparison tests
stepwise	run command with stepwise variable inclusion/exclusion
xi	run command after expanding factor variables and interactions
capture	run command and capture its return code
noisily	run command and show the output
quietly	run command and suppress the output
version	run command under specified version

The last group—capture, noisily, quietly, and version—have to do with programming Stata and, for historical reasons, capture, noisily, and quietly allow you to omit the colon, so one programmer might code

    quietly regress ...

and another

    quietly: regress ...

All the other prefix commands require the colon.

11.2 Abbreviation rules

Stata allows abbreviations. In this manual, we usually avoid abbreviating commands, variable names, and options to ensure readability:

    . summarize myvar, detail

Experienced Stata users, on the other hand, tend to abbreviate the same command as

    . sum myv, d

As a general rule, command, option, and variable names may be abbreviated to the shortest string of characters that uniquely identifies them.

This rule is violated if the command or option does something that cannot easily be undone; the command must then be spelled out in its entirety.

Also, a few common commands and options are allowed to have even shorter abbreviations than the general rule would allow.

The general rule is applied, without exception, to variable names.

11.2.1 Command abbreviation

The shortest allowed abbreviation for a command or option can be determined by looking at the command's syntax diagram. This minimal abbreviation is shown by underlining:

regress

rename

replace

rotate

run

If there is no underlining, no abbreviation is allowed. For example, replace may not be abbreviated, the underlying reason being that replace changes the data.

regress can be abbreviated reg, regr, regre, or regres, or it can be spelled out in its entirety.

Sometimes short abbreviations are also allowed. Commands that begin with the letter *d* include decode, describe, destring, dir, discard, display, do, and drop, which suggests that the shortest allowable abbreviation for describe is desc. However, since describe is such a commonly used command, you may abbreviate it with the single letter d. You may also abbreviate the list command with the single letter l.

The other exception to the general abbreviation rule is that commands that alter or destroy data must be spelled out completely. Two commands that begin with the letter *d*, discard and drop, are destructive in the sense that, once you give one of these commands, there is no way to undo the result. Therefore, both must be spelled out.

The final exceptions to the general rule are commands implemented as ado-files. Such commands may not be abbreviated. Ado-file commands are external, and their names correspond to the names of disk files.

11.2.2 Option abbreviation

Option abbreviation follows the same logic as command abbreviation: you determine the minimum acceptable abbreviation by examining the command's syntax diagram. The syntax diagram for summarize reads, in part,

summarize ... , detail format

Option detail may be abbreviated d, de, det, ..., detail. Similarly, option format may be abbreviated f, fo, ..., format.

Options clear and replace occur with many commands. The clear option indicates that even though completing this command will result in the loss of all data in memory, and even though the data in memory have changed since the data were last saved on disk, you want to continue. clear must be spelled out, as in use newdata, clear.

The replace option indicates that it's okay to save over an existing dataset. If you type save mydata and the file mydata.dta already exists, you will receive the message "file mydata.dta already exists", and Stata will refuse to overwrite it. To allow Stata to overwrite the dataset, you would type save mydata, replace. replace may not be abbreviated.

❏ Technical Note

replace is a stronger modifier than clear and is one you should think about before using. With a mistaken clear, you can lose hours of work, but with a mistaken replace, you can lose days of work.

❏

11.2.3 Variable-name abbreviation

1. Variable names may be abbreviated to the shortest string of characters that uniquely identifies them given the data currently loaded in memory.

   If your dataset contained four variables, state, mrgrate, dvcrate, and dthrate, you could refer to the variable dvcrate as dvcrat, dvcra, dvcr, dvc, or dv. You might type list dv to list the data on dvcrate. You could not refer to the variable dvcrate as d, however, since that abbreviation does not distinguish dvcrate from dthrate. If you were to type list d, Stata would respond with the message "ambiguous abbreviation". (If you wanted to refer to *all* variables that started with the letter *d*, you could type list d*; see [U] **11.4 varlists**.)

2. The character ~ may be used to mean that "zero or more characters go here". For instance, r~8 might refer to the variable rep78, or rep1978, or repair1978, or just r8. (The ~ character is similar to the * character in [U] **11.4 varlists**, except that it adds the restriction "and only one variable matches this specification".)

   In (1), we said that you could abbreviate variables. You could type dvcr to refer to dvcrate, but, if there were more than one variable that started with the letters dvcr, you would receive an error. Typing dvcr is the same as typing dvcr~.

11.3 Naming conventions

A name is a sequence of one to 32 letters (A–Z and a–z), digits (0–9), and underscores (_).

Programmers: local macro names can have no more than 31 characters in the name; see [U] **18.3.1 Local macros**.

Stata reserves the following names:

_all	double	long	_rc
_b	float	_n	_skip
byte	if	_N	using
_coef	in	_pi	with
_cons	int	_pred	

You may not use these reserved names for your variables.

The first character of a name must be a letter or an underscore. We recommend, however, that you not begin your variable names with an underscore. All of Stata's built-in variables begin with an underscore, and we reserve the right to incorporate new _*variables* freely.

Stata respects case; that is, myvar, Myvar, and MYVAR are three distinct names.

All objects in Stata—not just variables—follow this naming convention.

11.4   varlists

A *varlist* is a list of variable names. The variable names in a *varlist* refer either exclusively to new (not yet created) variables or exclusively to existing variables. A *newvarlist* always refers exclusively to new (not yet created) variables.

11.4.1   Lists of existing variables

In lists of existing variable names, variable names may be repeated.

▷ Example 11

If you type `list state mrgrate dvcrate state`, the variable `state` will be listed twice, once in the leftmost column and again in the rightmost column of the list.

◁

Existing variable names may be abbreviated as described in [U] **11.2 Abbreviation rules**. You may also use '∗' to indicate that "zero or more characters go here". For instance, if you suffix ∗ to a partial variable name (for example, sta∗), you are referring to all variable names that start with that letter combination. If you prefix ∗ to a letter combination (for example, ∗rate), you are referring to all variables that end in that letter combination. If you put ∗ in the middle (for example, m∗rate), you are referring to all variables that begin and end with the specified letters. You may put more than one ∗ in an abbreviation.

▷ Example 12

If the variables `poplt5`, `pop5to17`, and `pop18p` are in our dataset, we may type `pop∗` as a shorthand way to refer to all three variables. For instance, `list state pop∗` lists the variables `state`, `poplt5`, `pop5to17`, and `pop18p`.

If we had a dataset with variables `inc1990`, `inc1991`, ..., `inc1999` along with variables `incfarm1990`, ..., `incfarm1999`; `pop1990`, ..., `pop1999`; and `ms1990`, ..., `ms1999`, then `∗1995` would be a shorthand way of referring to `inc1995`, `incfarm1995`, `pop1995`, and `ms1995`. We could type, for instance, `list ∗1995`.

In that same dataset, typing `list i∗95` would be a shorthand way of listing `inc1995` and `incfarm1995`.

Typing `list i∗f∗95` would be a shorthand way of listing to `incfarm1995`.

◁

∼ is an alternative to ∗, and really, it means the same thing. The difference is that ∼ indicates that if more than one variable matches the specified pattern, Stata will complain rather than substituting all the variables that match the specification.

▷ Example 13

In the previous example, we could have typed `list i∼f∼95` to list `incfarm1995`. If, however, our dataset also included variable `infant1995`, then `list i∗f∗95` would list both variables and `list i∼f∼95` would complain that `i∼f∼95` is an ambiguous abbreviation.

◁

You may use ? to specify that one character goes here. Remember, * means zero or more characters go here, so ?* can be used to mean one or more characters goes here, ??* can be used to mean two or more characters go here, and so on.

▷ Example 14

In a dataset containing variables rep1, rep2, ..., rep78, rep? would refer to rep1, rep2, ..., rep9, and rep?? would refer to rep10, rep11, ..., rep78.

◁

You may place a dash (-) between two variable names to specify all the variables stored between the two listed variables, inclusive. You can determine storage order by using describe; it lists variables in the order in which they are stored.

▷ Example 15

If the dataset contains the variables state, mrgrate, dvcrate, and dthrate, in that order, typing list state-dvcrate is equivalent to typing list state mrgrate dvcrate. In both cases, three variables are listed.

◁

11.4.2   Lists of new variables

In lists of *new variables*, no variable names may be repeated or abbreviated.

You may specify a dash (-) between two variable names that have the same letter prefix and that end in numbers. This form of the dash notation indicates a range of variable names in ascending numerical order.

For example, typing input v1-v4 is equivalent to typing input v1 v2 v3 v4. Typing infile state v1-v3 ssn using rawdata is equivalent to typing infile state v1 v2 v3 ssn using rawdata.

You may specify the storage type before the variable name to force a storage type other than the default. The numeric storage types are byte, int, long, float (the default), and double. The string storage types are str#, where # is replaced with an integer between 1 and 244, inclusive, representing the maximum length of the string. See [U] **12 Data**.

For instance, the list var1 str8 var2 var3 specifies that var1 and var3 be given the default storage type and that var2 be stored as a str8—a string whose maximum length is eight characters.

The list var1 int var2 var3 specifies that var2 be stored as an int. You may use parentheses to bind a list of variable names. The list var1 int(var2 var3) specifies that both var2 and var3 be stored as ints. Similarly, the list var1 str20(var2 var3) specifies that both var2 and var3 be stored as str20s. The different storage types are listed in [U] **12.2.2 Numeric storage types** and [U] **12.4.4 String storage types**.

▷ Example 16

Typing infile str2 state str10 region v1-v5 using mydata reads the state and region strings from the file mydata.raw and stores them as str2 and str10, respectively, along with the variables v1 through v5, which are stored as the default storage type float (unless we have specified a different default with the set type command).

Typing `infile str10(state region) v1-v5 using mydata` would achieve almost the same result, except that the `state` and `region` values recorded in the data would both be assigned to `str10` variables. (We could then use the `compress` command to shorten the strings. See [D] **compress**; it is well worth reading.)

◁

❑ Technical Note

You may append a colon and a *value label name* to numeric variables. (See [U] **12.6 Dataset, variable, and value labels** for a description of value labels.) For instance, `var1 var2:myfmt` specifies that the variable `var2` be associated with the value label stored under the name `myfmt`. This has the same effect as typing the list `var1 var2` and then subsequently giving the command `label values var2 myfmt`.

The advantage of specifying the value label association with the colon notation is that value labels can then be assigned by the current command; see [D] **input** and [D] **infile (free format)**.

❑

▷ Example 17

Typing `infile int(state:stfmt region:regfmt) v1-v5 using mydata, automatic` reads the state and region data from the file `mydata.raw` and stores them as `int`s, along with the variables `v1` through `v5`, which are stored as the default storage type.

In our previous example, both state and region were strings, so how can strings be stored in a numeric variable? See [U] **12.6 Dataset, variable, and value labels** for the complete answer. The colon notation specifies the name of the value label, and the `automatic` option tells Stata to assign unique numeric codes to all character strings. The numeric code for state, which Stata will make up on the fly, will be stored in the `state` variable. The mapping from numeric codes to words will be stored in the value label named `stfmt`. Similarly, regions will be assigned numeric codes, which are stored in `region`, and the mapping will be stored in `regfmt`.

If we were to `list` the data, the `state` and `region` variables would look like strings. `state`, for instance, would appear to contain things like `AL`, `CA`, and `WA`, but actually it would contain only numbers like 1, 2, 3, and 4.

◁

11.4.3   Time-series varlists

Time-series varlists are a variation on varlists of existing variables. When a command allows a time-series varlist, you may include time-series operators. For instance, `L.gnp` refers to the lagged value of variable `gnp`. The time-series operators are

operator	meaning
L.	lag x_{t-1}
L2.	2-period lag x_{t-2}
...	
F.	lead x_{t+1}
F2.	2-period lead x_{t+2}
...	
D.	difference $x_t - x_{t-1}$
D2.	difference of difference $x_t - x_{t-1} - (x_{t-1} - x_{t-2}) = x_t - 2x_{t-1} + x_{t-2}$
...	
S.	"seasonal" difference $x_t - x_{t-1}$
S2.	lag-2 (seasonal) difference $x_t - x_{t-2}$
...	

Time-series operators may be repeated and combined. L3.gnp refers to the third lag of variable gnp. So do LLL.gnp, LL2.gnp, and L2L.gnp. LF.gnp is the same as gnp. DS12.gnp refers to the one-period difference of the 12-period difference. LDS12.gnp refers to the same concept, lagged once.

D1. = S1., but D2. ≠ S2., D3. ≠ S3., and so on. D2. refers to the difference of the difference. S2. refers to the two-period difference. If you wanted the difference of the difference of the 12-period difference of gnp, you would write D2S12.gnp.

Operators may be typed in uppercase or lowercase. Most users would type d2s12.gnp instead of D2S12.gnp.

You may type operators however you wish; Stata internally converts operators to their canonical form. If you typed ld2ls12d.gnp, Stata would present the operated variable as L2D3S12.gnp.

In addition to using *operator#*, Stata understands *operator*(*numlist*) to mean a set of operated variables. For instance, typing L(1/3).gnp in a varlist is the same as typing 'L.gnp L2.gnp L3.gnp'. The operators can also be applied to a list of variables by enclosing the variables in parentheses; e.g.,

```
. use http://www.stata-press.com/data/r10/gxmpl1
. list year L(1/3).(gnp cpi)
```

	year	L.gnp	L2.gnp	L3.gnp	L.cpi	L2.cpi	L3.cpi
1.	1989	.	.	.	.	.	.
2.	1990	5837.9	.	.	124	.	.
3.	1991	6026.3	5837.9	.	130.7	124	.
4.	1992	6367.4	6026.3	5837.9	136.2	130.7	124
5.	1993	6689.3	6367.4	6026.3	140.3	136.2	130.7
6.	1994	7098.4	6689.3	6367.4	144.5	140.3	136.2
7.	1995	7433.4	7098.4	6689.3	148.2	144.5	140.3
8.	1996	7851.9	7433.4	7098.4	152.4	148.2	144.5

The parentheses notation may be used with any operator. Typing D(1/3).gnp would return the first through third differences.

The parentheses notation may be used in operator lists with multiple operators, such as L(0/3)D2S12.gnp.

Operator lists may include up to one set of parentheses, which may enclose a *numlist*; see [U] **11.1.8 numlist**.

Before you can use time-series operators in varlists, you must set the time variable by using the `tsset` command:

```
. list l.gnp
time variable not set
r(111);
. tsset time
  (output omitted )
. list l.gnp
  (output omitted )
```

See [TS] **tsset**. The time variable must take on integer values. Also, the data must be sorted on the time variable. `tsset` handles this, but later you might encounter

```
. list l.mpg
not sorted
r(5);
```

Then type `sort time` or type `tsset` to reestablish the order.

The time-series operators respect the time variable. `L2.gnp` refers to gnp_{t-2}, regardless of missing observations in the dataset. In the following dataset, the observation for 1992 is missing:

```
. use http://www.stata-press.com/data/r10/gxmpl2
. list year gnp l2.gnp, separator(0)
```

	year	gnp	L2.gnp
1.	1989	5837.9	.
2.	1990	6026.3	.
3.	1991	6367.4	5837.9
4.	1993	7098.4	6367.4	← note, filled in correctly
5.	1994	7433.4	.
6.	1995	7851.9	7098.4

Operated variables may be used in expressions:

```
. generate gnplag2 = l2.gnp
(3 missing values generated)
```

Stata also understands cross-sectional time-series data. If you have cross sections of time series, you indicate this when you `tsset` the data:

```
. tsset country year
```

See [TS] **tsset**. In fact, you can type that, or you can type

```
. xtset country year
```

`xtset` is how you set panel data just as `tsset` is how you set time-series data and here the two commands do the same thing. Some panel datasets are not cross-sectional time series, however, in that the second variable is not time, so `xtset` also allows

```
. xtset country
```

See [XT] **xtset**.

11.5 by varlist: construct

> by *varlist*: *command*

The by prefix causes *command* to be repeated for each unique set of values of the variables in the *varlist*. *varlist* may contain numeric, string, or a mixture of numeric and string variables. (*varlist* may not contain time-series operators.)

by is an optional prefix to perform a Stata command separately for each group of observations where the values of the variables in the *varlist* are the same.

During each iteration, the values of the system variables _n and _N are set in relation to the first observation in the by-group; see [U] **13.7 Explicit subscripting**. The in *range* qualifier cannot be used with by *varlist*: because ranges specify absolute rather than relative observation numbers.

❏ Technical Note

The inability to combine in and by is not really a constraint since if provides all the functionality of in and a bit more. If you wanted to perform *command* for the first three observations in each of the by-groups, you could type

. by *varlist*: *command* if _n<=3

❏

The results of *command* would be the same as if you had formed separate datasets for each group of observations, saved them, used each separately, and issued *command*.

▷ Example 18

We provide some examples using by in [U] **11.1.2 by varlist:** above. We demonstrate the effect of by on _n, _N, and explicit subscripting in [U] **13.7 Explicit subscripting**.

by requires that the data first be sorted. For instance, if we had data on the average January and July temperatures in degrees Fahrenheit for 420 cities located in the Northeast and West and wanted to obtain the averages, by region, across those cities, we might type

```
. use http://www.stata-press.com/data/r10/citytemp3
(City Temperature Data)
. by region: summarize tempjan tempjuly
not sorted
r(5);
```

Stata refused to honor our request since the data are not sorted by region. We must either sort the data by region first (see [D] **sort**) or specify by's sort option (which has the same effect):

(Continued on next page)

```
. by region, sort: summarize tempjan tempjuly
```

-> region = NE

Variable	Obs	Mean	Std. Dev.	Min	Max
tempjan	164	27.88537	3.543096	16.6	31.8
tempjuly	164	73.35	2.361203	66.5	76.8

-> region = N Cntrl

Variable	Obs	Mean	Std. Dev.	Min	Max
tempjan	284	21.69437	5.725392	2.2	32.6
tempjuly	284	73.46725	3.103187	64.5	81.4

-> region = South

Variable	Obs	Mean	Std. Dev.	Min	Max
tempjan	250	46.1456	10.38646	28.9	68
tempjuly	250	80.9896	2.97537	71	87.4

-> region = West

Variable	Obs	Mean	Std. Dev.	Min	Max
tempjan	256	46.22539	11.25412	13	72.6
tempjuly	256	72.10859	6.483131	58.1	93.6

◁

▷ Example 19

Using the same data as in the example above, we estimate regressions, by region, of average January temperature on average July temperature. Both temperatures are specified in degrees Fahrenheit.

```
. by region: regress tempjan tempjuly
```

-> region = NE

Source	SS	df	MS		
Model	1529.74026	1	1529.74026		
Residual	516.484453	162	3.18817564		
Total	2046.22471	163	12.5535258		

Number of obs = 164
F(1, 162) = 479.82
Prob > F = 0.0000
R-squared = 0.7476
Adj R-squared = 0.7460
Root MSE = 1.7855

tempjan	Coef.	Std. Err.	t	P>|t|	[95% Conf. Interval]	
tempjuly	1.297424	.0592303	21.90	0.000	1.180461	1.414387
_cons	-67.28066	4.346781	-15.48	0.000	-75.86431	-58.697

```
-> region = N Cntrl
      Source |       SS          df       MS              Number of obs =      284
-------------+------------------------------             F(  1,    282) =   115.89
       Model | 2701.97917         1  2701.97917          Prob > F      =   0.0000
    Residual | 6574.79175       282  23.3148644          R-squared     =   0.2913
-------------+------------------------------             Adj R-squared =   0.2887
       Total | 9276.77092       283  32.7801093          Root MSE      =   4.8285

-------------+----------------------------------------------------------------
     tempjan |      Coef.   Std. Err.      t    P>|t|     [95% Conf. Interval]
-------------+----------------------------------------------------------------
    tempjuly |   .9957259   .0924944    10.77   0.000     .8136589    1.177793
       _cons |  -51.45888   6.801344    -7.57   0.000    -64.84673   -38.07103
-------------+----------------------------------------------------------------


-> region = South
      Source |       SS          df       MS              Number of obs =      250
-------------+------------------------------             F(  1,    248) =    95.17
       Model | 7449.51623         1  7449.51623          Prob > F      =   0.0000
    Residual | 19412.2231       248  78.2750933          R-squared     =   0.2773
-------------+------------------------------             Adj R-squared =   0.2744
       Total | 26861.7394       249  107.878471          Root MSE      =   8.8473

-------------+----------------------------------------------------------------
     tempjan |      Coef.   Std. Err.      t    P>|t|     [95% Conf. Interval]
-------------+----------------------------------------------------------------
    tempjuly |    1.83833   .1884392     9.76   0.000     1.467185    2.209475
       _cons |    -102.74   15.27187    -6.73   0.000    -132.8191   -72.66089
-------------+----------------------------------------------------------------


-> region = West
      Source |       SS          df       MS              Number of obs =      256
-------------+------------------------------             F(  1,    254) =     2.84
       Model | 357.161728         1  357.161728          Prob > F      =   0.0932
    Residual | 31939.9031       254  125.74765           R-squared     =   0.0111
-------------+------------------------------             Adj R-squared =   0.0072
       Total | 32297.0648       255  126.655156          Root MSE      =   11.214

-------------+----------------------------------------------------------------
     tempjan |      Coef.   Std. Err.      t    P>|t|     [95% Conf. Interval]
-------------+----------------------------------------------------------------
    tempjuly |   .1825482   .1083166     1.69   0.093    -.0307648    .3958613
       _cons |    33.0621    7.84194     4.22   0.000     17.61859     48.5056
-------------+----------------------------------------------------------------
```

The regressions show that a 1-degree increase in the average July temperature in the Northeast corresponds to a 1.3-degree increase in the average January temperature. In the West, however, it corresponds to a 0.18-degree increase, which is only marginally significant.

◁

❑ Technical Note

by has a second syntax that is especially useful when you want to play it safe:

by *varlist₁* (*varlist₂*) : *command*

This says that Stata is to verify that the data are sorted by *varlist₁* *varlist₂* and then, assuming that is true, perform *command* by *varlist₁*. For instance,

```
. by subject (time): gen finalval = val[_N]
```

By typing this, we want to create new variable `finalval`, which contains, in each observation, the final observed value of `val` for each subject in the data. The final value will be the last value if, within subject, the data are sorted by time. The above command verifies that the data are sorted by `subject` and `time` and then, if they are, performs

```
. by subject: gen finalval = val[_N]
```

If the data are not sorted properly, an error message will instead be issued. Of course, we could have just typed

```
. by subject: gen finalval = val[_N]
```

after verifying for ourselves that the data were sorted properly, as long as we were careful to look.

`by`'s second syntax can be used with `by`'s `sort` option, so we can also type

```
. by subject (time), sort: gen finalval = val[_N]
```

which is equivalent to

```
. sort subject time
. by subject: gen finalval = val[_N]
```

❑

11.6   File-naming conventions

Some commands require that you specify a *filename*. Filenames are specified in the way natural for your operating system:

Windows	Unix	Macintosh
mydata	mydata	mydata
mydata.dta	mydata.dta	mydata.dta
c:mydata.dta	~friend/mydata.dta	~friend/mydata.dta
"my data"	"my data"	"my data"
"my data.dta"	"my data.dta"	"my data.dta"
myproj\mydata	myproj/mydata	myproj/mydata
"my project\my data"	"my project/my data"	"my project/my data"
C:\analysis\data\mydata	~/analysis/data/mydata	~/analysis/data/mydata
"C:\my project\my data"	"~/my project/my data"	"~/my project/my data"
..\data\mydata	../data/mydata	../data/mydata
"..\my project\my data"	"../my project/my data"	"../my project/my data"

Usually (the exceptions being `copy`, `dir`, `ls`, `erase`, `rm`, and `type`), Stata automatically provides a file extension if you do not supply one. For instance, if you type `use mydata`, Stata assumes that you mean `use mydata.dta` since `.dta` is the file extension Stata normally uses for data files.

Stata provides 18 default file extensions that are used by various commands:

.ado	automatically loaded do-files
.dct	ASCII data dictionary
.do	do-file
.dta	Stata-format dataset
.dtasig	datasignature file
.gph	graph
.irf	impulse–response function datasets
.log	log file in text format
.mata	Mata source code
.mlib	Mata library
.mmat	Mata matrix
.mo	Mata object file
.out	file saved by outsheet
.raw	ASCII-format dataset
.smcl	log file in SMCL format
.sthlp	help files
.sum	checksum files to verify network transfers
.ster	saved estimates

You do not have to name your data files with the .dta extension—if you type an explicit file extension, it will override the default. For instance, if your dataset was stored as myfile.dat, you could type use myfile.dat. If your dataset was stored as simply myfile with no file extension, you could type the period at the end of the filename to indicate that you are explicitly specifying the null extension. You would type use myfile. to use this dataset.

All operating systems allow blanks in filenames, and so does Stata. However, if the filename includes a blank, you must enclose the filename in double quotes. Typing

```
. save "my data"
```

would create the file my data.dta. Typing

```
. save my data
```

would be an error.

❑ Technical Note

Stata also uses 12 other file extensions. These files are of interest only to advanced programmers or are for Stata's internal use. They are

.class	class file for object-oriented programming; see [P] **class**
.dlg	dialog resource file
.idlg	dialog resource include file
.ihlp	help include file
.key	search's keyword database file
.maint	maintenance file (for Stata's internal use only)
.mnu	menu file (for Stata's internal use only)
.pkg	user-site package file
.plugin	compiled addition (DLL)
.scheme	control file for a graph scheme
.style	graph style file
.toc	user-site description file

❑

11.6.1   A special note for Macintosh users

Have you seen the notation `myfolder/myfile` before? This notation is called a path and describes the location of a file or folder (also called a directory).

You do not have to use this notation if you do not like it. You could instead restrict yourself to using files only in the current folder. If that turns out to be too restricting, Stata for Macintosh provides enough menus and buttons that you can probably get by. You may, however, find the notation convenient. In case you do, here is the rest of the definition.

The character / is called a path delimiter and delimits folder names and filenames in a path. If the path starts with no path delimiter, the path is relative to the current folder.

For example, the path `myfolder/myfile` refers to the file `myfile` in the folder `myfolder`, which is contained in the current folder.

The characters .. refer to the folder containing the current folder. Thus, `../myfile` refers to `myfile` in the folder containing the current folder, and `../nextdoor/myfile` refers to `myfile` in the folder `nextdoor` in the folder containing the current folder.

If a path starts with a path delimiter, the path is called an absolute path and describes a fixed location of a file or folder name, regardless of what the current folder is. The leading / in an absolute path refers to the root directory, which is the main hard drive from which the operating system is booted. For example, the path `/myfolder/myfile` refers to the file `myfile` in the folder `myfolder`, which is contained in the main hard drive.

The character ~ refers to the user's home directory. Thus, the path `~/myfolder/myfile` refers to `myfile` in the folder `myfolder` in the user's home directory.

11.6.2   A special note for Unix users

Stata understands ~ to mean your home directory. Stata understands this, even if you do not use `csh(1)` as your shell.

11.7   References

Kolev, G. I. 2006. Stata tip 31: Scalar or variable? The problem of ambiguous names. *Stata Journal* 6: 279–280.

Ryan, P. 2005. Stata tip 22: Variable name abbreviation. *Stata Journal* 5: 465–466.

12 Data

Contents

12.1 Data and datasets

Data form a rectangular table of numeric and string values in which each row is an observation on all the variables and each column contains the observations on one variable. Variables are designated by *variable names*. Observations are numbered sequentially from 1 to _N. The following example of data contains the first five odd and first five even positive integers, along with a string variable:

```
        odd   even    name
   1.     1      2    Bill
   2.     3      4    Mary
   3.     5      6     Pat
   4.     7      8   Roger
   5.     9     10    Sean
```

The observations are numbered 1 to 5, and the variables are named odd, even, and name. Observations are referred to by number, and variables by name.

A *dataset* is *data* plus labelings, formats, notes, and characteristics.

All aspects of *data* and *datasets* are defined here.

12.2 Numbers

A *number* may contain a sign, an integer part, a decimal point, a fraction part, an e or E, and a signed integer exponent. Numbers may *not* contain commas; for example, the number 1,024 must be typed as 1024 (or 1024. or 1024.0). The following are examples of valid numbers:

```
5
-5
5.2
.5
5.2e+2
5.2e-2
```

❏ Technical Note

As a convenience for Fortran users, you may use d or D, as well as e, to indicate exponential notation. Thus, the number 520 may be written 5.2d+2, 5.2D+2, 5.2d+02, or 5.2D+02.

❏

❏ Technical Note

Stata also allows numbers to be represented in a hexadecimal/binary format, defined as

$$\left[+|-\right]0.0\left[\langle zeros\rangle\right]\{\text{X}|\text{x}\}\text{-3ff}$$

or

$$\left[+|-\right]1.\langle hexdigit\rangle\left[\langle hexdigits\rangle\right]\{\text{X}|\text{x}\}\{+|-\}\langle hexdigit\rangle\left[\langle hexdigits\rangle\right]$$

The lead digit is always 0 or 1; it is 0 only when the number being expressed is zero. A maximum of 13 digits to the right of the hexadecimal point are allowed. The power ranges from -3ff to +3ff. The number is expressed in hexadecimal (base 16) digits; the number aX+b means $a \times 2^b$. For instance, 1.0X+3 is 2^3 or 8. 1.8X+3 is 12 because 1.8_{16} is $1 + 8/16 = 1.5$ in decimal and the number is thus $1.5 \times 2^3 = 1.5 \times 8 = 12$.

Stata can also display numbers using this format; see [U] **12.5.1 Numeric formats**. For example,

```
. di 1.81x+2
6.015625

. di %21x 6.015625
+1.8100000000000X+002
```

This hexadecimal format is of special interest to numerical analysts.

❏

12.2.1 Missing values

A number may also take on the special value *missing*, denoted by a period (.). You specify a missing value anywhere that you may specify a number. Missing values differ from ordinary numbers in one respect: any arithmetic operation on a missing value yields a missing value.

In fact, there are 27 missing values in Stata: '.', the one just discussed, as well as .a, .b, ..., and .z, which are known as extended missing values. The missing value '.' is known as the default or system missing value. In any case, some people use extended missing values to indicate why a certain value is unknown—the question was not asked, the person refused to answer, etc. Other people have no use for extended missing values and just use '.'.

Stata's default or system missing value will be returned when you perform an arithmetic operation on missing values or when the arithmetic operation is not defined, such as division by zero, or the logarithm of a nonpositive number.

```
. display 2/0
.
. list
```

	a
1.	.b
2.	.
3.	.a
4.	3
5.	6

```
. generate x = a + 1
(3 missing values generated)
. list
```

	a	x
1.	.b	.
2.	.	.
3.	.a	.
4.	3	4
5.	6	7

Numeric missing values are represented by "large positive values". The ordering is

$$\text{all numbers} < . < .a < .b < \cdots < .z$$

Thus, the expression

$$\text{age} > 60$$

is true if variable age is greater than 60 or is missing. Similarly,

$$\text{gender} \, ! = 0$$

is true if gender is not zero or is missing.

The way to exclude missing values is to ask whether the value is less than '.', and the way to detect missing values is to ask whether the value is greater than or equal to '.'. For instance,

```
. list if age>60 & age<.
. generate agegt60 = 0 if age<=60
. replace agegt60 = 1 if age>60 & age<.
. generate agegt60 = (age>60) if age<.
```

❏ Technical Note

Before Stata 8, Stata only had one representation for missing values, the period (.).

To ensure that old programs and do-files continue to work properly, when version is set less than 8, all missing values are treated as being the same. Thus, . == .a == .b == .z, and so 'exp==.' and 'exp!=.' work just as they previously did.

❏

▷ Example 1

We have data on the income of husbands and wives recorded in the variables hincome and wincome, respectively. Typing the list command, we see that your data contain

```
. use http://www.stata-press.com/data/r10/gxmpl3
. list
```

	hincome	wincome
1.	32000	0
2.	35000	34000
3.	47000	.b
4.	.z	50000
5.	.a	.

The values of wincome in the third and fifth observations are *missing*, as distinct from the value of wincome in the first observation, which is known to be zero.

If we use the generate command to create a new variable, income, that is equal to the sum of hincome and wincome, three missing values would be produced:

```
. generate income = hincome + wincome
(3 missing values generated)
. list
```

	hincome	wincome	income
1.	32000	0	32000
2.	35000	34000	69000
3.	47000	.b	.
4.	.z	50000	.
5.	.a	.	.

generate produced a warning message that 3 missing values were created, and when we list the data, we see that 47,000 plus *missing* yields *missing*.

◁

❏ Technical Note

Stata stores numeric missing values as the largest 27 numbers allowed by the particular storage type; see [U] **12.2.2 Numeric storage types**. There are two important implications. First, if you sort on a variable that has missing values, the missing values will be placed last, and the sort order of any missing values will follow the rule regarding the properties of missing values stated above.

```
. sort wincome
. list wincome
```

	wincome
1.	0
2.	34000
3.	50000
4.	.
5.	.b

The second implication concerns relational operators and missing values. Do not forget that a missing value will be larger than any numeric value.

```
. list if wincome > 40000
```

	hincome	wincome	income
3.	.z	50000	.
4.	.a	.	.
5.	47000	.b	.

Observations 4 and 5 are listed because '.' and '.b' are both missing and thus are greater than 40,000. Relational operators are discussed in detail in [U] **13.2.3 Relational operators**.

❏

▷ Example 2

In producing statistical output, Stata ignores observations with missing values. Continuing with the example above, if we request summary statistics on hincome and wincome by using the summarize command, we obtain

```
. summarize hincome wincome
```

Variable	Obs	Mean	Std. Dev.	Min	Max
hincome	3	38000	7937.254	32000	47000
wincome	3	28000	25534.29	0	50000

Some commands discard the entire observation (known as *casewise deletion*) if one of the variables in the observation is missing. If we use the correlate command to obtain the correlation between hincome and wincome, for instance, we obtain

```
. correlate hincome wincome
(obs=2)
```

	hincome	wincome
hincome	1.0000	
wincome	1.0000	1.0000

The correlation coefficient is calculated over two observations.

◁

12.2.2 Numeric storage types

Numbers can be stored in one of five variable types: byte, int, long, float (the default), or double. bytes are, naturally, stored in 1 byte. ints are stored in 2 bytes, longs and floats in 4 bytes, and doubles in 8 bytes. The table below shows the minimum and maximum values for each storage type.

Storage type	Minimum	Maximum	Closest to 0 without being 0	Bytes
byte	-127	100	± 1	1
int	$-32,767$	32,740	± 1	2
long	$-2,147,483,647$	2,147,483,620	± 1	4
float	$-1.70141173319 \times 10^{38}$	$1.70141173319 \times 10^{38}$	$\pm 10^{-38}$	4
double	$-8.9884656743 \times 10^{307}$	$+8.9884656743 \times 10^{307}$	$\pm 10^{-323}$	8

Do not confuse the term *integer*, which is a characteristic of a number, with int, which is a storage type. For instance, the number 5 is an integer, no matter how it is stored; thus, if you read that an argument must be an integer, that does not mean that it must be stored as an int.

12.3 Dates and times

Stata has nine date, time, and date-and-time numeric encodings known collectively as %t variables or values. They are

%tC	calendar date and time, adjusted for leap seconds
%tc	calendar date and time, ignoring leap seconds
%td	calendar date
%tw	week
%tm	calendar month
%tq	financial quarter
%th	financial half-year
%ty	calendar year
%tg	generic—user defined

All except %ty and %tg are based on 0 = beginning of January 1960. %tc and %tC record the number of milliseconds since then. %td records the number of days. The others record the numbers of weeks, months, quarters, or half-years. %ty simply records the year and %tg records whatever you wish.

For a full discussion of working with dates and times, see [U] **24 Dealing with dates and times**.

12.4   Strings

A *string* is a sequence of printable characters and is typically enclosed in double quotes. The quotes are not considered a part of the string but delimit the beginning and end of the string. The following are examples of valid strings:

```
"Hello, world"
"String"
"string"
" string"
"string "
""
"x/y+3"
"1.2"
```

All the strings above are distinct; that is, "String" is different from "string", which is different from " string", which is different from "string ". Also, "1.2" is a string and not a number because it is enclosed in quotes.

All strings in Stata are of varying length, which means that Stata internally records the length of the string and never loses track. There is never a circumstance in which a string cannot be delimited by quotes, but there are rare instances where strings do not have to be delimited by quotes, such as during data input. In those cases, nondelimited strings are stripped of their leading and trailing blanks. Delimited strings are always accepted as is.

The special string "", often called *null string*, is considered by Stata to be a *missing*. No special meaning is given to the string containing one period, ".".

In addition to double quotes for enclosing strings, Stata also allows compound double quotes: `"` and `"'`. You can type "*string*" or you can type `"*string*"'`, although users seldom type `"*string*"'`. Compound double quotes are of special interest to programmers because they nest and provide a way for a quoted string to itself contain double quotes (either simple or compound). See [U] **18.3.5 Double quotes**.

12.4.1   Strings containing identifying data

String variables often contain identifying information, such as the patient's name or the name of the city or state. Such strings are typically listed but are not used directly in statistical analysis, although the data might be sorted on the string or datasets might be merged on the basis of one or more string variables.

12.4.2   Strings containing categorical data

Occasionally, strings contain information that is to be used directly in analysis, such as the patient's sex, which might be coded "male" or "female". Stata shows a decided preference for such information to be numerically encoded and stored in numeric variables. Stata's statistical routines treat string variables as if every observation records a numeric missing value.

Stata provides two commands for converting string variables into numeric codes and back again: encode and decode; see [U] **23.2 Categorical string variables**. Also see [D] **destring** for information on mapping string variables to numeric and vice versa.

12.4.3 Strings containing numeric data

If a string variable contains the character representation of a number—for instance, myvar contains "1", "1.2", and "-5.2"—you can convert it directly into a numeric variable by using the real() function or the destring command, e.g., generate newvar=real(myvar).

Similarly, if you want to convert a numeric variable to its string representation, you can use the string() function or the tostring command, e.g., generate as_str=string(numvar).

See [D] **functions**.

12.4.4 String storage types

Strings are stored in string variables with storage types str1, str2, ..., str244. The storage type merely sets the maximum length of the string, not its actual length; thus, "example" has length 7 whether it is stored as a str7, a str10, or even a str244. On the other hand, an attempt to assign the string "example" to a str6 would result in "exampl".

The maximum length of a string in Stata is 244. String literals may exceed 244 characters, but only the first 244 characters are significant.

12.5   Formats: Controlling how data are displayed

Formats describe how a number or string is to be presented. For instance, how is the number 325.24 to be presented? As 325.2, or 325.24, or 325.240, or 3.2524e+02, or 3.25e+02, or some other way? The *display format* tells Stata exactly how to present such data. You do not have to specify display formats since Stata always makes reasonable assumptions about how to display a variable, but you always have the option.

12.5.1   Numeric formats

A Stata numeric format is formed by

first type	%	to indicate the start of the format
then optionally type	–	if you want the result left-aligned
then optionally type	0	if you want to retain leading zeros (1)
then type	a number w	stating the width of the result
then type	.	
then type	a number d	stating the number of digits to follow the decimal point
then type		
either	e	for scientific notation, e.g., 1.00e+03
or	f	for fixed format, e.g., 1000.0
or	g	for general format; Stata chooses based on the number being displayed
then optionally type	c	to indicate comma format (not allowed with e)

(1) Specifying 0 to mean "include leading zeros" will be honored only with the f format.

For example,

```
%9.0g    general format, 9 columns wide
                    sqrt(2) =  1.414214
                      1,000 =       1000
                 10,000,000 =  1.00e+07
%9.0gc   general format, 9 columns wide, with commas
                    sqrt(2) =  1.414214
                      1,000 =      1,000
                 10,000,000 =  1.00e+07
%9.2f    fixed format, 9 columns wide, 2 decimal places
                    sqrt(2) =       1.41
                      1,000 =    1000.00
                 10,000,000 = 10000000.00
%9.2fc   fixed format, 9 columns wide, 2 decimal places, with commas
                    sqrt(2) =       1.41
                      1,000 =   1,000.00
                 10,000,000 = 10,000,000.00
%9.2e    exponential format, 9 columns wide
                    sqrt(2) =   1.41e+00
                      1,000 =   1.00e+03
                 10,000,000 =   1.00e+07
```

Stata has three numeric format types: e, f, and g. The formats are denoted by a leading percent sign (%) followed by the string $w.d$, where w and d stand for two integers. The first integer, w, specifies the width of the format. The second integer d specifies the number of digits that are to follow the decimal point. d must be less than w. Finally, a character denotes the format type (e, f, or g), and to that may optionally be appended a c indicating that commas are to be included in the result (c is not allowed with e.)

By default, every numeric variable is given a $%w.0g$ format, where w is large enough to display the largest number of the variable's type. The $%w.0g$ format is a set of formatting rules that present the values in as readable a fashion as possible without sacrificing precision. The g format changes the number of decimal places displayed whenever it improves the readability of the current value.

The default formats for each of the numeric variable types are

```
byte      %8.0g
int       %8.0g
long      %12.0g
float     %9.0g
double    %10.0g
```

You can change the format of a variable by using the `format` *varname* %*fmt* command.

In addition to $%w.0g$, allowed is $%w.0gc$ to display numbers with commas. "One thousand" is displayed as 1000 in $%9.0g$ format and as 1,000 in $%9.0gc$ format.

In addition to using $%w.0g$ and $%w.0gc$, you can use $%w.dg$ and $%w.dgc$, $d > 0$. For example, $%9.4g$ and $%9.4gc$. The 4 means to display approximately four significant digits. For instance, the number 3.14159265 in $%9.4g$ format is displayed as 3.142, 31.4159265 as 31.42, 314.159265 as 314.2, and 3141.59265 as 3142. The format is not exactly a significant-digit format because 31415.9265 is displayed as 31416, not as 3.142e+04.

Under the f format, values are always displayed with the same number of decimal places, even if this results in a loss in the displayed precision. Thus, the f format is similar to the C f format. Stata's f format is also similar to the Fortran F format, but, unlike the Fortran F format, it switches to g whenever a number is too large to be displayed in the specified f format.

In addition to $%w.df$, the format $%w.dfc$ can display numbers with commas.

The e format is similar to the C e and the Fortran E format. Every value is displayed as a leading digit (with a minus sign, if necessary), followed by a decimal point, the specified number of digits, the letter e, a plus sign or a minus sign, and the power of 10 (modified by the preceding sign) that multiplies the displayed value. When the e format is specified, the width must exceed the number of digits that follow the decimal point by at least seven to accommodate the leading sign and digit, the decimal point, the e, and the signed power of 10.

▷ Example 3

Below we have a 5-observation dataset with three variables: e_fmt, f_fmt, and g_fmt. All three variables have the same values stored in them; only the display format varies. describe shows the display format to the right of the variable type:

```
. use http://www.stata-press.com/data/r10/format
. describe
Contains data from http://www.stata-press.com/data/r10/format.dta
  obs:             5
  vars:            3                        12 Mar 2007 15:18
  size:           80 (99.9% of memory free)

              storage  display    value
variable name  type    format     label     variable label

e_fmt          float   %9.2e
f_fmt          float   %10.2f
g_fmt          float   %9.0g

Sorted by:
```

The formats for each of these variables were set by typing

```
. format e_fmt %9.2e
. format f_fmt %10.2f
```

It was not necessary to set the format for the g_fmt variable since Stata automatically assigned it the %9.0g format. Nevertheless, we could have typed format g_fmt %9.0g if we wished. Listing the data results in

```
. list
```

	e_fmt	f_fmt	g_fmt
1.	2.80e+00	2.80	2.801785
2.	3.96e+06	3962322.50	3962323
3.	4.85e+00	4.85	4.852834
4.	-5.60e-06	-0.00	-5.60e-06
5.	6.26e+00	6.26	6.264982

◁

❏ Technical Note

The discussion above is incomplete. There is one other format available that will be of interest to numerical analysts. The %21x format displays base 10 numbers in a hexadecimal (base 16) format. The number is expressed in hexadecimal (base 16) digits; the number $aX+b$ means $a \times 2^b$. For example,

```
. display %21x 1234.75
+1.34b0000000000X+00a
```

Thus, the base 10 number 1,234.75 has a base 16 representation of 1.34bX+0a, meaning

$$\left(1 + 3 \cdot 16^{-1} + 4 \cdot 16^{-2} + 11 \cdot 16^{-3}\right) \times 2^{10}$$

Remember, the hexadecimal–decimal equivalents are

hexadecimal	decimal
0	0
1	1
2	2
3	3
4	4
5	5
6	6
7	7
8	8
9	9
a	10
b	11
c	12
d	13
e	14
f	15

See [U] **12.2 Numbers**.

❏

12.5.2   European numeric formats

The three numeric formats e, f, and g will use ',' to indicate the decimal symbol if you specify their width and depth as w,d rather than $w.d$. For instance, the format %9,0g will display what Stata would usually display as 1.5 as 1,5.

If you use the European specification with fc or gc, the "comma" will be presented as a period. For instance, %9,0gc would display what Stata would usually display as 1,000.5 as 1.000,5.

If this way of presenting numbers appeals to you, consider using Stata's set dp comma command. set dp comma tells Stata to interpret nearly all $\%w.d\{g|f|e\}$ formats as $\%w,d\{g|f|e\}$ formats. Most of Stata is written using a period to represent the decimal symbol, and that means that, even if you set the appropriate $\%w,d\{g|f|e\}$ format for your data, it will affect only displays of the data. For instance, if you type summarize to obtain summary statistics or regress to obtain regression results, the decimal will still be shown as a period.

set dp comma changes that and affects all of Stata. With set dp comma, it does not matter whether your data are formatted $\%w.d\{g|f|e\}$ or $\%w,d\{g|f|e\}$. All results will be displayed using a comma as the decimal character:

```
. use http://www.stata-press.com/data/r10/auto
(1978 Automobile Data)

. set dp comma

. summarize mpg weight foreign
    Variable |       Obs        Mean    Std. Dev.       Min        Max
-------------+--------------------------------------------------------
         mpg |        74     21,2973    5,785503         12         41
      weight |        74    3019,459    777,1936       1760       4840
     foreign |        74     ,2972973    ,4601885          0          1

. regress mpg weight foreign
      Source |       SS       df       MS              Number of obs =      74
-------------+------------------------------           F(  2,    71) =   69,75
       Model | 1619,2877        2  809,643849           Prob > F      =  0,0000
    Residual | 824,171761       71  11,608053           R-squared     =  0,6627
-------------+------------------------------           Adj R-squared =  0,6532
       Total | 2443,45946       73  33,4720474          Root MSE      =  3,4071

------------------------------------------------------------------------------
         mpg |      Coef.   Std. Err.      t    P>|t|     [95% Conf. Interval]
-------------+----------------------------------------------------------------
      weight |  -,0065879    ,0006371   -10,34   0,000    -,0078583   -,0053175
     foreign |  -1,650029    1,075994    -1,53   0,130     -3,7955     ,4954422
       _cons |    41,6797    2,165547    19,25   0,000     37,36172    45,99768
------------------------------------------------------------------------------
```

You can switch the decimal character back to a period by typing `set dp period`.

❑ Technical Note

    `set dp comma` makes drastic changes inside Stata, and we mention this because some older, user-written programs may not be able to deal with those changes. If you are using an older user-written program, you might `set dp comma` and then find that the program does not work and instead presents some sort of syntax error.

    If, using any program, you do get an unanticipated error, try setting `dp` back to `period`. See [D] **format** for more information.

    Also understand that `set dp comma` affects how Stata outputs numbers, not how it inputs them. You must still use the period to indicate the decimal point on all input. Even with `set dp comma`, you type

```
. replace x=1.5 if x==2
```

                                                                                           ❑

12.5.3   Date and time formats

    Date and time formats are really a numeric format because Stata stores dates as the number of milliseconds, days, weeks, months, quarters, half-years, or years from 01jan1960; see [U] **24 Dealing with dates and times**.

    The syntax of the `%t` format is

first type	`%`	to indicate the start of the format
then optionally type	`-`	if you want the result left-aligned
then type	`t`	
then type	*character*	to indicate the units
then optionally type	*other characters*	to indicate how the date/time is to be displayed

The letter you type to specify the units is

C   milliseconds from 01jan1960, adjusted for leap seconds
c   milliseconds from 01jan1960, ignoring leap seconds
d   days from 01jan1960
w   weeks from 1960-w1
m   calendar months from jan1960
q   quarters from 1960-q1
h   half years from 1960-h1
g   user-defined units

There are many codes you can type after that to specify exactly how the date/time is to be displayed, but usually, you do not. Most users use the default %tc for date/times and %td for dates. See *Formatting date and time values* in [D] **dates and times** for details.

12.5.4   String formats

The syntax for a string format is

first type	%	to indicate the start of the format
then optionally type	–	if you want the result left-aligned
then type	a number	indicating the width of the result
then type	s	

For instance, %10s represents a string format of width 10.

For str$w$, the default format is %ws or %9s, whichever is wider. For example, a str10 variable receives a %10s format. Strings are displayed right-justified in the field, unless the minus sign is coded; %-10s would display the string left-aligned.

▷ Example 4

Our automobile data contain a string variable called make.

```
. use http://www.stata-press.com/data/r10/auto
(1978 Automobile Data)

. describe make
```

variable name	storage type	display format	value label	variable label
make	str18	%-18s		Make and Model

```
. list make in 63/67
```

```
         make

63.   Mazda GLC
64.   Peugeot 604
65.   Renault Le Car
66.   Subaru
67.   Toyota Celica
```

These values are left-aligned because make has a display format of %-18s. If we want to right-align the values, we could change the format:

```
. format %18s make

. list make in 63/67
```

	make
63.	Mazda GLC
64.	Peugeot 604
65.	Renault Le Car
66.	Subaru
67.	Toyota Celica

◁

12.6   Dataset, variable, and value labels

Labels are strings used to label elements in Stata, such as labels for datasets, variables, and values.

12.6.1   Dataset labels

Associated with every dataset is an 80-character *dataset label*, which is initially set to blanks. You can use the label data "*text*" command to define the dataset label.

▷ Example 5

We have just entered 1980 state data on marriage rates, divorce rates, and median ages. The describe command will describe the data in memory:

```
. describe
Contains data
   obs:          50
   vars:          4
   size:       1,200 (99.8% of memory free)
```

variable name	storage type	display format	value label	variable label
state	str8	%9s		
median_age	float	%9.0g		
marriage_rate	long	%12.0g		
divorce_rate	long	%12.0g		

```
Sorted by:
     Note:  dataset has changed since last saved
```

describe shows that there are 50 observations on four variables named state, median_age, marriage_rate, and divorce_rate. state is stored as a str8; median_age is stored as a float; and marriage_rate and divorce_rate are both stored as longs. Each variable's display format (see [U] **12.5 Formats: controlling how data are displayed**) is shown. Finally, the data are not in any particular sort order, and the dataset has changed since it was last saved on disk.

We can label the data by typing label data "1980 state data". We type this and then type describe again:

```
. label data "1980 state data"

. describe
Contains data
    obs:           50                              1980 state data
    vars:           4
    size:        1,200 (99.7% of memory free)

                 storage  display     value
variable name    type     format      label      variable label

state            str8     %9s
median_age       float    %9.0g
marriage_rate    long     %12.0g
divorce_rate     long     %12.0g

Sorted by:
     Note:  dataset has changed since last saved
```
◁

The dataset label is displayed by the describe and use commands.

12.6.2 Variable labels

In addition to the name, every variable has associated with it an 80-character *variable label*. The variable labels are initially set to blanks. You use the label variable *varname* "*text*" command to define a new variable label.

▷ Example 6

We have entered data on four variables: state, median_age, marriage_rate, and divorce_rate. describe portrays the data we entered:

```
. describe
Contains data from states.dta
    obs:           50                              1980 state data
    vars:           4
    size:        1,200 (99.7% of memory free)

                 storage  display     value
variable name    type     format      label      variable label

state            str8     %9s
median_age       float    %9.0g
marriage_rate    long     %12.0g
divorce_rate     long     %12.0g

Sorted by:
     Note:  dataset has changed since last saved
```

We can associate labels with the variables by typing

```
. label variable median_age "Median Age"

. label variable marriage_rate "Marriages per 100,000"

. label variable divorce_rate "Divorces per 100,000"
```

From then on, the result of describe will be

```
. describe
Contains data
    obs:            50                          1980 state data
    vars:            4
    size:        1,200 (99.7% of memory free)

              storage  display    value
variable name  type    format     label    variable label

state          str8    %9s
median_age     float   %9.0g               Median Age
marriage_rate  long    %12.0g              Marriages per 100,000
divorce_rate   long    %12.0g              Divorces per 100,000

Sorted by:
    Note:  dataset has changed since last saved
```

◁

Whenever Stata produces output, it will use the variable labels rather than the variable names to label the results if there is room.

12.6.3 Value labels

Value labels define a correspondence or mapping between numeric data and the words used to describe what those numeric values represent. Mappings are named and defined by the `label define` *lblname* # "*string*" # "*string*"... command. The maximum length for the *lblname* is 32 characters. # must be an integer or an extended missing value (.a, .b, ..., .z). The maximum length of *string* is 32,000 characters. Named mappings are associated with variables by the `label values` *varname lblname* command.

▷ Example 7

The definition makes value labels sound more complicated than they are in practice. We create a dataset on individuals in which we record a person's sex, coding 0 for males and 1 for females. If our dataset also contained an employee number and salary, it might resemble the following:

```
. use http://www.stata-press.com/data/r10/gxmpl4
(2007 Employee data)

. describe

Contains data from http://www.stata-press.com/data/r10/gxmpl4.dta
    obs:             7                          2007 Employee data
    vars:            3                          11 Feb 2007 15:31
    size:          112 (99.9% of memory free)

              storage  display    value
variable name  type    format     label    variable label

empno          float   %9.0g               Employee number
sex            float   %9.0g               Sex
salary         float   %8.0fc              Annual salary, exclusive of
                                             bonus

Sorted by:
```

```
. list
```

	empno	sex	salary
1.	57213	0	34,000
2.	47229	1	37,000
3.	57323	0	34,000
4.	57401	0	34,500
5.	57802	1	37,000
6.	57805	1	34,000
7.	57824	0	32,500

We could create a mapping called sexlabel defining 0 as "Male" and 1 as "Female", and then associate that mapping with the variable sex by typing

```
. label define sexlabel 0 "Male" 1 "Female"
. label values sex sexlabel
```

From then on, our data would appear as

```
. describe
Contains data from http://www.stata-press.com/data/r10/gxmpl4.dta
  obs:            7                          2007 Employee data
 vars:            3                          11 Feb 2007 15:31
 size:          112 (99.8% of memory free)
```

variable name	storage type	display format	value label	variable label
empno	float	%9.0g		Employee number
sex	float	%9.0g	sexlabel	Sex
salary	float	%8.0fc		Annual salary, exclusive of bonus

```
Sorted by:
. list
```

	empno	sex	salary
1.	57213	Male	34,000
2.	47229	Female	37,000
3.	57323	Male	34,000
4.	57401	Male	34,500
5.	57802	Female	37,000
6.	57805	Female	34,000
7.	57824	Male	32,500

Notice not only that the value label is used to produce words when we list the data but also that the association of the variable sex with the value label sexlabel is shown by the describe command.

◁

❑ Technical Note

Value labels and variables may share the same name. For instance, rather than calling the value label `sexlabel` in the example above, we could just as well have named it `sex`. We would then type `label values sex sex` to associate the value label named `sex` with the variable named `sex`.

❑

▷ Example 8

Stata's `encode` and `decode` commands provide a convenient way to go from string variables to numerically coded variables and back again. Let's pretend that, in the example above, rather than coding 0 for males and 1 for females, we created a string variable recording either `"male"` or `"female"`.

```
. use http://www.stata-press.com/data/r10/gxmpl5
(2007 Employee data)

. describe

Contains data from http://www.stata-press.com/data/r10/gxmpl5
  obs:                7                           2007 Employee data
  vars:               3                           11 Feb 2007 15:37
  size:             126 (99.8% of memory free)
```

variable name	storage type	display format	value label	variable label
empno	float	%9.0g		Employee number
sex	str6	%9s		Sex
salary	float	%8.0fc		Annual salary, exclusive of bonus

```
Sorted by:

. list
```

	empno	sex	salary
1.	57213	male	34,000
2.	47229	female	37,000
3.	57323	male	34,000
4.	57401	male	34,500
5.	57802	female	37,000
6.	57805	female	34,000
7.	57824	male	32,500

We now want to create a numerically encoded variable—we will call it `gender`—from the string variable. We want to do this, say, because we typed `anova salary sex` to perform a one-way ANOVA of salary on sex, and we were told that there were "no observations". We then remembered that all Stata's statistical commands treat string variables as if they contain nothing but missing values. The statistical commands work only with numerically coded data.

```
. encode sex, generate(gender)
. describe
Contains data from http://www.stata-press.com/data/r10/gxmpl5.dta
  obs:            7                          2007 Employee data
  vars:           4                          11 Feb 2007 15:37
  size:         154 (99.8% of memory free)
```

variable name	storage type	display format	value label	variable label
empno	float	%9.0g		Employee number
sex	str6	%9s		Sex
salary	float	%8.0fc		Annual salary, exclusive of bonus
gender	long	%8.0g	gender	Sex

```
Sorted by:
     Note:  dataset has changed since last saved
```

encode adds a new `long` variable called `gender` to the data and defines a new value label called gender. The value label `gender` maps 1 to the string `male` and 2 to `female`, so if we were to `list` the data, we could not tell the difference between the `gender` and `sex` variables. However, they are different. Stata's statistical commands know how to deal with `gender` but do not understand the `sex` variable. See [D] **encode**.

❑

❑ Technical Note

Perhaps rather than employee data, our data are on persons undergoing sex-change operations. There would therefore be two sex variables in our data, sex before the operation and sex after the operation. Assume that the variables are named `presex` and `postsex`. We can associate the *same* value label to each variable by typing

```
. label define sexlabel 0 "Male" 1 "Female"
. label values presex sexlabel
. label values postsex sexlabel
```

❑

❑ Technical Note

Stata's input commands (`input` and `infile`) can switch from the words in a value label back to the numeric codes. Remember that `encode` and `decode` can translate a string to a numeric mapping and vice versa, so we can map strings to numeric codes either at the time of input or later.

For example,

```
. label define sexlabel 0 "Male" 1 "Female"
. input empno sex:sexlabel salary, label
        empno        sex      salary
  1. 57213 Male 34000
  2. 47229 Female 37000
  3. 57323 0 34000
  4. 57401 Male 34500
  5. 57802 Female 37000
  6. 57805 Female 34000
  7. 57824 Male 32500
  8. end
```

The `label define` command defines the value label `sexlabel`. `input empno sex:sexlabel salary, label` tells Stata to input three variables from the keyboard (`empno`, `sex`, and `salary`), attach the value label `sexlabel` to the `sex` variable, and look up any words that are typed in the value label to try to convert them to numbers. To prove that it works, we `list` the data that we recently entered:

```
. list
```

	empno	sex	salary
1.	57213	Male	34000
2.	47229	Female	37000
3.	57323	Male	34000
4.	57401	Male	34500
5.	57802	Female	37000
6.	57805	Female	34000
7.	57824	Male	32500

Compare the information we typed for observation 3 with the result listed by Stata. We typed 57323 0 34000. Thus, the value of `sex` in the third observation is 0. When Stata listed the observation, it indicated the value is `Male` because we told Stata in our `label define` command that zero is equivalent to `Male`.

Let's now add one more observation to our data:

```
. input, label
          empno        sex      salary
  8. 67223 FEmale 33000
'FEmale' cannot be read as a number
  8. 67223 Female 33000
  9. end
```

At first we typed 67223 FEmale 33000, and Stata responded with "'FEmale' cannot be read as a number". Remember that Stata always respects case, so `FEmale` is not the same as `Female`. Stata prompted us to enter the line again, and we did so, this time correctly.

❑

❑ Technical Note

Coupled with the `automatic` option, Stata not only can go from words to numbers but also can create the mapping. Let's input the data again, but this time, rather than type the data in at the keyboard, let's read the data from a file. Assume that we have an ASCII file called `employee.raw` stored on our disk that contains

```
57213 Male 34000
47229 Female 37000
57323 Male 34000
57401 Male 34500
57802 Female 37000
57805 Female 34000
57824 Male 32500
```

The `infile` command can read these data and create the mapping automatically:

```
. label list sexlabel
value label sexlabel not found
r(111);
```

```
. infile empno sex:sexlabel salary using employee, automatic
(7 observations read)
```

Our first command, `label list sexlabel`, is only to prove that we had not previously defined the value label `sexlabel`. Stata `infiled` the data without complaint. We now have

```
. list
```

	empno	sex	salary
1.	57213	Male	34000
2.	47229	Female	37000
3.	57323	Male	34000
4.	57401	Male	34500
5.	57802	Female	37000
6.	57805	Female	34000
7.	57824	Male	32500

Of course, `sex` is just another numeric variable; it does not actually take on the values `Male` and `Female`—it takes on numeric codes that have been automatically mapped to `Male` and `Female`. We can find out what that mapping is by using the `label list` command:

```
. label list sexlabel
sexlabel:
            1 Male
            2 Female
```

We discover that Stata attached the codes 1 to `Male` and 2 to `Female`. Anytime we want to see what our data really look like, ignoring the value labels, we can use the `nolabel` option:

```
. list, nolabel
```

	empno	sex	salary
1.	57213	1	34000
2.	47229	2	37000
3.	57323	1	34000
4.	57401	1	34500
5.	57802	2	37000
6.	57805	2	34000
7.	57824	1	32500

❏

12.6.4 Labels in other languages

A dataset can contain labels—data, variable, and value—in up to 100 languages. To discover the languages available for the dataset in memory, type `label language`. You will see this

```
. label language
Language for variable and value labels

In this dataset, value and variable labels have been defined in only one
language:  default

To create new language:          . label language <name>, new
To rename current language:      . label language <name>, rename
```

or something like this:

```
. label language
```

Language for variable and value labels

```
    Available languages:
            de
            en
            sp
    Currently set is:               . label language sp
    To select different language:   . label language <name>
    To create new language:         . label language <name>, new
    To rename current language:     . label language <name>, rename
```

Right now, the example dataset is set with sp (Spanish) labels:

```
. describe

Contains data
  obs:            74                        Automóviles, 1978
  vars:           12                        3 Oct 2006 13:53
  size:         3,478 (99.7% of memory free)
```

	storage	display	value	
variable name	type	format	label	variable label
make	str18	%-18s		Marca y modelo
price	int	%8.0gc		Precio
mpg	int	%8.0g		Consumo de combustible
rep78	int	%8.0g		Historia de reparaciones
headroom	float	%6.1f		Cabeza adelante
trunk	int	%8.0g		Volumen del maletero
weight	int	%8.0gc		Peso
length	int	%8.0g		Longitud
turn	int	%8.0g		Radio de giro
displacement	int	%8.0g		Cilindrada
gear_ratio	float	%6.2f		Relación de cambio
foreign	byte	%8.0g		Extranjero

```
Sorted by:  foreign
```

To create labels in more than one language, you set the new language and then define the labels in the standard way; see [D] **label language**.

12.7 Notes attached to data

A dataset may contain notes, which are nothing more than little bits of text that you define and review with the notes command. Typing note, a colon, and the text defines a note:

```
. note:  Send copy to Bob once verified.
```

You can later display whatever notes you have previously defined by typing notes:

```
. notes
_dta:
  1.  Send copy to Bob once verified.
```

Notes are saved with the data, so once you save your dataset, you can replay this note in the future, too.

You can add more notes:

```
. note: Mary wants a copy, too.
```

```
. notes

_dta:
  1.  Send copy to Bob once verified.
  2.  Mary wants a copy, too.
```

The notes you have added so far are attached to the data generically, which is why Stata prefixes them with _dta when it lists them. You can attach notes to variables:

```
. note state: verify values for Nevada.

. note state: what about the two missing values?

. notes

_dta:
  1.  Send copy to Bob once verified.
  2.  Mary wants a copy, too.

state:
  1.  verify values for Nevada.
  2.  what about the two missing values?
```

When you describe your data, you can see whether notes are attached to the dataset or to any of the variables:

```
. describe

Contains data from states.dta
  obs:           50                          1980 state data
  vars:           4
  size:        1,200 (99.3% of memory free)  (_dta has notes)

               storage  display    value
variable name  type     format     label    variable label

state          str8     %9s                *
median_age     float    %9.0g               Median Age
marriage_rate  long     %12.0g              Marriages per 100,000
divorce_rate   long     %12.0g              Divorces per 100,000
                                            * indicated variables have notes

Sorted by:
    Note:  dataset has changed since last saved
```

See [D] **notes** for a complete description of this feature.

12.8   Characteristics

Characteristics are an arcane feature of Stata but are of great use to Stata programmers. In fact, the notes command described above was implemented using characteristics.

The dataset itself and each variable within the dataset have associated with them a set of characteristics. Characteristics are named and referred to as *varname*[*charname*], where *varname* is the name of a variable or _dta. The characteristics contain text and are stored with the data in the Stata-format .dta dataset, so they are recalled whenever the data are loaded.

How are characteristics used? The [XT] **xt** commands need to know the name of the panel variable, and some of these commands also need to know the name of the time variable. xtset is used to specify the panel variable and optionally the time variable. Users need xtset their data only once. Stata then remembers this information, even from a different Stata session. Stata does this with characteristics: _dta[iis] contains the name of the panel variable and _dta[tis] contains the name of the time variable. When an xt command is issued, the command checks these characteristics to obtain the panel and time variables' names. If this information is not found, then the data have not previously been `xtset` and an error message is issued. This use of characteristics is hidden from the user—no mention is made of how the commands remember the identity of the panel variable and the time variable.

Occasionally, commands identify their use of characteristics explicitly. The `xi` command (see [R] **xi**) states that it drops the first level of a categorical variable but that, if you wish to control which level is dropped, you can set the variable's `omit` characteristic. In the documentation, an example is provided where the user types

    . char agegrp[omit] 3

to set the default omission group to 3. As with the [XT] **xt** commands, if the user saves the data after setting the characteristic, the preferred omission group will be remembered from one session to the next.

As a Stata user, you need only understand how to set and clear a characteristic for the few commands that explicitly reveal their use of characteristics. You set a variable *varname*'s characteristic *charname* to *x* by typing

    . char varname[charname] x

You set the data's characteristic *charname* to be *x* by typing

    . char _dta[charname] x

You clear a characteristic by typing

    . char varname[charname]

where *varname* is either a variable name or _dta. You can clear a characteristic, even if it has never been set.

The most important feature of characteristics is that Stata remembers them from one session to the next; they are saved with the data.

❑ Technical Note

Programmers will want to know more. A technical description is found in [P] **char**, but for an overview, you may refer to *varname*'s *charname* characteristic by embedding its name in single quotes and typing '*varname*[*charname*]'; see [U] **18.3.13 Referencing characteristics**.

You can fetch the names of all characteristics associated with *varname* by typing

    . local macname : char varname[]

The maximum length of the contents of a characteristic is 8,681 characters for Small Stata and 67,784 characters for Stata/IC, Stata/SE, and Stata/MP. The association of names with characteristics is by convention. If you, as a programmer, wish to create new characteristics for use in your ado-files, do so, but include at least one capital letter in the characteristic name. The current convention reserves all lowercase names for "official" Stata.

❑

12.9 Reference

Cox, N. J. 2006. Stata tip 33: Sweet sixteen: Hexadecimal formats and precision problems. *Stata Journal* 6: 282–283.

13 Functions and expressions

Contents

If you have not read [U] **11 Language syntax**, please do so before reading this entry.

13.1 Overview

Examples of expressions include

```
2+2
miles/gallons
myv+2/oth
(myv+2)/oth
ln(income)
age<25 & income>50000
age<25 | income>50000
age==25
name=="M Brown"
fname + " " + lname
substr(name,1,10)
_pi
val[_n-1]
L.gnp
```

Expressions like those above are allowed anywhere *exp* appears in a syntax diagram. One example is [D] **generate**:

> generate *newvar* = *exp* $\big[$ *if* $\big]$ $\big[$ *in* $\big]$

The first *exp* specifies the contents of the new variable, and the optional second expression restricts the subsample over which it is to be defined. Another is [R] **summarize**:

> summarize $\big[$ *varlist* $\big]$ $\big[$ *if* $\big]$ $\big[$ *in* $\big]$

The optional expression restricts the sample over which summary statistics are calculated.

Algebraic and string expressions are specified in a natural way using the standard rules of hierarchy. You may use parentheses freely to force a different order of evaluation.

▷ Example 1

myv+2/oth is interpreted as myv+(2/oth). If you wanted to change the order of the evaluation, you could type (myv+2)/oth.

◁

13.2   Operators

Stata has four different classes of operators: arithmetic, string, relational, and logical. Each type is discussed below.

13.2.1   Arithmetic operators

The *arithmetic operators* in Stata are + (addition), − (subtraction), * (multiplication), / (division), ^ (raise to a power), and the prefix − (negation). Any arithmetic operation on a missing value or an impossible arithmetic operation (such as division by zero) yields a missing value.

▷ Example 2

The expression -(x+y^(x-y))/(x*y) denotes the formula

$$ -\frac{x + y^{x-y}}{x \cdot y} $$

and evaluates to *missing* if x or y is missing or zero.

◁

13.2.2   String operators

The + sign is also used as a string operator for the *concatenation* of two strings. Stata determines by context whether + means addition or concatenation. If + appears between two strings, Stata concatenates them. If + appears between two numeric values, Stata adds them.

▷ Example 3

The expression "this"+"that" results in the string "thisthat", whereas the expression 2+3 results in the number 5. Stata issues the error message "type mismatch" if the arguments on either side of the + sign are not of the same type. Thus, the expression 2+"this" is an error, as is 2+"3".

The expressions on either side of the + can be arbitrarily complex:

$$\text{substr(string(20+2),1,1) + upper(substr("rf",1+1,1))}$$

The result of the above expression is the string "2F". See [D] **functions** below for a description of the substr(), string(), and upper() functions.

◁

13.2.3   Relational operators

The *relational operators* are > (greater than), < (less than), >= (greater than or equal), <= (less than or equal), == (equal), and != (not equal). Observe that the relational operator for equality is a pair of equal signs. This convention distinguishes relational equality from the =*exp* assignment phrase.

❏ Technical Note

You may use ~ anywhere ! would be appropriate to represent the logical operator "not". Thus, the not-equal operator may also be written as ~=.

❏

Relational expressions are either *true* or *false*. Relational operators may be used on either numeric or string subexpressions; thus, the expression 3>2 is *true*, as is "zebra">"cat". In the latter case, the relation merely indicates that "zebra" comes after the word "cat" in the dictionary. All uppercase letters precede all lowercase letters in Stata's book, so "cat">"Zebra"sort order, strings is also *true*.

Missing values may appear in relational expressions. If x were a numeric variable, the expression x>=. is *true* if x is missing and *false* otherwise. A missing value is greater than any nonmissing value; see [U] **12.2.1 Missing values**.

▷ Example 4

You have data on age and income and wish to list the subset of the data for persons aged 25 years or less. You could type

    . list if age<=25

If you wanted to list the subset of data of persons aged exactly 25, you would type

    . list if age==25

Note the double equal sign. It would be an error to type list if age=25.

◁

Although it is convenient to think of relational expressions as evaluating to *true* or *false*, they actually evaluate to numbers. A result of *true* is defined as 1 and *false* is defined as 0.

▷ Example 5

The definition of *true* and *false* makes it easy to create indicator, or dummy, variables. For instance,

```
generate incgt10k=income>10000
```

creates a variable that takes on the value 0 when income is less than or equal to $10,000, and 1 when income is greater than $10,000. Since missing values are greater than all nonmissing values, the new variable incgt10k will also take on the value 1 when income is *missing*. It would be safer to type

```
generate incgt10k=income>10000 if income<.
```

Now, observations in which income is *missing* will also contain *missing* in incgt10k. See [U] **25 Dealing with categorical variables** for more examples.

◁

❏ Technical Note

Although you will rarely wish to do so, since arithmetic and relational operators both evaluate to numbers, there is no reason you cannot mix the two types of operators in one expression. For instance, (2==2)+1 evaluates to 2, since 2==2 evaluates to 1, and $1 + 1$ is 2.

Relational operators are evaluated after all arithmetic operations. Thus, the expression (3>2)+1 is equal to 2, whereas 3>2+1 is equal to 0. Evaluating relational operators last guarantees the *logical* (as opposed to the *numeric*) interpretation. It should make sense that 3>2+1 is *false*.

❏

13.2.4   Logical operators

The *logical operators* are & (and), | (or), and ! (not). The logical operators interpret any nonzero value (including *missing*) as *true* and zero as *false*.

▷ Example 6

If you have data on age and income and wish to list data for persons making more than $50,000 along with persons under the age of 25 making more than $30,000, you could type

```
list if income>50000 | income>30000 & age<25
```

The & takes precedence over the |. If you were unsure, however, you could have typed

```
list if income>50000 | (income>30000 & age<25)
```

In either case, the statement will also list all observations for which income is *missing*, since *missing* is greater than 50,000.

◁

❏ Technical Note

Like relational operators, logical operators return 1 for *true* and 0 for *false*. For example, the expression 5 & . evaluates to 1. Logical operations, except for !, are performed after all arithmetic and relational operations; the expression 3>2 & 5>4 is interpreted as (3>2) & (5>4) and evaluates to 1.

❏

13.2.5 Order of evaluation, all operators

The order of evaluation (from first to last) of all operators is ! (or ~), ^, - (negation), /, *, - (subtraction), +, != (or ≅), >, <, <=, >=, ==, &, and |.

13.3 Functions

Stata provides mathematical functions, probability and density functions, matrix functions, string functions, functions for dealing with dates and time series, and a set of special functions for programmers. You can find all of these documented in [D] **functions**. Stata's matrix programming language, Mata, provides more functions and those are documented in the *Mata Reference Manual*, or online (type `help mata functions`).

Functions are merely a set of rules; you supply the function with arguments, and the function evaluates the arguments according to the rules that define the function. Since functions are essentially subroutines that evaluate arguments and cause no action on their own, functions must be used in conjunction with a Stata command. Functions are indicated by the function name, an open parenthesis, an expression or expressions separated by commas, and a close parenthesis.

For example,

```
. display sqrt(4)
2
```

or

```
. display sqrt(2+2)
2
```

demonstrates the simplest use of a function. Here we have used the mathematical function, `sqrt()`, which takes one number (or expression) as its argument and returns its square root. The function was used with the Stata command `display`. If we had simply typed

```
. sqrt(4)
```

Stata would have returned the error message

```
unrecognized command:  sqrt
r(199);
```

Functions can operate on variables, as well. For example, suppose that you wanted to generate a random variable that has observations drawn from a log-normal distribution. You could type

```
. set obs 5
obs was 0, now 5
. generate y = uniform()
. replace y = invnormal(y)
(5 real changes made)
. replace y = exp(y)
(5 real changes made)
. list
```

	y
1.	.686471
2.	2.380994
3.	.2814537
4.	1.215575
5.	.2920268

You could have saved yourself some typing by typing just

```
. generate y = exp(invnormal(uniform()))
```

Functions accept expressions as arguments.

All functions are defined over a specified domain and return values within a specified range. Whenever an argument is outside a function's domain, the function will return a missing value or issue an error message, whichever is most appropriate. For example, if you supplied the log() function with an argument of zero, the log(0) would return a missing value because zero is outside the natural logarithm function's domain. If you supplied the log() function with a string argument, Stata would issue a "type mismatch" error because log() is a numerical function and is undefined for strings. If you supply an argument that evaluates to a value that is outside the function's range, the function will return a missing value. Whenever a function accepts a string as an argument, the string must be enclosed in double quotes, unless you provide the name of a variable that has a string storage type.

13.4   System variables (_variables)

Expressions may also contain _variables (pronounced "underscore variables"), which are built-in system variables that are created and updated by Stata. They are called _variables because their names all begin with the underscore character, '_'.

The _variables are

[eqno]_b[varname] (synonym: [eqno]_coef[varname]) contains the value (to machine precision) of the coefficient on varname from the most recently fitted model (such as ANOVA, regression, Cox, logit, probit, and multinomial logit). See [U] **13.5 Accessing coefficients and standard errors** below for a complete description.

_cons is always equal to the number 1 when used directly and refers to the intercept term when used indirectly, as in _b[_cons].

_n contains the number of the current observation.

_N contains the total number of observations in the dataset.

_pi contains the value of π to machine precision.

_rc contains the value of the return code from the most recent capture command.

[eqno]_se[varname] contains the value (to machine precision) of the standard error of the coefficient on varname from the most recently fitted model (such as ANOVA, regression, Cox, logit, probit, and multinomial logit). See [U] **13.5 Accessing coefficients and standard errors** below for a complete description.

13.5   Accessing coefficients and standard errors

After fitting a model, you can access the coefficients and standard errors and use them in subsequent expressions. Also see [R] **predict** (and [U] **20 Estimation and postestimation commands**) for an easier way to obtain predictions, residuals, and the like.

13.5.1 Simple models

First, let's consider estimation methods that yield one estimated equation with a one-to-one correspondence between coefficients and variables such as cnreg, logit, ologit, oprobit, probit, regress, and tobit. _b[*varname*] (synonym _coef[*varname*]) contains the coefficient on *varname* and _se[*varname*] contains its standard error, and both are recorded to machine precision. Thus, _b[age] refers to the calculated coefficient on the age variable after typing, say, regress response age sex, and _se[age] refers to the standard error on the coefficient. _b[_cons] refers to the constant and _se[_cons] to its standard error. Thus, you might type

```
. regress response age sex
. generate asif = _b[_cons] + _b[age]*age
```

13.5.2 ANOVA and MANOVA models

In ANOVA there is no simple relationship between the coefficients and the variables. For continuous variables in the model, _b[*varname*] refers to the coefficient. This works just as it does in simple models. For categorical variables, you must specify the level as well as the variable. _b[drug[2]] refers to the coefficient on the second level of drug. For interactions, _b[drug[2]*disease[1]] refers to the coefficient on the second level of drug and the first level of disease. Standard errors are obtained similarly using _se[]. Thus, you might type

```
. anova outcome sex age drug sex*age drug*age, continuous(age)
. generate age_effect = _b[age]*age
. replace age_effect = age_effect + _b[sex[1]*age] if sex==1
. replace age_effect = age_effect + _b[sex[2]*age] if sex==2
```

You access the coefficients and standard errors after manova in the same manner as with anova—using _b[] and _se[]—but, because manova is a multiple-equation estimator, the _b[] and _se[] must be preceded by an equation number in square brackets; see [U] **13.5.3 Multiple-equation models**.

13.5.3 Multiple-equation models

The syntax for referring to coefficients and standard errors in multiple-equation models is the same as in the simple-model case, except that _b[] and _se[] are preceded by an equation number in square brackets. There are, however, many alternatives in how you may type requests. The way that you are supposed to type requests is

$$[eqno]\_b[varname]$$
$$[eqno]\_se[varname]$$

but you may substitute _coef[] for _b[]. In fact, you may omit the _b[] altogether, and most Stata users do:

$$[eqno][varname]$$

You may also omit the second pair of square brackets:

$$[eqno]varname$$

There are two ways to specify the equation number *eqno*: either as an absolute equation number or as an "indirect" equation number. In the absolute form, the number is preceded by a '#' sign. Thus, [#1]displ refers to the coefficient on displ in the first equation (and [#1]_se[displ] refers to its standard error). You can even use this form for simple models, such as regress, if you prefer. regress estimates one equation, so [#1]displ refers to the coefficient on displ, just as _b[displ] does. Similarly, [#1]_se[displ] and _se[displ] are equivalent. The logic works both ways—in the multiple-equation context, _b[displ] refers to the coefficient on displ in the first equation and _se[displ] refers to its standard error. _b[*varname*] (_se[*varname*]) is just another way of saying [#1]*varname* ([#1]_se[*varname*]).

Equations may also be referred to indirectly. [res]displ refers to the coefficient on displ in the equation named res. Equations are often named after the corresponding dependent variable name if there is such a concept in the fitted model, so [res]displ might refer to the coefficient on displ in the equation for variable res.

For multinomial logit (mlogit), multinomial probit (mprobit), and similar commands, equations are named after the levels of the single dependent categorical variable. In these models, there is one dependent variable, and there is an equation corresponding to each of the outcomes (values taken on) recorded in that variable, except for the one that is taken to be the base outcome. [res]displ would be interpreted as the coefficient on displ in the equation corresponding to the outcome res. If outcome res is the base outcome, Stata treats [res]displ as zero (and Stata does the same for [res]_se[displ]).

Continuing with the multinomial outcome case: the outcome variable must be numeric. The syntax [res]displ would be understood only if there were a value label associated with the numeric outcome variable and res were one of the labels. If your data are not labeled, then you can use the usual multiple-equation syntax [##]*varname* and [##]_se[*varname*] to refer to the coefficient and standard error for variable *varname* in the #th equation.

For mlogit, if your data are not labeled, you can also use the syntax [#]*varname* and [#]_se[*varname*] (without the '#') to refer to the coefficient and standard error for *varname* in the equation for outcome #.

13.6   Accessing results from Stata commands

Most Stata commands—not just estimation commands—save results so that you can access them in subsequent expressions. You do that by referring to e(*name*), r(*name*), s(*name*), or c(*name*).

```
. summarize age
. generate agedev = age-r(mean)
. regress mpg weight
. display "The number of observations used is " e(N)
```

Most commands are categorized as r-class, meaning that they save results in r(). The returned results—such as r(mean)—are available immediately following the command, and if you are going to refer to them, you need to refer to them soon because the next command will probably replace what is in r().

e-class commands are Stata's estimation commands—commands that fit models. Results in e() remain available until the next model is fitted.

s-class commands are parsing commands—commands used by programmers to interpret commands you type. Few commands save anything in s().

There are no c-class commands. c() contains values that are always available, such as c(current_date) (today's date), c(pwd) (the current directory), C(N) (the number of observations), and so on. There are many c() values and they are documented in [P] **creturn**.

Every command of Stata is designated r-class, e-class, or s-class, or, if the command saves nothing, n-class. r stands for return as in returned results, e stands for estimation as in estimation results, s stands for string, and, admittedly, this last acronym is weak, n stands for null.

You can find out what is stored where by looking in the *Saved Results* section for the particular command in the *Reference* manual. If you know the class of a command—and it is easy enough to guess—you can also see what is stored by typing return list, ereturn list, or sreturn list:

See [R] **saved results** and [U] **18.8 Accessing results calculated by other programs**.

13.7   Explicit subscripting

Individual observations on variables can be referred to by subscripting the variables. Explicit subscripts are specified by following a variable name with square brackets that contain an expression. The result of the subscript expression is truncated to an integer, and the value of the variable for the indicated observation is returned. If the value of the subscript expression is less than 1 or greater than _N, a missing value is returned.

13.7.1   Generating lags and leads

When you type something like

```
. generate y = x
```

Stata interprets it as if you typed

```
. generate y = x[_n]
```

which means that the first observation of y is to be assigned the value from the first observation of x, the second observation of y is to be assigned the value from the second observation on x, and so on. If you instead typed

```
. generate y = x[1]
```

you would set each observation of y equal to the first observation on x. If you typed

```
. generate y = x[2]
```

you would set each observation of y equal to the second observation on x. If you typed

```
. generate y = x[0]
```

Stata would merely copy a missing value into every observation of y because observation 0 does not exist. The same would happen if you typed

```
. generate y = x[100]
```

and you had fewer than 100 observations in your data.

When you type the square brackets, you are specifying explicit subscripts. Explicit subscripting combined with the _*variable* _n can be used to create lagged values on a variable. The lagged value of a variable x can be obtained by typing

```
. generate xlag = x[_n-1]
```

If you are really interested in lags and leads, you probably have time-series data and would be better served by using the time-series operators, such as L.x. Time-series operators can be used with varlists and expressions and they are safer because they account for gaps in the data; see [U] **11.4.3 Time-series varlists** and [U] **13.8 Time-series operators**. Even so, it is important that you understand how the above works.

The built-in underscore variable _n is understood by Stata to mean the observation number of the current observation. That is why

```
. generate y = x[_n]
```

results in observation 1 of x being copied to observation 1 of y and similarly for the rest of the observations. Consider

```
. generate xlag = x[_n-1]
```

_n-1 evaluates to the observation number of the previous observation. For the first observation, _n-1 = 0 and therefore xlag[1] is set to missing. For the second observation, _n-1 = 1 and xlag[2] is set to the value of x[1], and so on.

Similarly, the lead of x can be created by

```
. generate xlead = x[_n+1]
```

Here the last observation on the new variable xlead will be *missing* because _n+1 will be greater than _N (_N is the total number of observations in the dataset).

13.7.2 Subscripting within groups

When a command is preceded by the by *varlist*: prefix, subscript expressions and the underscore variables _n and _N are evaluated relative to the subset of the data currently being processed. For example, consider the following (admittedly not very interesting) data:

```
. use http://www.stata-press.com/data/r10/gxmpl6
. list
```

	bvar	oldvar
1.	1	1.1
2.	1	2.1
3.	1	3.1
4.	2	4.1
5.	2	5.1

To see how _n, _N, and explicit subscripting work, let's create three new variables demonstrating each and then list their values:

```
. generate small_n = _n
. generate big_n = _N
. generate newvar = oldvar[1]
```

```
. list
```

	bvar	oldvar	small_n	big_n	newvar
1.	1	1.1	1	5	1.1
2.	1	2.1	2	5	1.1
3.	1	3.1	3	5	1.1
4.	2	4.1	4	5	1.1
5.	2	5.1	5	5	1.1

small_n (which is equal to _n) goes from 1 to 5, and big_n (which is equal to _N) is 5. This should not be surprising; there are 5 observations in the data, and _n is supposed to count observations, whereas _N is the total number. newvar, which we defined as oldvar[1], is 1.1. Indeed, we see that the first observation on oldvar is 1.1.

Now, let's repeat those same three steps, only this time preceding each step with the prefix by bvar:. First, we will drop the old values of small_n, big_n, and newvar so that we start fresh:

```
. drop small_n big_n newvar
. by bvar, sort: generate small_n=_n
. by bvar: generate big_n =_N
. by bvar: generate newvar=oldvar[1]
. list
```

	bvar	oldvar	small_n	big_n	newvar
1.	1	1.1	1	3	1.1
2.	1	2.1	2	3	1.1
3.	1	3.1	3	3	1.1
4.	2	4.1	1	2	4.1
5.	2	5.1	2	2	4.1

The results are different. Remember that we claimed that _n and _N are evaluated relative to the subset of data in the by-group. Thus, small_n (_n) goes from 1 to 3 for bvar = 1 and from 1 to 2 for bvar = 2. big_n (_N) is 3 for the first group and 2 for the second. Finally, newvar (oldvar[1]) is 1.1 and 4.1.

▷ Example 7

You now know enough to do some amazing things.

Suppose that you have data on individual states and you have another variable in your data called region that divides the states into the four census regions. You have a variable x in your data, and you want to make a new variable called avgx to include in your regressions. This new variable is to take on the average value of x for the region in which the state is located. Thus, for California you will have the observation on x and the observation on the average value in the region, avgx. Here's how:

```
. by region, sort: generate avgx=sum(x)/_n
. by region: replace avgx=avgx[_N]
```

First, by region, we generate avgx equal to the running sum of x divided by the number of observations so far. The , sort ensures that the data are in region order. We have, in effect, created the running average of x within region. It is the last observation of this running average, the overall average within the region, that interests us. So, by region, we replace every avgx observation in a region with the last observation within the region, avgx[_N].

Here is what we will see when we type these commands:

```
. use http://www.stata-press.com/data/r10/gxmpl7, clear
. by region, sort: generate avgx=sum(x)/_n
. by region: replace avgx=avgx[_N]
(46 real changes made)
```

In our example, there are no missing observations on x. If there had been, we would have obtained the wrong answer. When we created the running average, we typed

```
. by region, sort: generate avgx=sum(x)/_n
```

The problem is not with the sum() function. When sum() encounters a missing, it adds zero to the sum. The problem is with _n. Let's assume that the second observation in the first region has recorded a missing for x. When Stata processes the third observation in that region, it will calculate the sum of two elements (remember that one is missing) and then divide the sum by 3 when it should be divided by 2. There is an easy solution:

```
. by region: generate avgx=sum(x)/sum(x<.)
```

Rather than divide by _n, we divide by the total number of nonmissing observations seen on x so far, namely, the sum(x<.).

If our goal were simply to obtain the mean, we could have more easily accomplished it by typing egen avgx=mean(x), by(region); see [D] **egen**. egen, however, is written in Stata, and the above is how egen's mean() function works. The general principles are worth understanding.

◁

▷ Example 8

You have some patient data recording vital signs at various times during an experiment. The variables include patient, an ID number or name of the patient; time, a variable recording the date or time or epoch of the vital-sign reading; and vital, a vital sign. You probably have more than one vital sign, but one is enough to illustrate the concept. Each observation in your data represents a patient-time combination.

Let's assume that you have 1,000 patients and, for every observation on the same patient, you want to create a new variable called orig that records the patient's initial value of this vital sign.

```
. use http://www.stata-press.com/data/r10/gxmpl8, clear
. sort patient time
. by patient: generate orig=vital[1]
```

Observe that vital[1] refers not to the first reading on the first patient but to the first reading on the current patient, because we are performing the generate command by patient.

◁

▷ Example 9

Let's do one more example with these patient data. Suppose that we want to create a new dataset from our patient data that record not only the patient's identification, the time of the reading of the first vital sign, and the first vital sign reading itself, but also the time of the reading of the last vital sign and its value. We want 1 observation per patient. Here's how:

```
. sort patient time
. by patient: generate lasttime=time[_N]
. by patient: generate lastvital=vital[_N]
. by patient: drop if _n!=1
```

◁

13.8   Time-series operators

Time-series operators allow you to refer to the lag of gnp by typing L.gnp, the second lag by typing L2.gnp, etc. There are also operators for lead (F), difference D, and seasonal difference S.

Time-series operators can be used with varlists and with expressions. See [U] **11.4.3 Time-series varlists** if you have not read it already. This section has to do with using time-series operators in expressions such as with generate. You do not have to create new variables; you can use the time-series operated variables directly.

13.8.1   Generating lags and leads

In a time-series context, referring to L2.gnp is better than referring to gnp[_n-2] because there might be missing observations. Pretend that observation 4 contains data for $t = 25$ and observation 5 data for $t = 27$. L2.gnp will still produce correct answers; L2.gnp for observation 5 will be the value from observation 4 because the time-series operators look at t to find the relevant observation. The more mechanical gnp[_n-2] just goes 2 observations back, which, here, would not produce the desired result.

Time-series operators can be used with varlists or with expressions, so you can type

```
. regress val L.gnp r
```

or

```
. generate gnplagged = L.gnp
. regress val gnplagged
```

Before you can type either one, however, you must use the tsset command to tell Stata the identity of the time variable; see [TS] **tsset**. Once you have tsset the data, anyplace you see an *exp* in a syntax diagram, you may type time series–operated variables, so you can type

```
. summarize r if F.gnp < gnp
```

or

```
. generate grew = 1 if gnp > L.gnp & L.gnp < .
. replace grew = 0 if grew >= . & L.gnp < .
```

or

```
. generate grew = (gnp > L.gnp) if L.gnp < .
```

(Continued on next page)

13.8.2   Operators within groups

Stata also understands panel or cross-sectional time-series data. For instance, if you type

. tsset country time

you are declaring that you have time-series data. The time variable is time, and you have time-series data for separate countries.

Once you have tsset both cross-sectional and time identifiers, you proceed just as you would if you had a simple time series.

. generate grew = (gnp > L.gnp) if L.gnp < .

would produce correct results. The L. operator will not confuse the observation at the end of one panel with the beginning of the next.

13.9   Label values

If you have not read [U] **12.6 Dataset, variable, and value labels**, please do so. You may use labels in an expression in place of the numeric values with which they are associated. To use a label in this way, type the label in double quotes followed by a colon and the name of the value label.

▷ Example 10

If the value label yesno associates the label yes with 1 and no with 0, then "yes":yesno (said aloud as the value of yes under yesno) is evaluated as 1. If the double-quoted label is not defined in the indicated value label, or if the value label itself is not found, a missing value is returned. Thus, the expression "maybe":yesno is evaluated as *missing*.

. use http://www.stata-press.com/data/r10/gxmpl9, clear

. list

	name	answer
1.	Mikulin	no
2.	Gaines	no
3.	Hilbe	yes
4.	DeLeon	no
5.	Cain	no
6.	Wann	yes
7.	Schroeder	no
8.	Cox	no
9.	Bishop	no
10.	Hardin	yes
11.	Lancaster	yes
12.	Poole	no

. list if answer=="yes":yesno

	name	answer
3.	Hilbe	yes
6.	Wann	yes
10.	Hardin	yes
11.	Lancaster	yes

In the above example, the variable answer is not a string variable; it is a numeric variable that has the associated value label yesno. Since yesno associates yes with 1 and no with 0, we could have typed list if answer==1 instead of what we did type. We could not have typed list if answer=="yes" because answer is not a string variable. If we had, we would have received the error message "type mismatch".

◁

13.10   Precision and problems therein

Examine the following short Stata session:

```
. drop _all

. input x y

              x         y
  1. 1 1.1
  2. 2 1.2
  3. 3 1.3
  4. end

. count if x==1
       1

. count if y==1.1
       0

. list
```

	x	y
1.	1	1.1
2.	2	1.2
3.	3	1.3

We created a dataset containing two variables, x and y. The first observation has x equal to 1 and y equal to 1.1. When we asked Stata to count the number of times that the variable x took on the value 1, we were told that it occurred once. Yet when we asked Stata to count the number of times y took on the value 1.1, we were told zero—meaning that it never occurred. What has gone wrong? When we list the data, we see that the first observation has y equal to 1.1.

Despite appearances, Stata has not made a mistake. Stata stores numbers internally in binary form, and the number 1.1 has no exact binary representation—that is, there is no finite string of binary digits that is equal to 1.1.

❑ Technical Note

The number 1.1 in binary form is 1.0001100110011 ..., where the period represents the binary point. The problem binary computers have with storing numbers like 1/10 is much like the problem we base-10 users have in precisely writing 1/11, which is 0.0909090909 ....

❑

The number that appears as 1.1 in the listing above is actually 1.1000000238419, which is off by roughly 2 parts in 10^8. Unless we tell Stata otherwise, it stores all numbers as floats, which are also known as *single-precision* or *4-byte reals*. On the other hand, Stata performs all internal calculations in double, which is also known as *double-precision* or *8-byte reals*. This is what leads to the difficulty.

In the above example, we compared the number 1.1, stored as a `float`, with the number 1.1 stored as a `double`. The double-precision representation of 1.1 is more accurate than the single-precision representation, but it is also different. Those two numbers are not equal.

There are several ways around this problem. The problem with 1.1 apparently not equaling 1.1 would never arise if the storage precision and the precision of the internal calculations were the same. Thus, you could store all your data as `doubles`. This takes more computer memory, however, and it is unlikely that (1) your data are really that accurate and (2) the extra digits would meaningfully affect any calculated result, even if the data were that accurate.

❏ Technical Note

This is unlikely to affect any calculated result because Stata performs all internal calculations in double precision. This is all rather ironic, since the problem would also not arise if we had designed Stata to use single precision for its internal calculations. Stata would be less accurate, but the problem would have been completely disguised from the user, making this entry unnecessary.

❏

Another solution is to use the `float()` function. `float(x)` rounds x to its `float` representation. If we had typed `count if y==float(1.1)` in the above example, we would have been informed that there is one such value.

13.11   References

Cox, N. J. 2006. Stata tip 33: Sweet sixteen: Hexadecimal formats and precision problems. *Stata Journal* 6: 282–283.

Gould, W. 2006. Mata Matters: Precision. *Stata Journal* 6: 550–560.

14 Matrix expressions

Contents

14.1 Overview

Stata has two matrix programming languages, one that might be called Stata's older matrix language and another that is called Mata. Stata's Mata is the new one, and there is an uneasy relationship between the two.

Below we discuss Stata's older language and leave the newer one to another manual—the *Mata Reference Manual* ([M])—or you can learn about the newer one by typing `help mata`.

We admit that the newer language is better in almost every way than the older language, but the older one still has a use because it is the one that Stata truly and deeply understands. Even when Mata wants to talk to Stata, matrixwise, it is the older language that Mata must use, so you must learn to use the older language as well as the new.

This is not nearly as difficult, or messy, as you might imagine because Stata's older language is remarkably easy to use, and really, there is not much to learn. Just remember that for heavy-duty programming, it will be worth your time to learn Mata, too.

14.1.1 Definition of a matrix

Stata's definition of a matrix includes a few details that go beyond the mathematics. To Stata, a matrix is a named entity containing an $r \times c$ ($0 < r \leq$ matsize, $0 < c \leq$ matsize) rectangular array of double-precision numbers (including missing values) that is bordered by a row and a column of names.

```
. matrix list A

A[3,2]
    c1  c2
r1   1   2
r2   3   4
r3   5   6
```

149

Here we have a 3×2 matrix named A containing elements 1, 2, 3, 4, 5, and 6. Row 1, column 2 (written $A_{1,2}$ in math and A[1,2] in Stata) contains 2. The columns are named c1 and c2 and the rows, r1, r2, and r3. These are the default names Stata comes up with when it cannot do better. The names do not play a role in the mathematics, but they are of great help when it comes to labeling the output.

The names are operated on just as the numbers are. For instance,

```
. matrix B=A'*A
. matrix list B
symmetric B[2,2]
     c1  c2
c1   35
c2   44  56
```

We defined $\mathbf{B} = \mathbf{A}'\mathbf{A}$. The row and column names of $\mathbf{B}$ are the same. Multiplication is defined for any $a \times b$ and $b \times c$ matrices, the result being $a \times c$. Thus, the row and column names of the result are the row names of the first matrix and the column names of the second matrix. We formed $\mathbf{A}'\mathbf{A}$, using the transpose of $\mathbf{A}$ for the first matrix—which also interchanged the names—and so obtained the names shown.

14.1.2   matsize

Matrices are limited to being no larger than matsize $\times$ matsize. The default value of matsize is 400 for Stata/MP and Stata/SE and 200 for Stata/IC, but you can reset this with the set matsize command; see [R] **matsize**.

The maximum value of matsize is 800 for Stata/IC, so matrices are not suitable for holding many data. This restriction does not prove a limitation because terms that appear in statistical formulas are of the form $(\mathbf{X}'\mathbf{W}\mathbf{Z})$ and Stata provides a command, matrix accum, for efficiently forming such matrices; see [U] **14.6 Creating matrices by accumulating data** below. The maximum value of matsize is 11,000 for Stata/MP and Stata/SE, so performing matrix operations directly on many data is more feasible. The matsize limit does not apply to Mata matrices; see the *Mata Reference Manual*.

14.2   Row and column names

Matrix rows and columns always have names. Stata is smart about setting these names when the matrix is created, and the matrix commands and operators manipulate these names throughout calculations, so the names typically are set correctly at the conclusion of matrix calculations.

For instance, consider the matrix calculation $\mathbf{b} = (\mathbf{X}'\mathbf{X})^{-1}\mathbf{X}'\mathbf{y}$ performed on real data:

```
. use http://www.stata-press.com/data/r10/auto
(1978 Automobile Data)
. matrix accum XprimeX = weight foreign
(obs=74)
. matrix vecaccum yprimeX = mpg weight foreign
. matrix b = invsym(XprimeX)*yprimeX'
. matrix list b
b[3,1]
                 mpg
 weight   -.00658789
foreign   -1.6500291
  _cons    41.679702
```

These names were produced without our ever having given a special command to place the names on the result. When we formed matrix XprimeX, Stata produced the result

```
. matrix list XprimeX
symmetric XprimeX[3,3]
            weight    foreign      _cons
 weight  7.188e+08
foreign      50950         22
  _cons     223440         22         74
```

`matrix accum` forms $\mathbf{X'X}$ matrices from data and sets the row and column names to the variable names used. The names are correct in the sense that, for instance, the (1,1) element is the sum across the observations of squares of `weight` and the (2,1) element is the sum of the product of `weight` and `foreign`.

Similarly, `matrix vecaccum` forms $\mathbf{y'X}$ matrices, and it sets the row and column names to the variable names used, so `matrix vecaccum yprimeX = mpg weight foreign` resulted in

```
. matrix list yprimeX
yprimeX[1,3]
        weight  foreign     _cons
mpg    4493720      545      1576
```

The final step, `matrix b = invsym(XprimeX)*yprimeX'`, manipulated the names, and, if you think carefully, you can derive the rules for yourself. `invsym()` (inversion) is much like transposition, so row and column names must be swapped. Here, however, the matrix was symmetric, so that amounted to leaving the names as they were. Multiplication amounts to taking the column names of the first matrix and the row names of the second. The final result is

```
. matrix list b
b[3,1]
                   mpg
 weight   -.00658789
foreign   -1.6500291
  _cons    41.679702
```

and the interpretation is $\mathtt{mpg} = -.00659\,\mathtt{weight} - 1.65\,\mathtt{foreign} + 41.68 + e$.

Researchers realized long ago that using matrix notation simplifies the description of complex calculations. What they may not have realized is that, corresponding to each mathematical definition of a matrix operator, there is a definition of the operator's effect on the names that can be used to carry the names forward through long and complex matrix calculations.

14.2.1   The purpose of row and column names

Mostly, matrices in Stata are used in programming estimators, and Stata uses row and column names to produce pretty output. Say that we wrote code—interactively or in a program—that produced the following coefficient vector b and covariance matrix V:

```
. matrix list b
b[1,3]
           weight  displacement        _cons
y1     -.00656711     .00528078    40.084522
. matrix list V
symmetric V[3,3]
                    weight  displacement         _cons
      weight     1.360e-06
displacement     -.0000103     .00009741
      _cons     -.00207455     .01188356     4.0808455
```

We could now produce standard estimation output by coding two more lines:

```
. ereturn post b V
. ereturn display
```

	Coef.	Std. Err.	z	P>|z|	[95% Conf.	Interval]
weight	-.0065671	.0011662	-5.63	0.000	-.0088529	-.0042813
displacement	.0052808	.0098696	0.54	0.593	-.0140632	.0246248
_cons	40.08452	2.02011	19.84	0.000	36.12518	44.04387

Stata's `ereturn` command knew to produce this output because of the row and column names on the coefficient vector and variance matrix. Moreover, we usually do nothing special in our code that produces b and V to set the row and column names because, given how matrix names work, they work themselves out.

Also, sometimes row and column names help us detect programming errors. Assume that we wrote code to produce matrices b and V but made a mistake. Sometimes our mistake will result in the wrong row and column names. Rather than the b vector we previously showed you, we might produce

```
. matrix list b
b[1,3]
          weight          c2        _cons
y1   -.00656711       42.23    40.084522
```

If we posted our estimation results now, Stata would refuse because it can tell by the names that there is a problem:

```
. ereturn post b V
name conflict
r(507);
```

Understand, however, that Stata follows the standard rules of matrix algebra; the names are just along for the ride. Matrices are summed by position, meaning that a directive to form $C = A + B$ results in $C_{11} = A_{11} + B_{11}$, regardless of the names, and it is not an error to sum matrices with different names:

```
. matrix list a
symmetric a[3,3]
               c1          c2          c3
    mpg     14419
 weight   1221120   1.219e+08
  _cons       545       50950          22
. matrix list b
symmetric b[3,3]
                     c1          c2          c3
displacement   3211055
        mpg    227102       22249
      _cons     12153        1041          52
. matrix c = a + b
. matrix list c
symmetric c[3,3]
                     c1          c2          c3
displacement   3225474
        mpg   1448222   1.219e+08
      _cons     12698       51991          74
```

Matrix row and column names are used to label output; they do not affect how matrix algebra is performed.

14.2.2   Three-part names

Row and column names have three parts: *equation_name:ts_operator.subname.*

In the examples shown so far, the first two parts have been blank; the row and column names consisted of *subnames* only. This is typical. Run any single-equation model (such as those produced by `regress`, `probit`, and `logistic`), and if you fetch the resulting matrices, you will find that they have row and column names of the *subname* form.

Those who work with time-series data will find matrices with row and column names of the form *ts_operator.subname.* For example,

```
. matrix list example1
symmetric example1[3,3]
                           L.
              rate        rate        _cons
   rate    3.0952534
 L.rate     .0096504    .00007742
  _cons   -2.8413483   -.01821928   4.8578916
```

We obtained this matrix by running a linear regression on `rate` and `L.rate` and then fetching the covariance matrix. Think of the row and column name `L.rate` no differently from how you think of `rate` or, in the previous examples, `r1`, `r2`, `c1`, `c2`, `weight`, and `foreign`.

Equation names are used to label partitioned matrices and, in estimation, occur in the context of multiple equations. Here is a matrix with *equation_names* and *subnames*:

```
. matrix list example2
symmetric example2[5,5]
                        mpg:         mpg:         mpg:         mpg:         mpg:
                     foreign        displ        _cons      foreign        _cons
   mpg:foreign    1.6483972
     mpg:displ      .004747     .00003876
     mpg:_cons   -1.4266352    -.00905773   2.4341021
weight:foreign  -51.208454     -4.665e-19   15.224135    24997.727
  weight:_cons   15.224135      2.077e-17  -15.224135   -7431.7565    7431.7565
```

Here is an example with all three parts filled in:

```
. matrix list example3
symmetric example3[5,5]
                        val:         val:         val:       weight:      weight:
                                       L.
                        rate         rate        _cons      foreign        _cons
    val:rate       2.2947268
  val:L.rate        .00385216    .0000309
    val:_cons     -1.4533912    -.0072726    2.2583357
weight:foreign  -163.86684      7.796e-17   49.384526    25351.696
  weight:_cons    49.384526     -1.566e-16  -49.384526   -7640.237     7640.237
```

`val:L.rate` is a column name, just as, in the previous section, `c2` and `foreign` were names.

Say that this last matrix is the variance matrix produced by a program we wrote and that our program also produced a coefficient vector `b`:

```
. matrix list b
b[1,5]
                 val:         val:         val:       weight:      weight:
                                L.
                 rate         rate        _cons      foreign        _cons
 y1    4.5366753   -.00316923    20.68421   -1008.7968    3324.7059
```

Here is the result of posting and displaying the results:

```
. ereturn post b example3
. ereturn display
```

	Coef.	Std. Err.	z	P>|z|	[95% Conf.	Interval]
val						
rate						
—	4.536675	1.514836	2.995	0.003	1.567652	7.505698
L1	-.0031692	.0055591	-0.570	0.569	-.0140648	.0077264
_cons	20.68421	1.502776	13.764	0.000	17.73882	23.6296
weight						
foreign	-1008.797	159.2222	-6.336	0.000	-1320.866	-696.7271
_cons	3324.706	87.40845	38.036	0.000	3153.388	3496.023

The equation names are used to separate one equation from the next.

14.2.3 Setting row and column names

You reset row and column names by using the `matrix rownames` and `matrix colnames` commands.

Before resetting the names, use `matrix list` to verify that the names are not set correctly; often, they already are. When you enter a matrix by hand, however, the row names are unimaginatively set to r1, r2, ..., and the column names to c1, c2, ...

```
. matrix a = (1,2,3\4,5,6)
. matrix list a
a[2,3]
      c1  c2  c3
r1     1   2   3
r2     4   5   6
```

Regardless of the current row and column names, `matrix rownames` and `matrix colnames` reset them:

```
. matrix colnames a = foreign alpha _cons
. matrix rownames a = one two
. matrix list a
a[2,3]
       foreign   alpha   _cons
one          1       2       3
two          4       5       6
```

You may set the *ts_operator* as well as the *subname*,

```
. matrix colnames a = foreign l.rate _cons
. matrix list a
a[2,3]
                    L.
       foreign    rate   _cons
one          1       2       3
two          4       5       6
```

and you may set equation names:

```
. matrix colnames a = this:foreign this:l.rate that:_cons

. matrix list a

a[2,3]
            this:     this:     that:
                      L.
          foreign     rate     _cons
  one         1         2         3
  two         4         5         6
```

See [P] **matrix rownames** for more information.

14.2.4   Obtaining row and column names

matrix list displays the matrix with its row and column names. In a programming context, you can fetch the row and column names into a macro using

$$\text{local} \ldots : \text{rowfullnames } \textit{matname}$$
$$\text{local} \ldots : \text{colfullnames } \textit{matname}$$
$$\text{local} \ldots : \text{rownames } \textit{matname}$$
$$\text{local} \ldots : \text{colnames } \textit{matname}$$
$$\text{local} \ldots : \text{roweq } \textit{matname}$$
$$\text{local} \ldots : \text{coleq } \textit{matname}$$

rowfullnames and colfullnames return the full names (*equation_name*:*ts_operator*.*subname*) listed one after the other.

rownames and colnames omit the equations and return *ts_operator*.*subname*, listed one after the other.

roweq and coleq return the equation names, listed one after the other.

See [P] **macro** and [P] **matrix define** for more information.

14.3   Vectors and scalars

Stata does not have vectors as such—they are considered special cases of matrices and are handled by the matrix command.

Stata does have scalars, although they are not strictly necessary because they, too, could be handled as special cases. See [P] **scalar** for a description of scalars.

14.4   Inputting matrices by hand

You input matrices using

matrix input *matname* = (...)

or

matrix *matname* = (...)

In either case, you enter the matrices by row. You separate one element from the next by using commas (,) and one row from the next by using backslashes (\). If you omit the word input, you are using the expression parser to input the matrix:

```
. matrix a = (1,2\3,4)
. matrix list a
a[2,2]
     c1  c2
r1   1   2
r2   3   4
```

This has the advantage that you can use expressions for any of the elements:

```
. matrix b = (1, 2+3/2 \ cos(_pi), _pi)
. matrix list b
b[2,2]
            c1          c2
r1           1         3.5
r2          -1   3.1415927
```

The disadvantage is that the matrix must be small, say, no more than 50 elements (regardless of the value of `matsize`).

`matrix input` has no such restriction, but you may not use subexpressions for the elements:

```
. matrix input c  = (1,2\3,4)
. matrix input d = (1, 2+3/2 \ cos(_pi), _pi)
invalid syntax
r(198);
```

Either way, after inputting the matrix, you will probably want to set the row and column names; see [U] **14.2.3 Setting row and column names** above.

14.5   Accessing matrices created by Stata commands

Some Stata commands—including all estimation commands—leave behind matrices that you can subsequently use. After executing an estimation command, type `ereturn list` to see what is available:

```
. use http://www.stata-press.com/data/r10/auto
(1978 Automobile Data)
. probit foreign mpg weight
  (output omitted )
. ereturn list
scalars:
                e(N) =  74
             e(ll_0) =  -45.03320955699139
               e(ll) =  -26.8441890057987
             e(df_m) =  2
             e(chi2) =  36.37804110238538
             e(r2_p) =  .4039023807124769
            e(N_cdf) =  0
            e(N_cds) =  0

macros:
          e(cmdline) : "probit foreign mpg weight"
            e(title) : "Probit regression"
              e(vce) : "oim"
           e(depvar) : "foreign"
             e(cmd) : "probit"
         e(crittype) : "log likelihood"
          e(predict) : "probit_p"
       e(properties) : "b V"
        e(estat_cmd) : "probit_estat"
         e(chi2type) : "LR"
```

```
matrices:
                    e(b)  :   1 x 3
                    e(V)  :   3 x 3
                e(rules)  :   1 x 4
functions:
                e(sample)
```

Most estimation commands leave behind e(b) (the coefficient vector) and e(V) (the variance–covariance matrix of the estimator):

```
. matrix list e(b)

e(b)[1,3]
            mpg       weight        _cons
y1   -.10395033   -.00233554     8.275464
```

You can refer to e(b) and e(V) in any matrix expression:

```
. matrix myb = e(b)

. matrix list myb

myb[1,3]
            mpg       weight        _cons
y1   -.10395033   -.00233554     8.275464

. matrix c = e(b)*invsym(e(V))*e(b)'

. matrix list c

symmetric c[1,1]
             y1
y1   22.440544
```

14.6   Creating matrices by accumulating data

In programming estimators, matrices of the form $\mathbf{X'X}$, $\mathbf{X'Z}$, $\mathbf{X'WX}$, and $\mathbf{X'WZ}$ often occur, where $\mathbf{X}$ and $\mathbf{Z}$ are data matrices. matrix accum, matrix glsaccum, matrix vecaccum, and matrix opaccum produce such matrices; see [P] **matrix accum**.

We recommend that you not load the data into a matrix and use the expression parser directly to form such matrices, although see [P] **matrix mkmat** if that is your interest. If that is your interest, be sure to read the *Technical Note* at the end of [P] **matrix mkmat**. There is much to recommend learning how to use the matrix accum commands.

14.7   Matrix operators

You can create new matrices or replace existing matrices by typing

matrix *matname* = *matrix_expression*

For instance,

```
. matrix A = invsym(R*V*R')
. matrix IAR = I(rowsof(A)) - A*R
. matrix beta = b*IAR' + r*A'
. matrix C = -C'
. matrix D = (A, B \ B', A)
. matrix E = (A+B)*C'
. matrix S = (S+S')/2
```

The following operators are provided:

Operator	Symbol
Unary operators	
negation	−
transposition	,
Binary operators	
(lowest precedence)	
row join	\
column join	,
addition	+
subtraction	−
multiplication	*
division by scalar	/
Kronecker product	#
(highest precedence)	

Parentheses may be used to change the order of evaluation.

Note in particular that , and \ are operators; (1,2) creates a 1×2 matrix (vector), and (A,B) creates a rowsof(A) $\times$ colsof(A)+colsof(B) matrix, where rowsof(A) = rowsof(B). (1\2) creates a 2×1 matrix (vector), and (A\B) creates a rowsof(A)+rowsof(B) $\times$ colsof(A) matrix, where colsof(A) = colsof(B). Thus, expressions of the form

```
matrix R = (A,B)*Vinv*(A,B)'
```

are allowed.

14.8 Matrix functions

In addition to the functions listed below, see [P] **matrix svd** for singular value decomposition, [P] **matrix symeigen** for eigenvalues and eigenvectors of symmetric matrices, and see [P] **matrix eigenvalues** for eigenvalues of nonsymmetric matrices. For a full description of the matrix functions, see [D] **functions**.

Matrix functions returning matrices:

cholesky(M)	I(n)	nullmat(*matname*)
corr(M)	inv(M)	sweep(M,i)
diag(v)	invsym(M)	vec(M)
get(*systemname*)	J(r,c,z)	vecdiag(M)
hadamard(M,N)	matuniform(r,c)	

Matrix functions returning scalars:

colnumb(M,s)	el(M,i,j)	rownumb(M,s)
colsof(M)	issymmetric(M)	rowsof(M)
det(M)	matmissing(M)	trace(M)
diag0cnt(M)	mreldif(X,Y)	

14.9 Subscripting

1. In matrix and scalar expressions, you may refer to *matname*[r,c], where r and c are scalar expressions, to obtain one element of *matname* as a scalar.

   Examples:
   ```
   matrix A = A / A[1,1]
   generate newvar = oldvar / A[2,2]
   ```

2. In matrix expressions, you may refer to *matname*[s_r,s_c], where s_r and s_c are string expressions, to obtain a submatrix with one element. The element returned is based on searching the row and column names.

   Examples:
   ```
   matrix B = V["price","price"]
   generate sdif = dif / sqrt(V["price","price"])
   ```

3. In matrix expressions, you may mix these two syntaxes and refer to *matname*[r,s_c] or to *matname*[s_r,c].

   Example:
   ```
   matrix b = b * R[1,"price"]
   ```

4. In matrix expressions, you may use *matname*[r_1..r_2,c_1..c_2] to refer to submatrices; r_1, r_2, c_1, and c_2 may be scalar expressions. If r_2 evaluates to missing, it is taken as referring to the last row of *matname*; if c_2 evaluates to missing, it is taken as referring to the last column of *matname*. Thus, *matname*[r_1...,c_1...] is allowed.

   Examples:
   ```
   matrix S = Z[1..4, 1..4]
   matrix R = Z[5..., 5...]
   ```

5. In matrix expressions, you may refer to *matname*[s_{r1}..s_{r2},s_{c1}..s_{c2}] to refer to submatrices where s_{r1}, s_{r2}, s_{c1}, and s_{c2}, are string expressions. The matrix returned is based on looking up the row and column names.

   If the string evaluates to an equation name only, all the rows or columns for the equation are returned.

   Examples:
   ```
   matrix S = Z["price".."weight", "price".."weight"]
   matrix L = D["mpg:price".."mpg:weight", "mpg:price".."mpg:weight"]
   matrix T1 = C["mpg:", "mpg:"]
   matrix T2 = C["mpg:", "price:"]
   ```

6. In matrix expressions, any of the above syntaxes may be combined.

   Examples:
   ```
   matrix T1 = C["mpg:", "price:weight".."price:displ"]
   matrix T2 = C["mpg:", "price:weight"...]
   matrix T3 = C["mpg:price", 2..5]
   matrix T4 = C["mpg:price", 2]
   ```

7. When defining an element of a matrix, use

$$\texttt{matrix } \textit{matname}\,[i,j] \;=\; \textit{expression}$$

where i and j are scalar expressions. The matrix *matname* must already exist.

Example:
```
matrix A = J(2,2,0)
matrix A[1,2] = sqrt(2)
```

8. To replace a submatrix within a matrix, use the same syntax. If the expression on the right evaluates to a scalar or 1×1 matrix, the element is replaced. If it evaluates to a matrix, the submatrix with top-left element at (i, j) is replaced. The matrix *matname* must already exist.

Example:
```
matrix A = J(4,4,0)
matrix A[2,2] = C'*C
```

14.10   Using matrices in scalar expressions

Scalar expressions are documented as *exp* in the Stata manuals:

$$\begin{aligned}
&\texttt{generate } \textit{newvar} = \textit{exp} \texttt{ if } \textit{exp} \ldots \\
&\texttt{replace } \textit{newvar} = \textit{exp} \texttt{ if } \textit{exp} \ldots \\
&\texttt{regress } \ldots \texttt{ if } \textit{exp} \ldots \\
&\texttt{if } \textit{exp} \; \{\ldots\} \\
&\texttt{while } \textit{exp} \; \{\ldots\}
\end{aligned}$$

Most importantly, scalar expressions occur in `generate` and `replace`, in the `if` *exp* modifier allowed on the end of many commands, and in the `if` and `while` commands for program control.

You will rarely need to refer to a matrix in any of these situations except when using the `if` and `while` commands.

In any case, you may refer to matrices in any of these situations, but the expression cannot require evaluation of matrix expressions returning matrices. Thus, you could refer to `trace(A)` but not to `trace(A+B)`.

It can be difficult to predict when an evaluation of an expression requires evaluating a matrix; even experienced users can be surprised. If you get the error message "matrix operators that return matrices not allowed in this context", r(509), you have encountered such a situation.

The solution is to split the line in two. For instance, you would change

```
if trace(A+B)==0 {
        . . .
}
```

to

```
matrix AplusB = A+B
if trace(AplusB)==0 {
        . . .
}
```

or even to

```
matrix Trace = trace(A+B)
if Trace[1,1]==0 {
        . . .
}
```

15 Printing and preserving output

15.1   Overview

Stata can record your session into a file called a log file but does not start a log automatically; you must tell Stata to record your session. By default, the resulting log file contains what you type and what Stata produces in response, recorded in a format called Stata Markup and Control Language (SMCL); see [P] **smcl**. The file can be printed or converted to ASCII text for incorporation into documents you create with your word processor.

To start a log:
    Your session is now being recorded
    in file *filename*.`smcl`.
	. `log using` *filename*

To temporarily stop logging:
    Temporarily stop:
    Resume:
	
	. `log off`
	. `log on`

To stop logging and close the file:
    You can now print *filename*.`smcl` or type:
    to create *filename*.`log` that you can
    load into your word processor.
	. `log close`
	. `translate` *filename*.`smcl` *filename*.`log`

Alternative ways to start logging:
    append to an existing log:
    replace an existing log:
	
	. `log using` *filename*, `append`
	. `log using` *filename*, `replace`

Using the GUI:
    To start a log:
    To temporarily stop logging:
    To resume:
    To stop logging and close the file:
    To print previous or current log:
	
	click the **Log** button
	click the **Log** button, and choose **Suspend**
	click the **Log** button, and choose **Resume**
	click the **Log** button, and choose **Close**
	select **File > View**, choose file,
	    right-click on the Viewer, and select **Print**

Also, `cmdlog` will produce logs containing solely what you typed—logs that, although not containing your results, are sufficient to recreate the session.

To start a command-only log:	. `cmdlog using` *filename*
To stop logging and close the file:	. `cmdlog close`
To recreate your session:	. `do` *filename*.`txt`

15.1.1   Starting and closing logs

With great foresight, you begin working in Stata and type `log using session` (or click the **Log** button) before starting your work:

```
. log using session

      log:  C:\example\session.smcl
 log type:  smcl
opened on:  17 Mar 2007, 12:35:08

. use http://www.stata-press.com/data/r10/census
(1980 Census data by state)

. tabulate reg [freq=pop]

  Census |
  region |      Freq.       Percent        Cum.
---------+-----------------------------------------
      NE |  49,135,283        21.75       21.75
 N Cntrl |  58,865,670        26.06       47.81
   South |  74,734,029        33.08       80.89
    West |  43,172,490        19.11      100.00
---------+-----------------------------------------
   Total | 225,907,472       100.00

. summarize medage

 Variable |       Obs        Mean    Std. Dev.       Min        Max
----------+--------------------------------------------------------
   medage |        50       29.54    1.693445       24.2       34.7

. log close
      log:  C:\example\session.smcl
 log type:  smcl
closed on:  17 Mar 2007, 12:35:38
```

There is now a file named `session.smcl` on your disk. If you were to look at it in a text editor or word processor, you would see something like this:

```
{smcl}
{com}{sf}{ul off}{txt}{.-}
      log:  {res}C:\example\session.smcl
  {txt}log type:  {res}smcl
  {txt}opened on:  {res}17 Mar 2007, 12:35:08
{txt}
{com}. use http://www.stata-press.com/data/r10/census
{txt}(1980 Census data by state)

{com}. tabulate reg [freq=pop]

     {txt}Census {c |}
     region {c |}      Freq.       Percent        Cum.
{hline 12}{c +}{hline 35}
         NE {c |}{res} 49,135,283        21.75       21.75
  {txt}   N Cntrl {c |}{res} 58,865,670        26.06       47.81
    (output omitted )
```

What you are seeing is SMCL, which Stata understands. Here is the result of typing the file using Stata's `type` command:

```
. type session.smcl
```

```
        log:  C:\example\session.smcl
   log type:  smcl
  opened on:  17 Mar 2007, 12:35:08
. use http://www.stata-press.com/data/r10/census
(1980 Census data by state)
. tabulate reg [freq=pop]
```

Census region	Freq.	Percent	Cum.
NE	49,135,283	21.75	21.75
N Cntrl	58,865,670	26.06	47.81
South	74,734,029	33.08	80.89
West	43,172,490	19.11	100.00
Total	225,907,472	100.00	

```
. summarize medage
```

Variable	Obs	Mean	Std. Dev.	Min	Max
medage	50	29.54	1.693445	24.2	34.7

```
. log close
        log:  C:\example\session.smcl
   log type:  smcl
  closed on:  17 Mar 2007, 12:35:38
```

```
. —
```

What you will see is a perfect copy of what you previously saw. If you use Stata to print the file, you will get a perfect printed copy, too.

SMCL files can be translated to ASCII text, which is a format more useful for inclusion into a word processing document. If you type translate *filename*.smcl *filename*.log, Stata will translate *filename*.smcl to ASCII and store the result in *filename*.log:

```
. translate session.smcl session.log
```

The resulting file session.log looks like this:

```
---------------------------------------------------------------------------
        log:  C:\example\session.smcl
   log type:  smcl
  opened on:  17 Mar 2007, 12:35:08
. use http://www.stata-press.com/data/r10/census
(1980 Census data by state)
. tabulate reg [freq=pop]
```

Census region	Freq.	Percent	Cum.
NE	49,135,283	21.75	21.75
N Cntrl	58,865,670	26.06	47.81
South	74,734,029	33.08	80.89

```
(output omitted )
```

When you use translate to create *filename*.log from *filename*.smcl, *filename*.log must not already exist:

```
. translate session.smcl session.log
file session.log already exists
r(602);
```

If the file does already exist and you wish to overwrite the existing copy, you can specify the `replace` option:

```
. translate session.smcl session.log, replace
```

See [R] **translate** for more information.

If you prefer, you can skip the SMCL and create ASCII text logs directly, either by specifying that you want the log in `text` format,

```
. log using session, text
```

or by specifying that the file to be created be a `.log` file:

```
. log using session.log
```

15.1.2   Appending to an existing log

Stata never lets you accidentally write over an existing log file. If you have an existing log file and you want to continue logging, you have three choices:

- create a new log file
- append the new log onto the existing log file by typing `log using` *logname*, `append`
- replace the existing log file by typing `log using` *logname*, `replace`

    For example, if you have an existing log file named `session.smcl`, you might type

```
. log using session, append
```

to append the new log to the end of the existing log file, `session.smcl`.

15.1.3   Temporarily suspending and resuming logging

Once you have started logging your session, you can turn logging on and off. When you turn logging off, Stata temporarily stops recording your session but leaves the log file open. When you turn logging back on, Stata continues to record your session, appending the additional record to the end of the file.

Say that the first time something interesting happens, you type `log using results` (or click on **Log** and open `results.smcl`). You then retype the command that produced the interesting result (or double-click the command in the Review window, or use the PageUp key to retrieve the command; see [U] **10 Keyboard use**). You now have a copy of the interesting result saved in the log file.

You are now reasonably sure that nothing interesting will occur, at least for a while. Rather than type `log close`, however, you type `log off`, or you click on **Log** and choose **Suspend**. From now on, nothing goes into the file. The next time something interesting happens, you type `log on` (or click on **Log** and choose **Resume**) and reissue the (interesting) command. After that, you type `log off`. You keep working like this—toggling the log on and off.

15.2   Placing comments in logs

Stata treats lines starting with a '*' as comments and ignores them. Thus, if you are working interactively and wish to make a comment, you can type '*' followed by your comment:

```
. * check that all the spells are completed
. _
```

Stata ignores your comment, but if you have a log going the comment now appears in the file.

❑ Technical Note

log can be combined with #review (see [U] **10 Keyboard use**) to bail you out when you have not adequately planned ahead. Say that you have been working in front of your computer, and you now realize that you have done what you wanted to do. Unfortunately, you are not sure exactly what it is you have done. Did you make a mistake? Could you reproduce the result? Unfortunately, you have not been logging your output. Typing #review will allow you to look over what commands you have issued, and, combined with log, will allow you to make a record. You can also see the commands that you have issued in the Review window. You can save those commands to a file by right-clicking on the Review window and selecting **Save Review Contents...**.

Type log using *filename*. Type #review 100. Stata will list the last 100 commands you gave, or however many it has stored. Since log is making a record, that list will also be stored in the file. Finally, type log close.

❑

15.3   Logging only what you type

Log files record everything that happens during a session, both what you type and what Stata produces in response.

Stata can also produce command log files—files that contain only what you type. These files are perfect for later going back and creating a Stata do-file.

cmdlog creates command log files, and its basic syntax is

cmdlog using *filename* [, replace]	creates *filename*.txt
cmdlog off	temporarily suspends command logging
cmdlog on	resumes command logging
cmdlog close	closes the command log file

See [R] **log** for all the details.

Command logs are plain ASCII text files. If you typed

```
. cmdlog using session
(cmdlog C:\example\session.txt opened)
. use http://www.stata-press.com/data/r10/census
(Census Data)
. tabulate reg [freq=pop]
  (output omitted )
. summarize medage
  (output omitted )
. cmdlog close
(cmdlog C:\example\session.txt closed)
```

file `mycmds.txt` would contain

```
use census
tabulate reg [freq=pop]
summarize medage
```

You can create both kinds of logs—full session logs and command logs—simultaneously, if you wish. A command log file can later be used as a do-file; see [R] **do**.

15.4 The log-button alternative

The capabilities of the `log` command (but not the `cmdlog` command) are available from Stata's GUI interface; just click on the **Log** button or select **Log** from the **File** menu.

You can use the Viewer to view logs, even logs that are in the process of being created. Just select **File > View**. If you are currently logging, the filename to view will already be filled in with the current log file, and all you need to do is click **OK**. Periodically, you can click the **Refresh** button to bring the Viewer up to date.

You can also use the Viewer to view previous logs.

You can access the Viewer by selecting **File > View**, or you can use the `view` command:

```
. view myoldlog.smcl
```

15.5 Printing logs

You print logs from the Viewer. Select **File > View**, or type 'view *logfilename*' from the command line to load the log into the Viewer, and then right-click on the Viewer and select **Print**.

You can also print logs by other means; see [R] **translate**.

15.6 Creating multiple log files simultaneously

Programmers or advanced users may wish to create more than one log file simultaneously. For example, you may wish to create a log file of your whole session but also create a separate log file for part of your session.

You can create multiple logs by using `log`'s `name()` option; see [R] **log**.

16 Do-files

Contents

16.1 Description

Rather than typing commands at the keyboard, you can create a text file containing commands and instruct Stata to execute the commands stored in that file. Such files are called *do-files* since the command that causes them to be executed is do.

A do-file is a standard ASCII text file that is executed by Stata when you type do *filename*. You can use any text editor or the built-in Do-file Editor to create do-files; see [GS] **14 Using the Do-file Editor—automating Stata**.

▷ Example 1

You can use do-files to create a batchlike environment in which you place all the commands you want to perform in a file and then instruct Stata to do that file. Assume that you use your text editor or word processor to create a file called `myjob.do` that contains these three lines:

```
─────────────────────────────────────────── top of myjob.do ───────────
    use http://www.stata-press.com/data/r10/census5
    tabulate region
    summarize marriage_rate divorce_rate median_age if state!="Nevada"
─────────────────────────────────────────── end of myjob.do ───────────
```

You then enter Stata and instruct Stata to do the file:

```
. do myjob

. use http://www.stata-press.com/data/r10/census5
(1980 Census data by state)
```

```
. tabulate region
```

Census region	Freq.	Percent	Cum.
NE	9	18.00	18.00
N Cntrl	12	24.00	42.00
South	16	32.00	74.00
West	13	26.00	100.00
Total	50	100.00	

```
. summarize marriage_rate divorce_rate median_age if state !="Nevada"
```

Variable	Obs	Mean	Std. Dev.	Min	Max
marriage_r~e	49	.0106791	.0021746	.0074654	.0172704
divorce_rate	49	.0054268	.0015104	.0029436	.008752
median_age	49	29.52653	1.708286	24.2	34.7

You typed only do myjob to produce this output. Since you did not specify the file extension, Stata assumed you meant do myjob.do; see [U] **11.6 File-naming conventions**.

◁

16.1.1   Version

We recommend that the first line in your do-file declare the Stata release you used when you wrote the do-file; myjob.do would read better as

———————————————————————————————— top of myjob.do ————————

```
version 10
use http://www.stata-press.com/data/r10/census
tabulate region
summarize marriage_rate divorce_rate median_age if state!="Nevada"
```
———————————————————————————————— end of myjob.do ————————

We admit that we do not always follow our own advice, as you will see many examples in this manual that do not include the version 10 line.

If you intend to keep the do-file, however, you should include this line since it ensures that your do-file will continue to work with future versions of Stata. Stata is under continual development, and sometimes things change in surprising ways.

For instance, in Stata 3.0, a new syntax for specifying the weights was introduced. If you had an old do-file written for Stata 2.1 that analyzed weighted data and did not have version 2.1 at the top, you would find that today's Stata would flag some of the file's lines as syntax errors. If you had the version 2.1 line, it would work just as it used to.

In Stata 4.0, we updated the random-number generator uniform() — the new one is better in that it has a longer period. If you wrote a do-file back in the days of Stata 3.1 that made a bootstrap calculation of variance and did not include version 3.1 at the top, it would now produce different (but equivalent) results. If you had included the line, it would produce the same results that it used to.

Skipping ahead to Stata 10, we introduced xtset and declared that, to use the xt commands, you must xtset your data first. Previously, you specified options on the end of each xt command that identified the group and, optionally, the time variables. Despite this change, if you include version 9 or earlier at the top of your do-file, the xt commands will continue to work the old way.

When running an old do-file that includes a `version` statement, you need not worry about setting the version back after it has completed. Stata automatically restores the previous value of `version` when the do-file completes.

16.1.2   Comments and blank lines in do-files

You may freely include blank lines in your do-file. In the previous example, the do-file could just as well have read

```
───────────────────────────────────── top of myjob.do ─────────
version 10
use http://www.stata-press.com/data/r10/census
tabulate region
summarize marriage_rate divorce_rate median_age if state!="Nevada"
───────────────────────────────────── end of myjob.do ─────────
```

There are four ways to include comments in a do-file.

1. Begin the line with a '`*`'; Stata ignores such lines. `*` cannot be used within Mata.

2. Place the comment in `/* */` delimiters.

3. Place the comment after two forward slashes, i.e., `//`. Everything after the `//` to the end of the current line is considered a comment (unless the `//` is part of `http://...`).

4. Place the comment after three forward slashes, i.e., `///`. Everything after the `///` to the end of the current line is considered a comment. However, when you use `///`, the next line joins with the current line. `///` lets you split long lines across multiple lines in the do-file.

❏ Technical Note

The `/* */`, `//`, and `///` comment indicators can be used in do-files and ado-files only; you may not use them interactively. You can, however, use the '`*`' comment indicator interactively.

❏

`myjob.do` then might read

```
───────────────────────────────────── top of myjob.do ─────────
* a sample analysis job
version 10
use http://www.stata-press.com/data/r10/census
/* obtain the summary statistics: */
tabulate region
summarize marriage_rate divorce_rate median_age if state!="Nevada"
───────────────────────────────────── end of myjob.do ─────────
```

or equivalently,

```
───────────────────────────────────── top of myjob.do ─────────
// a sample analysis job
version 10
use http://www.stata-press.com/data/r10/census
// obtain the summary statistics:
tabulate region
summarize marriage_rate divorce_rate median_age if state!="Nevada"
───────────────────────────────────── end of myjob.do ─────────
```

The style of comment indicator you use is up to you. One advantage of the `/* */` method is that it can be put at the end of lines:

```
——————————————————————————————————————————— top of myjob.do ———————————
* a sample analysis job
version 10

use http://www.stata-press.com/data/r10/census

tabulate region                  /* obtain summary statistics */
summarize marriage_rate divorce_rate median_age if state!="Nevada"
——————————————————————————————————————————— end of myjob.do ———————————
```

In fact, /* */ can be put anywhere, even in the middle of a line:

```
——————————————————————————————————————————— top of myjob.do ———————————
* a sample analysis job
version 10

use /* confirm this is latest */ http://www.stata-press.com/data/r10/census

tabulate region                  /* obtain summary statistics */
summarize marriage_rate divorce_rate median_age if state!="Nevada"
——————————————————————————————————————————— end of myjob.do ———————————
```

You can achieve the same results with the // and /// methods:

```
——————————————————————————————————————————— top of myjob.do ———————————
// a sample analysis job
version 10

use http://www.stata-press.com/data/r10/census

tabulate region                  // obtain summary statistics
summarize marriage_rate divorce_rate median_age if state!="Nevada"
——————————————————————————————————————————— end of myjob.do ———————————
```

or

```
——————————————————————————————————————————— top of myjob.do ———————————
// a sample analysis job
version 10

use /// confirm this is latest
http://www.stata-press.com/data/r10/census

tabulate region                  // obtain summary statistics
summarize marriage_rate divorce_rate median_age if state!="Nevada"
——————————————————————————————————————————— end of myjob.do ———————————
```

16.1.3 Long lines in do-files

When you use Stata interactively, you press *Enter* to end a line and tell Stata to execute it. If you need to type a line that is wider than the screen, you simply do it, letting it wrap or scroll.

You can follow the same procedure in do-files—if your editor or word processor will let you—but you can do better. You can change the end-of-line delimiter to ';' by using #delimit, you can comment out the line break by using /* */ comment delimiters, or you can use the /// line-join indicator.

▷ Example 2

In the following fragment of a do-file, we temporarily change the end-of-line delimiter:

── fragment of example.do ──────────

```
use mydata
#delimit ;
summarize weight price displ headroom rep78 length turn gear_ratio
          if substr(company,1,4)=="Ford" |
              substr(company,1,2)=="GM", detail ;
gen byte ford = substr(company,1,4)=="Ford" ;
#delimit cr
gen byte gm = substr(company,1,2)=="GM"
```

── fragment of example.do ──────────

Once we change the line delimiter to semicolon, all lines, even short ones, must end in semicolons. Stata treats carriage returns as no different from blanks. We can change the delimiter back to carriage return by typing #delimit cr.

The #delimit command is allowed only in do-files—it is not allowed interactively. You need not remember to set the delimiter back to carriage return at the end of a do-file because Stata will reset it automatically.

◁

▷ Example 3

The other way around long lines is to comment out the carriage return by using /* */ comment brackets or to use the /// line-join indicator. Thus, our code fragment could also read

── fragment of example.do ──────────

```
use mydata
summarize weight price displ headroom rep78 length turn gear_ratio /*
     */  if substr(company,1,4)=="Ford" |     /*
     */     substr(company,1,2)=="GM", detail
gen byte ford = substr(company,1,4)=="Ford"
gen byte gm = substr(company,1,2)=="GM"
```

── fragment of example.do ──────────

or

── fragment of example.do ──────────

```
use mydata
summarize weight price displ headroom rep78 length turn gear_ratio ///
          if substr(company,1,4)=="Ford" |     ///
              substr(company,1,2)=="GM", detail
gen byte ford = substr(company,1,4)=="Ford"
gen byte gm = substr(company,1,2)=="GM"
```

── fragment of example.do ──────────

◁

(Continued on next page)

16.1.4   Error handling in do-files

A do-file stops executing when (1) the end of the file is reached, (2) an `exit` is executed, or (3) an error (nonzero *return code*) occurs. If an error occurs, the remaining commands in the do-file are not executed.

If you press *Break* while executing a do-file, Stata responds as though an error has occurred, stopping the do-file. This happens because the return code is nonzero; see [U] **8 Error messages and return codes** for an explanation of return codes.

▷ Example 4

Here is what happens when we execute a do-file and then press *Break*:

```
. do myjob2
. version 10
. use census
(Census data)
. tabulate region
    Census
    region │    Freq.      Percent        Cum.
—Break—
r(1);

end of do-file
—Break—
r(1);

. _
```

When we pressed *Break*, Stata responded by typing —`Break`— and showed a return code of 1. Stata seemingly repeated itself, typing first "end of do-file", and then —`Break`— and the return code of 1 again. Do not worry about the repeated messages. The first message indicates that Stata was stopping the `tabulate` because you pressed *Break*, and the second message indicates that Stata is stopping the do-file for the same reason.

◁

▷ Example 5

Let's try our example again, but this time, let's introduce an error. We change the file `myjob2.do` to read

```
──────────────────────────────────────────────── top of myjob2.do ────────
version 10
use censas
tabulate region
summarize marriage_rate divorce_rate median_age if state!="Nevada"
──────────────────────────────────────────────── end of myjob2.do ────────
```

To introduce a subtle typographical error, we typed `use censas` when we meant `use census`. We assume that there is no file called `censas.dta`, so now we have an error. Here is what happens when you instruct Stata to `do` the file:

```
. do myjob2
. version 10
. use censas
```

```
file censas.dta not found
r(601);

end of do-file
r(601);

. _
```

When Stata was told to use censas, it responded with "file censas.dta not found" and a return code of 601. Stata then typed "end of do-file" and repeated the return code of 601. The repeated message occurred for the same reason it did when we pressed *Break* in the previous example. The use resulted in a return code of 601, so the do-file itself resulted in the same return code. The important thing to understand is that Stata stopped executing the file because there was an error.

◁

❑ Technical Note

We can tell Stata to continue executing the file even if there are errors by typing do *filename*, nostop. Here is the result:

```
. do myjob2, nostop

. version 10

. use censas
file censas.dta not found
r(601);

. tabulate region
no variables defined
r(111);

summarize marriage_rate divorce_rate median_age if state!="Nevada"
no variables defined
r(111);

end of do-file

. _
```

None of the commands worked because the do-file's first command failed. That is why Stata ordinarily stops. However, if our file had contained anything that could work, it would have worked. In general, we do not recommend coding in this manner, as unintended consequences can result when errors do not stop execution.

❑

16.1.5 Logging the output of do-files

You log the output of do-files just as you would an interactive session; see [U] **15 Printing and preserving output**.

Many users include the commands to start and stop the logging in the do-file itself:

```
──────────────────────────────── top of myjob3.do ───────────
version 10
log using myjob3, replace
* a sample analysis job
use census
tabulate region                      // obtain summary statistics
summarize marriage_rate divorce_rate median_age if state!="Nevada"
log close
──────────────────────────────── end of myjob3.do ───────────
```

We chose to open with `log using myjob3, replace`, the important part being the `replace` option. Had we omitted the option, we could not easily rerun our do-file. If `myjob3.smcl` had already existed and `log` was not told that it is okay to replace the file, the do-file would have stopped and instead reported that "file myjob3.smcl already exists". We could get around that, of course, by erasing the log file before running the do-file.

16.1.6   Preventing –more– conditions

Assume that you are running a do-file and logging the output so that you can look at it later. Then Stata's feature of pausing every time the screen is full is just an irritation: it means that you have to sit and watch the do-file run so you can clear the —more—.

The way around this is to include the line `set more off` in your do-file. Setting more to `off`, as explained in [U] **7 —more— conditions**, prevents Stata from ever issuing a —more—.

16.2   Calling other do-files

Do-files may call other do-files. Say that you wrote `makedata.do`, which infiles your data, generates a few variables, and saves `step1.dta`. Say that you wrote `anlstep1.do`, which performed a little analysis on `step1.dta`. You could then create a third do-file,

```
─────────────────────────────────────────────── top of master.do ───────────
    version 10
    do makedata
    do anlstep1
─────────────────────────────────────────────── end of master.do ───────────
```

and so in effect combine the two do-files.

Do-files may call other do-files, which, in turn, call other do-files, and so on. Stata allows do-files to be nested 64 deep.

Be not confused: `master.do` above could call 1,000 do-files one after the other, and still the level of nesting would be only two.

16.3   Running a do-file (Stata for Windows)

1. You can execute do-files by typing `do` followed by the filename, as we did above.

2. You can execute do-files by selecting **File > Do...**.

3. You can use the Do-file Editor to compose, save, and execute do-files; see [GSW] **14 Using the Do-file Editor—automating Stata**. To use the Do-file Editor, click on the **Do-file Editor** button, or type `doedit` in the Command window.

4. You can double-click the icon for the do-file to launch Stata and run the do-file. When the do-file completes, Stata will prompt you for the next command just as if you had started Stata the normal way. If you want Stata to exit instead, include `exit, STATA clear` as the last line of your do-file.

5. You can run the do-file in batch mode. See [GSW] **C.7 Stata batch mode** for details, but the short explanation is that you open a DOS window and type

```
C:\data> "C:\Program Files\Stata10\wstata" /s do myjob
```

or

```
C:\data> "C:\Program Files\Stata10\wstata" /b do myjob
```

to run in batch mode, assuming that you have installed Stata in the folder `C:\Program Files\Stata10`. `/b` and `/s` determine the kind of log produced, but put that aside for a second. When you start Stata in these ways, Stata will run in the background. When the do-file completes, the Stata icon on the taskbar will flash. You can then click on it to close Stata. If you want to stop the do-file before it completes, click on the Stata icon on the taskbar, and Stata will ask you if you want to cancel the job. If you want Stata to exit when the do-file is complete rather than flashing on the taskbar, also specify `/e` on the command line.

To log the output, you can start the log before executing the do-file or you can include the `log using` and `log close` in your do-file.

When you run Stata in these ways, Stata takes the following actions:

a. Stata automatically opens a log. If you specified `/s`, Stata will open a SMCL log; if you specified `/b`, Stata will open an ASCII text log. If your do-file is named *xyz*.`do`, the log will be called *xyz*.`smcl` (`/s`) or *xyz*.`log` (`/b`) in the same directory.

b. If your do-file explicitly opens another log, Stata will save two copies of the output.

c. Stata ignores —more— conditions and anything else that would cause the do-file to stop were it running interactively.

16.4   Running a do-file (Stata for Macintosh)

1. You can execute do-files by typing `do` followed by the filename, as we did above.

2. You can execute do-files by selecting **File > Do...**.

3. With Stata running, you can go to the Desktop and double-click the do-file.

4. You can use the Do-file Editor to compose, save, and execute do-files; see [GSM] **14 Using the Do-file Editor—automating Stata**. Click the **Do-file Editor** button, or type `doedit` in the Command window.

5. If Stata is not running, you can double-click on the icon for the do-file to launch Stata and run the do-file. When the do-file completes, Stata will prompt you for the next command just as if you had started Stata the normal way. If you want Stata to exit instead, include `exit, STATA clear` as the last line of your do-file.

6. You can run the do-file in batch mode. See [GSM] **C.4 Stata batch** for details, but the short explanation is that you open a Terminal window and type

```
% /Applications/Stata/Stata.app/Contents/MacOS/Stata -s do myjob
```

or

```
% /Applications/Stata/Stata.app/Contents/MacOS/Stata -b do myjob
```

to run in batch mode, assuming that you have installed Stata/IC in the folder `/Applications/Stata`. `-b` and `-s` determine the kind of log produced, but put that aside for a second. When you start Stata in these ways, Stata will run in the background. When the do-file completes, the Stata icon on the Dock will bounce until you put Stata into the foreground. You can then exit Stata. If you want to stop the do-file before it completes, right-click on the Stata icon on the Dock, and select **Quit**.

To log the output, you can start the log before executing the do-file or you can include the `log using` and `log close` in your do-file.

When you run Stata in these ways, Stata takes the following actions:

    a. Stata automatically opens a log. If you specified −s, Stata will open a SMCL log; if you specified −b, Stata will open an ASCII text log. If your do-file is named *xyz*.do, the log will be called *xyz*.smcl (−s) or *xyz*.log (−b) in the same directory.

    b. If your do-file explicitly opens another log, Stata will save two copies of the output.

    c. Stata ignores —more— conditions and anything else that would cause the do-file to stop were it running interactively.

16.5   Running a do-file (Stata for Unix)

1. You can execute do-files by typing do followed by the filename, as we did above.

2. You can execute do-files by selecting **File > Do...**.

3. You can use the Do-file Editor to compose, save, and execute do-files; see [GSW] **14 Using the Do-file Editor—automating Stata**. Click on the **Do-file Editor** button or type doedit in the Command window.

4. At the Unix prompt, you can type
    $ xstata do *filename*
or
    $ stata do *filename*
to launch Stata and run the do-file. When the do-file completes, Stata will prompt you for the next command just as if you had started Stata the normal way. If you want Stata to exit instead, include exit, STATA clear as the last line of your do-file.

To log the output, you can start the log before executing the do-file or you can include the log using and log close in your do-file.

5. At the Unix prompt, you can type
    $ stata −s do *filename* &
or
    $ stata −b do *filename* &
to run the do-file in the background. The above two examples both involve the use of stata, not xstata. Type stata, even if you usually use the GUI version of Stata, xstata. The examples differ only in that one specifies the −s option and the other, the −b option, which determines the kind of log that will be produced. In the above examples, Stata takes the following actions:

    a. Stata automatically opens a log. If you specified −s, Stata will open a SMCL log; if you specified −b, Stata will open an ASCII text log. If your do-file is named *xyz*.do, the log will be called *xyz*.smcl (−s) or *xyz*.log (−b) in the current directory (the directory from which you issued the stata command).

    b. If your do-file explicitly opens another log, Stata will save two copies of the output.

    c. Stata ignores —more— conditions and anything else that would cause the do-file to stop were it running interactively.

To reiterate: one way to run a do-file in the background and obtain an ASCII text log is by typing

    $ stata −b do myfile &

Another way uses standard redirection:

    $ stata < myfile.do > myfile.log &

The first way is slightly more efficient. Either way, Stata knows it is in the background and ignores —more— conditions and anything else that would cause the do-file to stop if it were running interactively. However, if your do-file contains either the #delimit command or the comment characters (/* at the end of one line and */ at the beginning of the next), the second method will not work. We recommend that you use the first method: stata -b do myfile &.

The choice between stata -b do myfile & and stata -s do myfile & is more personal. We prefer obtaining SMCL logs (-s) because they look better when printed, and, in any case, they can always be converted to ASCII text format with translate; see [R] **translate**.

16.6   Programming with do-files

This is an advanced topic, and we are going to refer to concepts not yet explained; see [U] **18 Programming Stata** for more information.

16.6.1   Argument passing

Do-files accept arguments, just as Stata programs do; this is described in [U] **18 Programming Stata** and [U] **18.4 Program arguments**. In fact, the logic Stata follows when invoking a do-file is the same as when invoking a program: the local macros are saved, and new ones are defined. Arguments are stored in the local macros '1', '2', and so on. When the do-file completes, the previous definitions are restored, just as with programs.

Thus, if you wanted your do-file to

1. use a dataset of your choosing,

2. tabulate a variable named region, and

3. summarize variables marriage_rate and divorce_rate,

you could write the do-file

```
────────────────────────────────────────────────── top of myxmpl.do ───────────
    use '1'
    tabulate region
    summarize marriage_rate divorce_rate
────────────────────────────────────────────────── end of myxmpl.do ───────────
```

and you could run this do-file by typing, for instance,

```
    . do myxmpl census
    (output omitted )
```

The first command—use '1'—would be interpreted as use census because census was the first argument you typed after do myxmpl.

An even better version of the do-file would read

```
────────────────────────────────────────────────── top of myxmpl.do ───────────
    args dsname
    use 'dsname'
    tabulate region
    summarize marriage_rate divorce_rate
────────────────────────────────────────────────── end of myxmpl.do ───────────
```

The `args` command merely assigns a better name to the argument passed. `args dsname` does not verify that what we type following `do myxmpl` is a filename—we would have to use the `syntax` command if we wanted to do that—but substituting 'dsname' for '1' does make the code more readable.

If our program were to receive two arguments, we could refer to them as '1' and '2', or we could put an 'args dsname other' at the top of our do-file and then refer to 'dsname' and 'other'.

To learn more about argument passing, see [U] **18.4 Program arguments**.

16.6.2   Suppressing output

There is an alternative to typing do *filename*; it is run *filename*. `run` works in the same way as do, except that neither the instructions in the file nor any of the output caused by those instructions is shown on the screen or in the log file.

For instance, with the above `myxmpl.do`, typing `run myxmpl census` results in

```
. run myxmpl census

.  _
```

All the instructions were executed, but none of the output was shown.

This is not useful here, but if the do-file contained only the definitions of Stata programs—see [U] **18 Programming Stata**—and you merely wanted to load the programs without seeing the code, run would be useful.

17 Ado-files

17.1 Description

Stata is programmable, and even if you never write a Stata program, Stata's programmability is still important. Many of Stata's features are implemented as Stata programs, and new features are implemented every day, both by StataCorp and by others.

1. You can obtain additions from the *Stata Journal.* You subscribe to the printed journal, but the software additions are available free over the Internet.

2. You can obtain additions from the Stata listserver, Statalist, where an active group of users advise each other on how to use Stata, and often, in the process, trade programs. Visit the Stata web site, http://www.stata.com, for instructions on how to subscribe; subscribing to the listserver is free.

3. The Boston College Statistical Software Components Archive (SSC) is a distributed database making available a large and constantly growing number of Stata programs. You can browse and search the archive, and you can find links to the archive from http://www.stata.com. Importantly, Stata knows how to access the archive and other places, as well. You can search for additions by using Stata's search, net command; see [R] **search**. You can immediately install materials you find with search, net by using the hyperlinks that will be displayed by search in the Results window or by using the net command. A specialized command, ssc, has several options available to help you find and install the user-written commands that are available from this site; see [R] **ssc**.

4. You can write your own additions to Stata.

This chapter is written for people who want to use ado-files. All users should read it. If you later decide you want to write ado-files, see [U] **18.11 Ado-files**.

17.2   What is an ado-file?

An ado-file defines a Stata command, but not all Stata commands are defined by ado-files.

When you type `summarize` to obtain summary statistics, you are using a command built into Stata.

When you type `ci` to obtain confidence intervals, you are running an ado-file. The results of using a built-in command or an ado-file are indistinguishable.

An ado-file is an ASCII text file that contains a Stata program. When you type a command that Stata does not know, it looks in certain places for an ado-file of that name. If Stata finds it, Stata loads and executes it, so it appears to you as if the ado-command is just another command built into Stata.

We just told you that Stata's `ci` command is implemented as an ado-file. That means that, somewhere, there is a file named `ci.ado`.

Ado-files usually come with help files. When you type `help ci` (or select **Help > Stata Command...**, and type ci), Stata looks for `ci.sthlp`, just as it looks for `ci.ado` when you use the `ci` command. A help file is also an ASCII text file that tells Stata's help system what to display.

17.3   How can I tell if a command is built in or an ado-file?

You can use the `which` command to determine whether a file is built in or implemented as an ado-file. For instance, `logistic` is an ado-file, and here is what happens when you type `which logistic`:

```
. which logistic
C:\Program Files\Stata10\ado\base\l\logistic.ado
*! version 3.3.0  26feb2007
```

`summarize` is a built-in command:

```
. which summarize
built-in command:   summarize
```

17.4   Can I look at an ado-file?

Certainly. When you type `which` followed by an ado-command, Stata reports where the file is stored:

```
. which logistic
C:\Program Files\Stata10\ado\base\l\logistic.ado
*! version 3.3.0  26feb2007
```

Ado-files are just ASCII text files containing the Stata program, so you can type them or view them in Stata's Viewer (or even look at them in your editor or word processor):

```
. type "C:\Program Files\Stata10\ado\base\l\logistic.ado"
*! version 3.3.0  26feb2007
program define logistic, eclass prop(or svyb svyj svyr swml) byable(onecall)
        version 6.0, missing
  (output omitted )
end
```

or

```
. viewsource logistic.ado
*! version 3.3.0  26feb2007
program define logistic, eclass prop(or svyb svyj svyr swml) byable(onecall)
        version 6.0, missing
  (output omitted )
end
```

The `type` command displays the contents of a file. The `viewsource` command searches for a file along the ado directories and displays the file in the Viewer. You can also look at the corresponding help file in raw form if you wish. If there is a help file, it is stored in the same place as the ado-file:

```
. type "C:\Program Files\Stata10\ado\base\l\logistic.sthlp", asis
{smcl}
{* *! version 1.1.15  23apr2007}{...}
{cmd:help logistic}{right:dialogs:  {bf:{dialog logistic}}  {dialog logistic,
> message(-svy-) name(svy_logistic):svy: logistic}}
{right:also see:  {help logistic postestimation}}
{hline}
  (output omitted )
```

or

```
. viewsource logistic.sthlp
{smcl}
{* *! version 1.1.15  23apr2007}{...}
{cmd:help logistic}{right:dialogs:  {bf:{dialog logistic}}  {dialog logistic,
> message(-svy-) name(svy_logistic):svy: logistic}}
{right:also see:  {help logistic postestimation}}
{hline}
  (output omitted )
```

17.5   Where does Stata look for ado-files?

Stata looks for ado-files in seven places, which can be categorized in three ways:

  I. The official ado directories:
   1. (UPDATES), the official updates directory containing updated ado-files from StataCorp
   2. (BASE), the official base directory containing the ado-files shipped with your version of Stata

 II. Your personal ado-directories:
   3. (SITE), the directory for ado-files your site might have installed
   4. (PLUS), the directory for ado-files you personally might have installed
   5. (PERSONAL), the directory for ado-files you might have written
   6. (OLDPLACE), the directory where Stata users used to save their personally written ado-files

III. The current directory:
   7. (.), the ado-files you have written just this instant or for just this project

The location of these directories varies from computer to computer, but Stata's `sysdir` command will tell you where they are on your computer:

```
. sysdir
    STATA:  C:\Program Files\Stata10\
  UPDATES:  C:\Program Files\Stata10\ado\updates\
     BASE:  C:\Program Files\Stata10\ado\base\
     SITE:  C:\Program Files\Stata10\ado\site\
     PLUS:  C:\ado\plus\
 PERSONAL:  C:\ado\personal\
 OLDPLACE:  C:\ado\
```

17.5.1   Where are the official ado-directories?

These are the directories listed as BASE and UPDATES by sysdir:

```
. sysdir
   STATA:  C:\Program Files\Stata10\
 UPDATES:  C:\Program Files\Stata10\ado\updates\
    BASE:  C:\Program Files\Stata10\ado\base\
    SITE:  C:\Program Files\Stata10\ado\site\
    PLUS:  C:\ado\plus\
PERSONAL:  C:\ado\personal\
OLDPLACE:  C:\ado\
```

1. BASE contains the ado-files we originally shipped to you.

2. UPDATES contains any updates you might have installed since then. You can install these updates by using the update command or by selecting **Help > Official Updates**; see [U] **17.8 How do I install official updates?**.

17.5.2   Where is my personal ado-directory?

These are the directories listed as PERSONAL, PLUS, SITE, and OLDPLACE by sysdir:

```
. sysdir
   STATA:  C:\Program Files\Stata10\
 UPDATES:  C:\Program Files\Stata10\ado\updates\
    BASE:  C:\Program Files\Stata10\ado\base\
    SITE:  C:\Program Files\Stata10\ado\site\
    PLUS:  C:\ado\plus\
PERSONAL:  C:\ado\personal\
OLDPLACE:  C:\ado\
```

1. PERSONAL is for ado-files you have written. Store your private ado-files here; see [U] **17.7 How do I add my own ado-files?**.

2. PLUS is for ado-files you personally installed but did not write. Such ado-files are usually obtained from the SJ, but they are sometimes found in other places, too. You find and install such files by using Stata's net command, or you can select **Help > SJ and User-written Programs**; see [U] **17.6 How do I install an addition?**.

3. SITE is really the opposite of a personal ado directory—it is a public directory corresponding to PLUS. If you are on a networked computer, the site administrator can install ado-files here, and all Stata users will then be able to use them just as if they all found and installed them in their PLUS directory for themselves. Site administrators find and install the ado-files just as you would, using Stata's net command, but they specify an option when they install something that tells Stata to write the files into SITE rather than PLUS; see [R] **net**.

4. OLDPLACE is for old-time Stata users. Prior to Stata 6, all "personal" ado-files, whether personally written or just personally installed, were written in the same directory—OLDPLACE. So that the old-time Stata users do not have to go back and rearrange what they have already done, Stata still looks in OLDPLACE.

17.6   How do I install an addition?

Additions come in four types:

1. User-written additions, which you might find in the SJ, etc.

2. Updates to user-written additions

> See [U] **17.9 How do I install updates to user-written additions?**.

3. Ado-files you have written

> See [U] **17.7 How do I add my own ado-files?** If you have an ado-file obtained from the Stata listserver or a friend, treat it as belonging to this case.

4. Official updates provided by StataCorp

> See [U] **17.8 How do I install official updates?**.

User-written additions you might find in the *Stata Journal* (SJ), etc., are obtained over the Internet. To access them on the Internet,

1. select **Help > SJ and User-written Programs**, and click on one of the links

or

2. type `net from http://www.stata.com`.

What to do next will be obvious, but, in case it is not, see [GSW] **20 Updating and extending Stata—Internet functionality**, [GSM] **20 Updating and extending Stata—Internet functionality**, or [GSU] **20 Updating and extending Stata—Internet functionality**. Also see [U] **28 Using the Internet to keep up to date**, [R] **net**, and [R] **adoupdate**.

17.7   How do I add my own ado-files?

You write a Stata program (see [U] **18 Programming Stata**), store it in a file ending in `.ado`, perhaps write a help file, and copy everything to the directory `sysdir` lists as PERSONAL:

```
. sysdir
   STATA:  C:\Program Files\Stata10\
 UPDATES:  C:\Program Files\Stata10\ado\updates\
    BASE:  C:\Program Files\Stata10\ado\base\
    SITE:  C:\Program Files\Stata10\ado\site\
    PLUS:  C:\ado\plus\
PERSONAL:  C:\ado\personal\
OLDPLACE:  C:\ado\
```

Here we would copy the files to `C:\ado\personal`.

While you are writing your ado-file, it is sometimes convenient to store the pieces in the current directory. Do that if you wish; you can move them to your personal ado-directory when the program is debugged.

17.8   How do I install official updates?

Updates are available over the Internet:

1. select **Help > Official Updates**, and then click on **http://www.stata.com**

or

2. type `update query`.

What to do next should be obvious, but in case it is not, see [GSW] **20 Updating and extending Stata—Internet functionality**, [GSM] **20 Updating and extending Stata—Internet functionality**, or [GSU] **20 Updating and extending Stata—Internet functionality**. Also see [U] **28 Using the Internet to keep up to date** and [R] **net**.

The official updates include bug fixes and new features but do not change the syntax of an existing command or change the way Stata works.

Once you have installed the updates, you can enter Stata and type `help whatsnew` (or select **Help > What's New?**) to learn about what has changed.

17.9   How do I install updates to user-written additions?

If you have previously installed user-written additions, you can check for updates to them by typing `adoupate`. If updates are available, you can install them by typing `adoupdate, update`. See [R] **adoupdate**.

17.10   Reference

Cox, N. J. 2006. Stata tip 30: May the source be with you. *Stata Journal* 6: 149–150.

18 Programming Stata

Contents

Stata programming is an advanced topic. Some Stata users live productive lives without ever programming Stata. After all, you do not need to know how to program Stata to input data, create new variables, and fit models. On the other hand, programming Stata is not difficult—at least if the problem is not difficult—and Stata's programmability is one of its best features. The real power of Stata is not revealed until you program it.

If you are uncertain whether to read this chapter, we recommend that you start reading and then bail out when it gets too arcane for you. You will learn things about Stata that you may find useful even if you never write a Stata program.

If you want even more, we offer courses over the Internet on Stata programming; see [U] **3.7 Net-Courses**.

18.1 Description

When you type a command that Stata does not recognize, Stata first looks in its memory for a program of that name. If Stata finds it, Stata executes the program.

There is no Stata command named `hello`,

```
. hello
unrecognized command
r(199);
```

but there could be if you defined a program named `hello`, and after that, the following might happen when you typed `hello`:

```
. hello
hi there
. _
```

This would happen if, beforehand, you had typed

```
. program hello
  1. display "hi there"
  2. end
. _
```

That is how programming works in Stata. A program is defined by

$$\text{program } \textit{progname}$$
$$\textit{Stata commands}$$
$$\text{end}$$

and it is executed by typing *progname* at Stata's dot prompt.

18.2 Relationship between a program and a do-file

Stata treats programs the same way it treats do-files. Below we will discuss passing arguments, consuming results from Stata commands, and other topics, but everything we say applies equally to do-files and programs.

Programs and do-files differ in the following ways:

1. You invoke a do-file by typing do *filename*. You invoke a program by simply typing the program's name.

2. Programs must be defined (loaded) before they are used, whereas all that is required to run a do-file is that the file exist. There are ways to make programs load automatically, however, so this difference is of little importance.

3. When you type do *filename*, Stata displays the commands it is executing and the results. When you type *progname*, Stata shows only the results, not the display of the underlying commands. This is an important difference in outlook: in a do-file, how it does something is as important as what it does. In a program, the how is no longer important. You might think of a program as a new feature of Stata.

Let us now mention some of the similarities:

1. Arguments are passed to programs and do-files in the same way.

2. Programs and do-files both contain Stata commands. Any Stata command you put in a do-file can be put in a program.

3. Programs may call other programs. Do-files may call other do-files. Programs may call do-files (this rarely happens), and do-files may call programs (this often happens). Stata allows programs (and do-files) to be nested up to 64 deep.

Now, here is the interesting thing: programs are typically defined in do-files (or in a variant of do-files called ado-files; we will get to that later).

You can define a program interactively, and that is useful for pedagogical purposes, but in real applications, you will compose your program in a text editor and store its definition in a do-file.

You have already seen your first program:

```
program hello
        display "hi there"
end
```

You could type those commands interactively, but if the body of the program were more complicated, that would be inconvenient. So instead, suppose that you typed the commands into a do-file:

```
─────────────────────────────────────────── top of hello.do ───────────
program hello
        display "hi there"
end
─────────────────────────────────────────── end of hello.do ───────────
```

Now, returning to Stata, you type

```
. do hello
. program hello
  1.        display "hi there"
  2. end
  .
end of do-file
```

Do you see that typing do hello did nothing but load the program? Typing do hello is the same as typing out the program's definition because that is all the do-file contains. The do-file was executed, but the statements in the do-file only defined the program hello; they did not execute it. Now that the program is loaded, we can execute it interactively:

```
. hello
hi there
```

So, that is one way you could use do-files and programs together. If you wanted to create new commands for interactive use, you could

1. Write the command as a `program` ... `end` in a do-file.

2. `do` the do-file before you use the new command.

3. Use the new command during the rest of the session.

There are more convenient ways to do this that would automatically load the do-file, but put that aside. The above method would work.

Another way we could use do-files and programs together is to put the definition of the program and its execution together into a do-file:

```
──────────────────────────────────────────── top of hello.do ──────────
program hello
        display "hi there"
end
hello
──────────────────────────────────────────── end of hello.do ──────────
```

Here is what would happen if we executed this do-file:

```
. do hello

. program hello
  1.          display "hi there"
  2. end

. hello
hi there

.
end of do-file
```

Do-files and programs are often used in such combinations. Why? Say that program `hello` is long and complicated and you have a problem where you need to do it twice. That would be a good reason to write a program. Moreover, you may wish to carry forth this procedure as a step of your analysis and, being cautious, do not want to perform this analysis interactively. You never intended program `hello` to be used interactively—it was just something you needed in the midst of a do-file—so you defined the program and used it there.

Anyway, there are many variations on this theme, but few people actually sit in front of Stata and interactively type `program` and then compose a program. They instead do that in front of their text editor. They compose the program in a do-file and then execute the do-file.

There is one other (minor) thing to know: once a program is defined, Stata does not allow you to redefine it:

```
. program hello
hello already defined
r(110);
```

Thus, in our most recent do-file that defines and executes `hello`, we could not rerun it in the same Stata session:

```
. do hello

. program hello
hello already defined
r(110);

end of do-file
r(110);
```

That problem is solved by typing `program drop hello` before redefining it. We could do that interactively, or we could modify our do-file:

```
──────────────────────────────── top of hello.do ────────────
    program drop hello
    program hello
            display "hi there"
    end
    hello
──────────────────────────────── end of hello.do ────────────
```

There is a problem with this solution. We can now rerun our do-file, but the first time we tried to run it in a Stata session, it would fail:

```
. do hello

. program drop hello
hello not found
r(111);

end of do-file
r(111);
```

The way around this conundrum is to modify the do-file:

```
──────────────────────────────── top of hello.do ────────────
    capture program drop hello
    program hello
            display "hi there"
    end
    hello
──────────────────────────────── end of hello.do ────────────
```

`capture` in front of a command makes Stata indifferent to whether the command works; see [P] **capture**. In real do-files containing programs, you will often see `capture program drop` before the program's definition.

To learn about the `program` command itself, see [P] **program**. It manipulates programs. `program` can define programs, drop programs, and show you a directory of programs that you have defined.

A program can contain any Stata command, but certain Stata commands are of special interest to program writers; see the *Programming* heading in the subject table of contents in the *Quick Reference and Index*.

18.3 Macros

Before we can begin programming, we must discuss macros, which are the variables of Stata programs.

A *macro* is a string of characters, called the *macroname*, that stands for another string of characters, called the *macro contents*.

Macros can be local or global. We will start with local macros because they are the most commonly used, but nothing really distinguishes one from the other at this stage.

18.3.1   Local macros

Local macro names can be up to 31 (not 32) characters long.

One sets the contents of a local macro with the `local` command. In fact, we can do this interactively. We will begin by experimenting with macros in this way to learn about them. If we type

```
. local shortcut "myvar thisvar thatvar"
```

then 'shortcut' is a synonym for "myvar thisvar thatvar". Note the single quotes around shortcut. We said that sentence exactly the way we meant to because

if you type	'shortcut',
i.e.,	left-single-quote shortcut right-single-quote,
Stata hears	myvar thisvar thatvar.

To access the contents of the macro, we use a left single quote (located at the upper left on most keyboards), the macro name, and a right single quote (located under the " on the right side of most keyboards).

The single quotes bracketing the macroname shortcut are called the macro-substitution characters. shortcut means shortcut. 'shortcut' means myvar thisvar thatvar.

So, if you typed

```
. list 'shortcut'
```

the effect would be exactly as if you typed

```
. list myvar thisvar thatvar
```

Macros can be used anywhere in Stata. For instance, if we also defined

```
. local cmd "list"
```

we could type

```
. 'cmd' 'shortcut'
```

to mean list myvar thisvar thatvar.

For another example, consider the definitions

```
. local prefix "my"
. local suffix "var"
```

Then

```
. 'cmd' 'prefix''suffix'
```

would mean list myvar.

18.3.2   Global macros

Let's put aside why Stata has two kinds of macros—local and global—and focus right now on how global macros work.

Global macros can have names that are up to 32 (not 31) characters long. You set the contents of a global macro by using the global rather than the local command:

```
. global shortcut "alpha beta"
```

You obtain the contents of a global macro by prefixing its name with a dollar sign: $shortcut is equivalent to "alpha beta".

In the previous section, we defined a local macro named shortcut, which is a different macro. 'shortcut' is still "myvar thisvar thatvar".

Local and global macros may have the same names, but even if they do, they are unrelated and are still distinguishable.

Global macros are just like local macros except that you set their contents with global rather than local, and you substitute their contents by prefixing them with a $ rather than enclosing them in ' '.

18.3.3 The difference between local and global macros

The difference between local and global macros is that local macros are private and global macros are public.

Say that you have written a program

```
program myprog
        code using local macro alpha
end
```

The local macro alpha in myprog is private in that no other program can modify or even look at alpha's contents. To make this point absolutely clear, assume that your program looks like this:

```
program myprog
        code using local macro alpha
        mysub
        more code using local macro alpha
end
program mysub
        code using local macro alpha
end
```

myprog calls mysub, and both programs use a local macro named alpha. Even so, the local macros in each program are different. mysub's alpha macro may contain one thing, but that has nothing to do with what myprog's alpha macro contains. Even when mysub begins execution, its alpha macro is different from myprog's. It is not that mysub's inherits myprog's alpha macro contents but is then free to change it. It is that myprog's alpha and mysub's alpha are entirely different things.

When you write a program using local macros, you need not worry that some other program has been written using local macros with the same names. Local macros are just that: local to your program.

Global macros, on the other hand, are available to all programs. If both myprog and mysub use the global macro beta, they are using the same macro. Whatever the contents of $beta are when mysub is invoked, those are the contents when mysub begins execution, and, whatever the contents of $beta are when mysub completes, those are the contents when myprog regains control.

18.3.4 Macros and expressions

From now on, we are going to use local and global macros according to whichever is convenient; whatever is said about one applies to the other.

Consider the definitions

```
. local one 2+2
. local two = 2+2
```

(which we could just as well have illustrated using the global command). In any case, note the equal sign in the second macro definition and the lack of the equal sign in the first. Formally, the first should be

```
. local one "2+2"
```

but Stata does not mind if we omit the double quotes in the local (global) statement.

local one 2+2 (with or without double quotes) copies the string 2+2 into the macro named one.

local two = 2+2 evaluates the expression 2+2, producing 4, and stores 4 in the macro named two.

That is, you type

local *macname* *contents*

if you want to copy *contents* to *macname*, and you type

local *macname* = *expression*

if you want to evaluate *expression* and store the result in *macname*.

In the second form, *expression* can be numeric or string. 2+2 is a numeric expression. As an example of a string expression,

```
. local res = substr("this",1,2) + "at"
```

stores that in res.

Since the expression can be either numeric or string, what is the difference between the following statements?

```
. local a "example"
. local b = "example"
```

Both statements store example in their respective macros. The first does so by a simple copy operation, whereas the second evaluates the expression "example", which is a string expression because of the double quotes that, here, evaluates to itself.

There is, however, a difference. Stata's expression parser is limited to handling strings of 244 characters. Strings longer than that are truncated.

The copy operation of the first syntax is not limited—it can copy up to the maximum length of a macro, which is currently 165,200 characters for Stata/IC and 8,681 for Small Stata. For Stata/MP and Stata/SE, the limit is $33 * c(\text{max\_k\_theory}) + 200$ characters, which for the default setting of 5,000 is 165,200 characters.

To a programmer, the length limit for string expressions may seem restrictive, but it is not, because of another feature discussed in [U] **18.3.6 Extended macro functions**.

There are some other issues of using macros and expressions that look a little strange to programmers coming from other languages, at least the first time they see them. Say that the macro 'i' contains 5. How would you increment i so that it contains $5 + 1 = 6$? The answer is

```
local i = 'i' + 1
```

Do you see why the single quotes are on the right but not the left? Remember, 'i' refers to the contents of the local macro named i, which, we just said, is 5. Thus, after expansion, the line reads

```
local i = 5 + 1
```

which is the desired result.

There is a another way to increment local macros that will be more familiar to some programmers, especially C programmers:

```
local ++i
```

As C programmers would expect, `local ++i` is more efficient (executes more quickly) than `local i = i+1`, but in terms of outcome, it is equivalent. You can decrement a local macro by using

```
local --i
```

`local --i` is equivalent to `local i = i-1` but executes more quickly. Finally,

```
local i++
```

will *not* increment the local macro i but instead redefines the local macro i to contain ++. There is, however, a context in which i++ (and i--) do work as expected; see [U] **18.3.7 Macro increment and decrement functions**.

18.3.5 Double quotes

Consider another local macro, 'answ', which might contain yes or no. In a program that was supposed to do something different on the basis of answ's content, you might code

```
if "'answ'" == "yes" {
        ...
}
else {
        ...
}
```

Note the odd-looking "'answ'", and now think about the line after substitution. The line reads either

```
if "yes" == "yes" {
```

or

```
if "no" == "yes" {
```

either of which is the desired result. Had we omitted the double quotes, the line would have read

```
if no == "yes" {
```

(assuming 'answ' contains no), and that is not at all the desired result. As the line reads now, no would not be a string but would be interpreted as a variable in the data.

The key to all this is to think of the line after substitution.

Double quotes are used to enclose strings: "yes", "no", "my dir\my file", "'answ'" (meaning that the contents of local macro answ, treated as a string), and so on. Double quotes are used with macros,

```
local a "example"
if "'answ'" == "yes" {
        ...
}
```

and double quotes are used by many Stata commands:

```
. regress lnwage age ed if sex=="female"
. gen outa = outcome if drug=="A"
. use "person file"
```

Do not omit the double quotes just because you are using a "quoted" macro:

```
. regress lnwage age ed if sex==""`x'"
. gen outa = outcome if drug==""`firstdrug'"
. use "`filename'"
```

Stata has two sets of double-quote characters, of which "" is one. The other is ‘"'’. They both work the same way:

```
. regress lnwage age ed if sex==`"female"'
. gen outa = outcome if drug==`"A"'
. use `"person file"'
```

No rational user would use ‘"'’ (called compound double quotes) instead of "" (called simple double quotes), but smart programmers do use them:

```
local a `"example"'
if `""`answ'""' == `"yes"' {
       ...
}
```

Why is ‘"example"'’ better than "example", ‘""`answ'""'’ better than "`answ'", and ‘"yes"'’ better than "yes"? The answer is that only ‘""`answ'""'’ is better than "`answ'"; ‘"example"'’ and ‘"yes"'’ are no better—and no worse—than "example" and "yes".

‘""`answ'""'’ is better than "`answ'" because the macro answ might itself contain (simple or compound) double quotes. The really great thing about compound double quotes is that they nest. Say that ‘answ' contained the string "I "think" so". Then,

Stata would find	if "`answ'"=="yes"
confusing because it would expand to	if "I "think" so"=="yes"
Stata would not find	if `""`answ'""'==`"yes"'
confusing because it would expand to	if `""I "think" so"'==`"yes"'

Open and close double quote in the simple form look the same; open quote is " and so is close quote. Open and close double quote in the compound form are distinguishable; open quote is ‘" and close quote is "', and so Stata can pair the close with the corresponding open double quote. ‘"I "think" so"'’ is easy for Stata to understand, whereas "I "think" so" is a hopeless mishmash. (If you disagree, consider what "A"B"C" might mean. Is it the quoted string A"B"C, or is it quoted string A, followed by B, followed by quoted string C?)

Since Stata can distinguish open from close quotes, even nested compound double quotes are understandable: ‘"I `"think"' so"'’. (What does "A"B"C" mean? Either it means ‘"A`"B"'C"'’ or it means ‘"A"'B`"C"'’.)

Yes, compound double quotes make you think that your vision is stuttering, especially when combined with the macro substitution ‘' characters. That is why we rarely use them, even when writing programs. You do not have to use exclusively one or the other style of quotes. It is perfectly acceptable to code

```
local a "example"
if `""`answ'""' == "yes" {
       ...
}
```

using compound double quotes where it might be necessary (‘""`answ'""'’) and using simple double quotes in other places (such as "yes"). It is also acceptable to use simple double quotes around macros (e.g., "`answ'") if you are certain that the macros themselves do not contain double quotes or (more likely) if you do not care what happens if they do.

Sometimes careful programmers should use compound double quotes. Later you will learn that Stata's `syntax` command interprets standard Stata syntax and so makes it easy to write programs that understand things like

```
. myprog mpg weight if strpos(make,"VW")!=0
```

`syntax` works—we are getting ahead of ourselves—by placing the if *exp* typed by the user in the local macro if. Thus, 'if' will contain "if strpos(make,"VW")!=0" here. Now, say that you are at a point in your program where you want to know whether the user specified an if *exp*. It would be natural to code

```
if '"'if'"' != "" {
        // the if exp was specified
        ...
}
else {
        // it was not
        ...
}
```

We used compound double quotes around the macro 'if'. The local macro 'if' might contain double quotes, so we placed compound double quotes around it.

18.3.6   Extended macro functions

In addition to allowing =*exp*, `local` and `global` provide *extended functions*. The use of an extended function is denoted by a colon (:) following the macro name, as in

```
local       lbl : variable label myvar
local filenames : "." files "*.dta"
local       xi : word 'i' of 'list'
```

Some macro extended functions access a piece of information. In the first example, the variable label associated with variable myvar will be stored in macro lbl. Other macro extended functions perform operations to gather the information. In the second example, macro filenames will contain the names of all the .dta datasets in the current directory. Still other macro extended functions perform an operation on their arguments and return the result. In the third example, xi will contain the 'i'th word (element) of 'list'. See [P] **macro** for a list of the macro extended functions.

Another useful source of information is c(), documented in [P] **creturn**:

```
local  today "'c(current_date)'"
local curdir "'c(pwd)'"
local  newn = c(N)+1
```

c() refers to a prerecorded list of values, which may be used directly in expressions or which may be quoted and the result substituted anywhere. c(current_date) returns today's date in the form *"dd MON yyyy"*. Thus, the first example stores in macro today that date. c(pwd) returns the current directory, such as C:\data\proj. Thus, the second example stores in macro curdir the current directory. c(N) returns the number of observations of the data in memory. Thus, the third example stores in macro newn that number, plus one.

Note the use of quotes with c(). We could just as well have coded the first two examples as

```
local  today = c(current_date)
local curdir = c(pwd)
```

c() is a Stata function in the same sense that sqrt() is a Stata function. Thus, we can use c() directly in expressions. It is a special property of macro expansion, however, that you may use the c() function inside macro-expansion quotes. The same is not true of sqrt().

In any case, whenever you need a piece of information, whether it be about the dataset or about the environment, look in [P] **macro** and [P] **creturn**. It is likely to be in one place or the other, and sometimes, it is in both. You can obtain the current directory by using

```
local curdir = c(pwd)
```

or by using

```
local curdir : pwd
```

When information is in both, it does not matter which source you use.

18.3.7   Macro increment and decrement functions

We mentioned incrementing macros in [U] **18.3.4 Macros and expressions**. The construct

```
command that makes reference to 'i'
local ++i
```

occurs so commonly in Stata programs that it is convenient (and faster when executed) to collapse both lines of code into one and to increment (or decrement) i at the same time that it is referred to. Stata allows this:

```
while ('++i' < 1000) {
        . . .
}
while ('i++' < 1000) {
        . . .
}
while ('--i' > 0) {
        . . .
}
while ('i--' > 0) {
        . . .
}
```

Above we have chosen to illustrate this by using Stata's while command, but ++ and -- can be used anyplace in any context, just so long as it is enclosed in macro-substitution quotes.

When the ++ or -- appears before the name, the macro is first incremented or decremented, and then the result is substituted.

When the ++ or -- appears after the name, the current value of the macro is substituted and then the macro is incremented or decremented.

❑ Technical Note

Do not use the inline ++ or -- operators when a part of the line might not be executed. Consider

```
if ('i'==0) local j = 'k++'
```

versus

```
if (`i'==0) {
        local j = `k++'
}
```

The first will not do what you expect because macros are expanded before the line is interpreted. Thus, the first will result in k always being incremented, whereas the second increments k only when `i'==0.

❑

18.3.8 Macro expressions

Typing

```
command that makes reference to `=exp'
```

is equivalent to

```
local macroname = exp
command that makes reference to `macroname'
```

although the former runs faster and is easier to type. When you use `=exp' within some larger command, *exp* is evaluated by Stata's expression evaluator, and the results are inserted as a literal string into the larger command. Then the command is executed. For example,

```
summarize u4
summarize u`=2+2'
summarize u`=4*(cos(0)==1)'
```

all do the same thing. *exp* can be any valid Stata expression and thus may include references to variables, matrices, scalars, or even other macros. In the last case, just remember to enclose the submacros in quotes:

```
replace `var' = `group'[`=`j'+1']
```

Also, typing

```
command that makes reference to `:extended macro function'
```

is equivalent to

```
local macroname : extended macro function
command that makes reference to `macroname'
```

Thus, one might code

```
format y `:format x'
```

to assign to variable y the same format as the variable x.

❑ **Technical Note**

There is another macro expansion operator, . (called dot), which is used in conjunction with Stata's class system; see [P] **class** for more information.

There is also a macro expansion function, macval(), which is for use when expanding a macro—`macval(name)'—which confines the macro expansion to the first level of *name*, thereby suppressing the expansion of any embedded references to macros within *name*. Only two or three Stata users have or will ever need this, but, if you suspect you are one of them, see [P] **macro** and then see [P] **file** for an example.

❑

18.3.9 Advanced local macro manipulation

This section is really an aside to help test your understanding of macro substitution. The tricky examples illustrated below sometimes occur in real programs.

1. Say that you have macros x1, x2, x3, and so on. Obviously, ʻx1ʼ refers to the contents of x1, ʻx2ʼ to the contents of x2, etc. What does ʻxʻiʼʼ refer to? Suppose that ʻiʼ contains 6.

   The rule is to expand the inside first:

   $$\text{ʻxʻiʼʼ}\quad\text{expands to}\quad\text{ʻx6ʼ}$$
   $$\text{ʻx6ʼ}\quad\text{expands to}\quad\text{the contents of local macro x6}$$

   So, there you have a vector of macros.

2. We have already shown adjoining expansions: ʻalphaʼʻbetaʼ expands to myvar if ʻalphaʼ contains my and ʻbetaʼ contains var. What does ʻalphaʼʻgammaʼʻbetaʼ expand to when gamma is undefined?

   Stata does not mind if you refer to a nonexistent macro. A nonexistent macro is treated as a macro with no contents. If local macro gamma does not exist, then

   $$\text{ʻgammaʼ}\quad\text{expands to}\quad\text{nothing}$$

   It is not an error. Thus, ʻalphaʼʻgammaʼʻbetaʼ expands to myvar.

3. You clear a local macro by setting its contents to nothing:

   ```
        local macname
   or   local macname ""
   or   local macname = ""
   ```

18.3.10 Advanced global macro manipulation

Global macros are rarely used, and when they are used, it is typically for communication between programs. You should never use a global macro where a local macro would suffice.

1. Constructions like xi are expanded sequentially. If $x contained this and $i 6, then xi expands to this6. If $x was undefined, then xi is just 6 because undefined global macros, like undefined local macros, are treated as containing nothing.

2. You can nest macro expansion by including braces, so if $i contains 6, ${x$i} expands to ${x6}, which expands to the contents of $x6 (which would be nothing if $x6 is undefined).

3. You can mix global and local macros. Assume that local macro j contains 7. Then, ${xʻjʼ} expands to the contents of $x7.

4. You also use braces to force the contents of global macros to run up against the succeeding text. For instance, assume that the macro drive contains "b:". If drive were a local macro, you could type

   ```
   ʻdriveʼmyfile.dta
   ```

   to obtain b:myfile.dta. Because drive is a global macro, however, you must type

   ```
   ${drive}myfile.dta
   ```

You could not type

    $drive myfile.dta

because that would expand to b: myfile.dta. You could not type

    $drivemyfile.dta

because that would expand to .dta.

5. Because Stata uses $ to mark global-macro expansion, printing a real $ is sometimes tricky. To display the string $22.15 with the display command, you can type display "\$22.15", although you can get away with display "$22.15" because Stata is rather smart. Stata would not be smart about display "$this" if you really wanted to display $this and not the contents of the macro this. You would have to type display "\$this". Another alternative would be to use the SMCL code for a dollar sign when you wanted to display it: display "{c S|}this"; see [P] **smcl**.

6. Real dollar signs can also be placed into the contents of macros, thus postponing substitution. First, let's understand what happens when we do not postpone substitution; consider the following definitions:

    global baseset "myvar thatvar"
    global bigset "$baseset thisvar"

$bigset is equivalent to "myvar thatvar thisvar". Now, say that we redefine the macro baseset:

    global baseset "myvar thatvar othvar"

The definition of bigset has not changed—it is still equivalent to "myvar thatvar thisvar". It has not changed because bigset used the definition of baseset that was current at the time it was defined. bigset no longer knows that its contents are supposed to have any relation to baseset.

Instead, let us assume that we had defined bigset as

    global bigset "\$baseset thisvar"

at the outset. Then $bigset is equivalent to "$baseset thisvar", which in turn is equivalent to "myvar thatvar othvar thisvar". Since bigset explicitly depends upon baseset, anytime we change the definition of baseset, we will automatically change the definition of bigset as well.

18.3.11   Constructing Windows filenames by using macros

Stata uses the \ character to tell its parser not to expand macros.

Windows uses the \ character as the directory path separator.

Mostly, there is no problem using a \ in a filename. However, if you are writing a program that contains a Windows path in macro path and a filename in fname, do not assemble the final result as

    'path'\'fname'

because Stata will interpret the \ as an instruction to not expand 'fname'. Instead, assemble the final result as

    'path'/'fname'

Stata understands / as a directory separator on all platforms.

18.3.12   Accessing system values

Stata programs often need access to system parameters and settings, such as the value of π, the current date and time, or the current working directory.

System values are accessed via Stata's c-class values. The syntax works much the same as if you were referring to a local macro. For example, a reference to the c-class value for π, `c(pi)`, will expand to a literal string containing 3.141592653589793 and could be used to do

```
. display sqrt(2*`c(pi)')
2.5066283
```

You could also access the current time

```
. display "`c(current_time)'"
11:34:57
```

C-class values are designed to provide one all-encompassing way to access system parameters and settings, including system directories, system limits, string limits, memory settings, properties of the data currently in memory, output settings, efficiency settings, network settings, and debugging settings.

See [P] **creturn** for a detailed list of what is available. Typing

```
. creturn list
```

will give you the list of current settings.

18.3.13   Referring to characteristics

Characteristics—see [U] **12.8 Characteristics**—are like macros associated with variables. They have names of the form *varname*[*charname*]—such as mpg[comment]—and you quote their names just as you do macro names to obtain their contents:

> To substitute the value of *varname*[*charname*], type   `*varname*[*charname*]`
> For example,   `mpg[comment]`

You set the contents using the `char` command:

> char *varname*[*charname*]   [["]*text*["]]

This is similar to the `local` and `global` commands, except that there is no =*exp* variation. You clear a characteristic by setting its contents to nothing just as you would with a macro:

> Type   char *varname*[*charname*]
> or   char *varname*[*charname*] ""

What is unique about characteristics is that they are saved with the data, meaning that their contents survive from one session to the next, and they are associated with variables in the data, so if you ever drop a variable, the associated characteristics disappear, too. (Also, _dta[*charname*] is associated with the data but not with any variable in particular.)

All the standard rules apply: characteristics may be referred to by quotation in any context, and the characteristic's contents are substituted for the quoted characteristic name. As with macros, referring to a nonexistent characteristic is not an error; it merely substitutes to nothing.

18.4   Program arguments

When you invoke a program or do-file, what you type following the program or do-file name are the arguments. For instance, if you have a program called xyz and type

. xyz mpg weight

then mpg and weight are the program's arguments, mpg being the first argument and weight the second.

Program arguments are passed to programs via local macros:

Macro	Contents
'0'	what the user typed exactly as the user typed it, odd spacing, double quotes, and all
'1'	the first argument (first word of '0')
'2'	the second argument (second word of '0')
'3'	the third argument (third word of '0')
...	...
'*'	the arguments '1', '2', '3', ..., listed one after the other and with one blank in between; similar to but different from '0' because odd spacing and double quotes are removed

That is, what the user types is passed to you in three different ways:

1. It is passed in '0' exactly as the user typed it, meaning quotes, odd spacing, and all.

2. It is passed in '1', '2', ... broken out into arguments on the basis of blanks (but with quotes used to force binding; we will get to that).

3. It is passed in '*' as "'1' '2' '3' ...", which is a crudely cleaned up version of '0'.

You will probably not use all three forms in one program.

We recommend that you ignore '*', at least for receiving arguments; it is included so that old Stata programs will continue to work.

Operating directly with '0' takes considerable programming sophistication, although Stata's syntax command makes interpreting '0' according to standard Stata syntax easy. That will be covered in [U] **18.4.4 Parsing standard Stata syntax** below.

The easiest way to receive arguments, however, is to deal with the positional macros '1', '2', ...

At the start of this section, we imagined an xyz program invoked by typing xyz mpg weight. Then '1' would contain mpg, '2' would contain weight, and '3' would contain nothing.

Let's write a program to report the correlation between two variables. Of course, Stata already has a command that can do this—correlate—and, in fact, we will implement our program in terms of correlate. It is silly, but all we want to accomplish right now is to show how Stata passes arguments to a program.

Here is our program:

```
program xyz
        correlate '1' '2'
end
```

Once the program is defined, we can try it:

```
. use http://www.stata-press.com/data/r10/auto
(1978 Automobile Data)

. xyz mpg weight
(obs=74)
```

	mpg	weight
mpg	1.0000	
weight	-0.8072	1.0000

See how this works? We typed `xyz mpg weight`, which invoked our `xyz` program with '1' being `mpg` and '2' being `weight`. Our program gave the command `correlate '1' '2'`, and that expanded to `correlate mpg weight`.

Stylistically, this is not a good example of the use of positional arguments, but realistically, there is nothing wrong with it. The stylistic problem is that if `xyz` is really to report the correlation between two variables, it ought to allow standard Stata syntax, and that is not a difficult thing to do. Realistically, the program works.

Positional arguments, however, play an important role, even for programmers who care about style. When we write a subroutine—a program to be called by another program and not intended for direct human use—we often pass information by using positional arguments.

Stata forms the positional arguments '1', '2', ... by taking what the user typed following the command (or do-file), parsing it on white space with double quotes used to force binding, and stripping the quotes. The arguments are formed on the basis of words, but double-quoted strings are kept together as one argument but with the quotes removed.

Let's create a program to illustrate these concepts. Although we would not normally define programs interactively, this program is short enough that we will:

```
. program listargs
  1.    display "The 1st argument you typed is:  '1'"
  2.    display "The 2nd argument you typed is:  '2'"
  3.    display "The 3rd argument you typed is:  '3'"
  4.    display "The 4th argument you typed is:  '4'"
  5.  end
```

The `display` command simply types the double-quoted string following it; see [P] **display**.

Let's try our program:

```
. listargs
The 1st argument you typed is:
The 2nd argument you typed is:
The 3rd argument you typed is:
The 4th argument you typed is:
```

We type `listargs`, and the result shows us what we already know—we typed nothing after the word `listargs`. There are no arguments. Let's try it again, this time adding `this is a test`:

```
. listargs this is a test
The 1st argument you typed is:  this
The 2nd argument you typed is:  is
The 3rd argument you typed is:  a
The 4th argument you typed is:  test
```

We learn that the first argument is 'this', the second is 'is', and so on. Blanks always separate arguments. You can, however, override this feature by placing double quotes around what you type:

```
. listargs "this is a test"
The 1st argument you typed is:  this is a test
The 2nd argument you typed is:
The 3rd argument you typed is:
The 4th argument you typed is:
```

This time we typed only one argument, 'this is a test'. When we place double quotes around what we type, Stata interprets whatever we type inside the quotes to be one argument. Here '1' contains 'this is a test' (the double quotes were removed).

We can use double quotes more than once:

```
. listargs "this is" "a test"
The 1st argument you typed is:  this is
The 2nd argument you typed is:  a test
The 3rd argument you typed is:
The 4th argument you typed is:
```

The first argument is 'this is' and the second argument is 'a test'.

18.4.1 Named positional arguments

Positional arguments can be named: in your code, you do not have to refer to '1', '2', '3', ...; you can instead refer to more meaningful names, such as n, a, and b; numb, alpha, and beta; or whatever else you find convenient. You want to do this because programs coded in terms of '1', '2', ... are hard to read and therefore are more likely to contain errors.

You obtain better-named positional arguments by using the args command:

```
program progname
        args argnames
        . . .
end
```

For instance, if your program received four positional arguments and you wanted to call them varname, n, oldval, and newval, you would code

```
program progname
        args varname n oldval newval
        . . .
end
```

varname, n, oldval, and newval become new local macros, and args simply copies '1', '2', '3', and '4' to them. It does not change '1', '2', '3', and '4'—you can still refer to the numbered macros if you wish—and it does not verify that your program receives the right number of arguments. If our example above were invoked with just two arguments, 'oldval' and 'newval' would contain nothing. If it were invoked with five arguments, the fifth argument would still be out there, stored in local macro '5'.

Let's make a command to create a dataset containing n observations on x ranging from a to b. Such a command would be useful, for instance, if we wanted to graph some complicated mathematical function and experiment with different ranges. It is convenient if we can type the range of x over which we wish to make the graph rather than concocting the range by hand. (In fact, Stata already has such a command—range—but it will be instructive to write our own.)

Before writing this program, we had better know how to proceed, so here is how you could create a dataset containing n observations with x ranging from a to b:

1. `drop _all`
   to clear whatever data are in memory.

2. `set obs` n
   to make a dataset of n observations on no variables; if n were 100, we would type `set obs`
   100.

3. `gen x = (_n-1)/(n-1)*(`b`-`a`)+`a
   because the built-in variable `_n` is 1 in the first observation, 2 in the second, and so on; see
   [U] **13.4 System variables (_variables)**.

So, the first version of our program might read

```
program rng                              // arguments are n a b
        drop _all
        set obs `1'
        generate x = (_n-1)/(_N-1)*(`3'-`2')+`2'
end
```

The above is just a direct translation of what we just said. `1` corresponds to n, `2` corresponds
to a, and `3` corresponds to b. This program, however, would be far more understandable if we
changed it to read

```
program rng
        args n a b
        drop _all
        set obs `n'
        generate x = (_n-1)/(_N-1)*(`b'-`a')+`a'
end
```

18.4.2 Incrementing through positional arguments

Some programs contain k arguments, where k varies, but it does not much matter because the
same thing is done to each argument. One such program is `summarize`: type `summarize mpg` to
obtain summary statistics on `mpg`, and type `summarize mpg weight` to obtain first summary statistics
on `mpg` and then summary statistics on `weight`.

```
program ...
        local i = 1
        while "``i''" != "" {
                logic stated in terms of ``i''
                local ++i
        }
end
```

Equivalently, if the logic that uses ``i'' contains only one reference to ``i'',

```
program ...
        local i = 1
        while "``i''" != "" {
                logic stated in terms of ``i++''
        }
end
```

Note the tricky construction ``i'', which then itself is placed in double quotes—"``i''"—for the
`while` loop. To understand it, say that i contains 1 or, equivalently, `i` is 1. Then ``i'' is `1` is
the name of the first variable. "``i''" is the name of the first variable in quotes. The `while` asks
if the name of the variable is nothing and, if it is not, executes. Now `i` is 2, and "``i''" is the
name of the second variable, in quotes. If that name is not "", we continue. If the name is "", we
are done.

Say that you were writing a subroutine that was to receive k variables, but the code that processes each variable needs to know (while it is processing) how many variables were passed to the subroutine. You need first to count the variables (and so derive k) and then, knowing k, pass through the list again.

```
program progname
        local k = 1                     // count the number of arguments
        while "`k'" != "" {
                local ++k
        }
        local --k                       // k contains one too many
                                        // now pass through again
        local i = 1
        while `i' <= `k' {
                code in terms of `i' and `k'
                local ++i
        }
end
```

In the above example, we have used `while`, Stata's all-purpose looping command. Stata has two other looping commands, `foreach` and `forvalues`, and they sometimes produce code that is more readable and executes more quickly. We direct you to read [P] **foreach** and [P] **forvalues**, but at this point, there is nothing they can do that `while` cannot do. Above we coded

```
local i = 1
while `i' <= `k' {
        code in terms of `i' and `k'
        local ++i
}
```

to produce logic that looped over the values `i' = 1 to `k'. We could have instead coded

```
forvalues i = 1(1)`k' {
        code in terms of `i' and `k'
}
```

Similarly, at the beginning of this subsection, we said that you could use the following code in terms of `while` to loop over the arguments received:

```
program ...
        local i = 1
        while "`i'" != "" {
                logic stated in terms of `i'
                local ++i
        }
end
```

Equivalent to the above would be

```
program ...
        foreach x of local 0 {
                logic stated in terms of `x'
        }
end
```

See [P] **foreach** and [P] **forvalues**.

You can combine `args` and incrementing through an unknown number of positional arguments. Say that you were writing a subroutine that was to receive (1) `varname`, the name of some variable; (2) `n`, which is some sort of count; and (3) at least one and maybe 20 variable names. Perhaps you are to sum the variables, divide by `n`, and store the result in the first variable. What the program does is irrelevant; here is how we could receive the arguments:

```
program progname
        args varname n
        local i 3
        while "''i''" != "" {
                logic stated in terms of ''i''
                local ++i
        }
end
```

18.4.3   Using macro shift

Another way to code the repeat-the-same-process problem for each argument is

```
program ...
        while "'1'" != "" {
                logic stated in terms of '1'
                macro shift
        }
end
```

macro shift shifts '1', '2', '3', ..., one to the left: what was '1' disappears, what was '2' becomes '1', what was '3' becomes '2', and so on.

The outside while loop continues the process until macro '1' contains nothing.

macro shift is an older construct that we no longer advocate using. Instead, we recommend that you use the techniques described in the previous subsection, that is, references to ''i'' and foreach/forvalues.

There are two reasons we make this recommendation: (1) macro shift destroys the positional macros '1', '2', which must then be reset using tokenize should you wish to pass through the argument list again, and (more importantly) (2) if the number of arguments is large (which in Stata/MP and Stata/SE is more likely), macro shift can be incredibly slow.

❑ Technical Note

macro shift can do one thing that would be difficult to do by other means.

'*', the result of listing the contents of the numbered macros one after the other with one blank between, changes with macro shift. Say that your program received a list of variables and that the first variable was the dependent variable and the rest were independent variables. You want to save the first variable name in 'lhsvar' and all the rest in 'rhsvars'. You could code

```
program progname
        local lhsvar "'1'"
        macro shift 1
        local rhsvars "'*'"
        ...
end
```

Now suppose that one macro contains a list of variables and you want to split the contents of the macro in two. Perhaps 'varlist' is the result of a syntax command (see [U] **18.4.4 Parsing standard Stata syntax**), and you now wish to split 'varlist' into 'lhsvar' and 'rhsvars'. tokenize will reset the numbered macros:

```
program progname
        ...
        tokenize 'varlist'
        local lhsvar "'1'"
        macro shift 1
        local rhsvars "'*'"
        ...
end
```

❏

18.4.4   Parsing standard Stata syntax

Let us now switch to '0' from the positional arguments '1', '2', ...

You can parse '0' (what the user typed) according to standard Stata syntax with one command. Remember that standard Stata syntax is

$$\left[\text{by } \textit{varlist:}\right] \quad \textit{command} \quad \left[\textit{varlist}\right] \quad \left[\text{=}\textit{exp}\right] \quad \left[\text{using } \textit{filename}\right] \quad \left[\textit{if}\right] \quad \left[\textit{in}\right] \quad \left[\textit{weight}\right]$$

$$\left[, \textit{ options}\right]$$

See [U] **11 Language syntax**.

The syntax command parses standard syntax. You code what amounts to the syntax diagram of your command in your program, and then syntax looks at '0' (it knows to look there) and compares what the user typed with what you are willing to accept. Then one of two things happens: either syntax stores the pieces in an easily processable way or, if what the user typed does not match what you specified, syntax issues the appropriate error message and stops your program.

Consider a program that is to take two or more variable names along with an optional if *exp* and in *range*. The program would read

```
program ...
        syntax varlist(min=2) [if] [in]
        ...
end
```

You will have to read [P] **syntax** to learn how to specify the syntactical elements, but the command is certainly readable, and it will not be long until you are guessing correctly about how to fill it in. And yes, the square brackets really do indicate optional elements, and you just use them with syntax in the natural way.

The one syntax command you code encompasses the parsing process. Here, if what the user typed matches "two or more variables and an optional if and in", syntax defines new local macros:

'varlist'	the two or more variable names
'if'	the if *exp* specified by the user (or nothing)
'in'	the in *range* specified by the user (or nothing)

To see that this works, experiment with the following program:

```
program tryit
        syntax varlist(min=2) [if] [in]
        display "varlist now contains |'varlist'|"
        display '"if now contains |'if'|"'
        display "in now contains |'in'|"
end
```

Below we experiment:

```
. tryit mpg weight
varlist now contains |mpg weight|
if now contains ||
in now contains ||

. tryit mpg weight displ if foreign==1
varlist now contains |mpg weight displ|
if now contains |if foreign==1|
in now contains ||

. tryit mpg wei in 1/10
varlist now contains |mpg weight|
if now contains ||
in now contains |in 1/10|

. tryit mpg
too few variables specified
r(102);
```

In our third try we abbreviated the weight variable as wei, yet, after parsing, syntax unabbreviated the variable for us.

If this program were next going to step through the variables in the varlist, the positional macros '1', '2', ... could be reset by coding

```
tokenize 'varlist'
```

See [P] **tokenize**. tokenize 'varlist' resets '1' to be the first word of 'varlist', '2' to be the second word, and so on.

18.4.5 Parsing immediate commands

Immediate commands are described in [U] **19 Immediate commands**—they take numbers as arguments. By convention, when you name immediate commands, you should make the last letter of the name *i*. Assume that mycmdi takes as arguments two numbers, the first of which must be a positive integer, and allows the options alpha and beta. The basic structure is

```
program mycmdi
        gettoken n 0 : 0, parse(" ,")        /* get first number */
        gettoken x 0 : 0, parse(" ,")        /* get second number */
        confirm integer number 'n'           /* verify first is integer */
        confirm number 'x'                   /* verify second is number */

        if 'n'<=0 error 2001                 /* check that n is positive */
        place any other checks here

        syntax [, Alpha Beta]                /* parse remaining syntax */
        make calculation and display output
end
```

See [P] **gettoken**.

18.4.6 Parsing nonstandard syntax

If you wish to interpret nonstandard syntax and positional arguments are not adequate for you, you know that you face a formidable programming task. The key to the solution is the gettoken command.

gettoken can pull one token from the front of a macro according to the parsing characters you specify and, optionally, define another macro or redefine the initial macro to contain the remaining (unparsed) characters. That is,

Say that '0' contains	"this is what the user typed"
After gettoken,	
new macro 'token' could contain	"this"
and '0' could still contain	"this is what the user typed"
or	
new macro 'token' could contain	"this"
and new macro 'rest' could contain	" is what the user typed"
and '0' could still contain	"this is what the user typed"
or	
new macro 'token' could contain	"this"
and '0' could contain	" is what the user typed"

A simplified syntax of gettoken is

> gettoken *emname1* [*emname2*] : *emname3* [, <u>p</u>arse(*pchars*) <u>q</u>uotes
>
> <u>m</u>atch(*lmacname*) bind]

where *emname1*, *emname2*, *emname3*, and *lmacname* are the names of local macros. (Stata provides a way to work with global macros, but in practice that is seldom necessary; see [P] **gettoken**.)

gettoken pulls the first token from *emname3* and stores it in *emname1*, and if *emname2* is specified, stores the remaining characters from *emname3* in *emname2*. Any of *emname1*, *emname2*, and *emname3* may be the same macro. Typically, gettoken is coded

> gettoken *emname1*   : 0 [, *options*]
>
> gettoken *emname1* 0 : 0 [, *options*]

since '0' is the macro containing what the user typed. The first coding is used for token lookahead, should that be necessary, and the second is used for committing to taking the token.

gettoken's options are

parse("*string*")   for specifying parsing characters
the default is parse(" "), meaning to parse on white space
it is common to specify parse(`"" "'), meaning to parse on white space
and double quote
(`"" "' is the string double-quote-space in compound double quotes)

quotes   to specify that outer double quotes *not* be stripped

match(*lmacname*)   to bind on parentheses and square brackets
lmacname will be set to contain "(", "[", or nothing, depending on
whether *emname1* was bound on parentheses or brackets or if match()
turned out to be irrelevant
emname1 will have the outside parentheses or brackets removed

gettoken binds on double quotes whenever a (simple or compound) double quote is encountered at the beginning of *emname3*. Specifying parse(`"" "') ensures that double-quoted strings are isolated.

quote specifies that double quotes not be removed from the source in defining the token. For instance, in parsing ""this is" a test", the next token is "this is" if quote is not specified and is ""this is"" if quote is specified.

match() specifies that parentheses and square brackets be matched in defining tokens. The outside level of parentheses or brackets is stripped. In parsing "(2+3)/2", the next token is "2+3" if match() is specified. In practice, match() might be used with expressions, but it is more likely to be used to isolate bound varlists and time-series varlists.

18.5   Scalars and matrices

In addition to macros, scalars and matrices are provided for programmers; see [U] **14 Matrix expressions**, [P] **scalar** and [P] **matrix**.

As far as scalar calculations go, you can use macros or scalars. Remember, macros can hold numbers. Stata's scalars are, however, slightly faster and are a little more accurate than macros. The speed issue is so slight as to be nearly immeasurable. Macros are accurate to a minimum of 12 decimal digits, and scalars are accurate to roughly 16 decimal digits. Which you use makes little difference except in iterative calculations.

Stata also has a serious matrix programming language called Mata, which is the subject of another manual. Mata can be used to write subroutines that are called by Stata programs. See the *Mata Reference Manual*, and in particular, [M-1] **ado**.

18.6   Temporarily destroying the data in memory

It is sometimes necessary to modify the data in memory to accomplish a particular task. A well-behaved program, however, ensures that the user's data are always restored. The preserve command makes this easy:

```
code before the data need changing
preserve
code that changes data freely
```

When you give the preserve command, Stata makes a copy of the user's data on disk. When your program terminates—no matter how—Stata restores the data and erases the temporary file; see [P] **preserve**.

18.7   Temporary objects

If you write a substantial program, it will invariably require the use of temporary variables in the data, or temporary scalars, matrices, or files. Temporary objects are necessary while the program is making its calculations, and once the program completes they are discarded.

Stata provides three commands to create temporary objects: tempvar creates names for variables in the dataset, tempname creates names for scalars and matrices, and tempfile creates names for files. All are described in [P] **macro**, and all have the same syntax:

{ tempvar | tempname | tempfile } *macname* [*macname* . . .]

The commands create local macros containing names you may use.

18.7.1   Temporary variables

Say that, in making a calculation, you need to add variables sum_y and sum_z to the data. You might be tempted to code

```
. . .
gen sum_y = . . .
gen sum_z = . . .
. . .
```

but that would be poor because (1) the dataset might already have variables named sum_y and sum_z in it and (2) you will have to remember to drop the variables before your program concludes. Better is

```
. . .
tempvar sum_y
gen 'sum_y' = . . .
tempvar sum_z
gen 'sum_z' = . . .
. . .
```

or

```
. . .
tempvar sum_y sum_z
gen 'sum_y' = . . .
gen 'sum_z' = . . .
. . .
```

It is not necessary to explicitly drop 'sum_y' and 'sum_z' when you are finished, although you may if you wish. Stata will automatically drop any variables with names assigned by tempvar. After issuing the tempvar command, you must refer to the names with the enclosing quotes, which signifies macro expansion. Thus, after typing tempvar sum_y—the one case where you do not put single quotes around the name—refer thereafter to the variable 'sum_y', with quotes. tempvar does not create temporary variables. Instead tempvar creates names that may later be used to create new variables that will be temporary, and tempvar stores that name in the local macro whose name you provide.

A full description of tempvar can be found in [P] **macro**.

18.7.2   Temporary scalars and matrices

tempname works just like tempvar. For instance, a piece of your code might read

```
tempname YXX XXinv
matrix accum 'YXX' = price weight mpg
matrix 'XXinv' = invsym('YXX'[2..., 2...])
tempname b
matrix 'b' = 'XXinv'*'YXX'[1..., 1]
```

The above code solves for the coefficients of a regression on price on weight and mpg; see [U] **14 Matrix expressions** and [P] **matrix** for more information on the matrix commands.

As with temporary variables, temporary scalars and matrices are automatically dropped at the conclusion of your program.

18.7.3 Temporary files

In cases where you ordinarily might think you need temporary files, you may not because of Stata's ability to preserve and automatically restore the data in memory; see [U] **18.6 Temporarily destroying the data in memory** above.

For more complicated programs, Stata does provide temporary files. A code fragment might read

```
preserve                          /* save original data */
tempfile males females
keep if sex==1
save "'males'"
restore, preserve                 /* get back original data */
keep if sex==0
save "'females'"
```

As with temporary variables, scalars, and matrices, it is not necessary to delete the temporary files when you are through with them; Stata automatically erases them when your program ends.

18.8 Accessing results calculated by other programs

Stata commands that report results also save the results where they can be subsequently used by other commands or programs. This is documented in the *Saved Results* section of the particular command in the *Reference* manuals. Commands save results in one of three places:

1. r-class commands, such as summarize, save their results in r(); most commands are r-class.

2. e-class commands, such as regress, save their results in e(); e-class commands are Stata's model estimation commands.

3. s-class commands (there are no good examples) save their results in s(); this is a rarely used class that programmers sometimes find useful to help parse input.

Commands that do not save results are called n-class commands. More correctly, these commands require that you state where the result is to be saved, as in generate *newvar* = ...

▷ Example 1

You wish to write a program to calculate the standard error of the mean, which is given by the formula $\sqrt{s^2/n}$, where s^2 is the calculated variance. (You could obtain this statistic by using the ci command, but we will pretend that is not true.) You look at [R] **summarize** and learn that the mean is stored in r(mean), the variance in r(Var), and the number of observations in r(N). With that knowledge, you write the following program:

```
program meanse
        quietly summarize '1'
        display "     mean = " r(mean)
        display "SE of mean = " sqrt(r(Var)/r(N))
end
```

The result of executing this program is

```
. meanse mpg
      mean = 21.297297
SE of mean = .67255109
```

◁

If you run an r-class command and type `return list` or run an e-class command and type `ereturn list`, Stata will summarize what was saved:

```
. use http://www.stata-press.com/data/r10/auto
(1978 Automobile Data)

. regress mpg weight displ
  (output omitted )

. ereturn list

scalars:
                  e(N) =  74
               e(df_m) =  2
               e(df_r) =  71
                  e(F) =  66.78504752026517
                 e(r2) =  .6529306984682528
               e(rmse) =  3.45606176570828
                e(mss) =  1595.409691543724
                e(rss) =  848.0497679157352
               e(r2_a) =  .643154098425105
                 e(ll) =  -195.2397979466294
               e(ll_0) =  -234.3943376482347

macros:
            e(cmdline) : "regress mpg weight displ"
              e(title) : "Linear regression"
                e(vce) : "ols"
             e(depvar) : "mpg"
                e(cmd) : "regress"
         e(properties) : "b V"
            e(predict) : "regres_p"
              e(model) : "ols"
          e(estat_cmd) : "regress_estat"

matrices:
                  e(b) :  1 x 3
                  e(V) :  3 x 3

functions:
             e(sample)

. summarize mpg if foreign
```

Variable	Obs	Mean	Std. Dev.	Min	Max
mpg	22	24.77273	6.611187	14	41

```
. return list

scalars:
                  r(N) =  22
              r(sum_w) =  22
               r(mean) =  24.77272727272727
                r(Var) =  43.70779220779221
                 r(sd) =  6.611186898567625
                r(min) =  14
                r(max) =  41
                r(sum) =  545
```

In the example above, we ran `regress` followed by `summarize`. As a result, `e(N)` records the number of observations used by `regress` (equal to 74), and `r(N)` records the number of observations used by `summarize` (equal to 22). `r(N)` and `e(N)` are not the same.

If we now ran another r-class command—say, `tabulate`—the contents of `r()` would change, but those in `e()` would remain unchanged. You might, therefore, think that if we then ran another e-class command, say, `probit`, the contents of `e()` would change, but `r()` would remain unchanged. Although it is true that `e()` results remain in place until the next e-class command is executed, do not depend on r() remaining unchanged. If an e-class or n-class command were to use an r-class command as a subroutine, that would cause r() to change. Anyway, most commands are r-class, so the contents of r() change often.

❑ Technical Note

It is, therefore, of great importance that you access results stored in r() immediately after the command that sets them. If you need the mean and variance of the variable '1' for subsequent calculation, do *not* code

```
summarize '1'
...
... r(mean) ... r(Var) ...
```

Instead, code

```
summarize '1'
local mean = r(mean)
local var = r(Var)
...
... 'mean' ... 'var' ...
```

or

```
tempname mean var
summarize '1'
scalar 'mean' = r(mean)
scalar 'var' = r(Var)
...
... 'mean' ... 'var' ...
```

❑

Saved results, whether in r() or e(), come in three types: scalars, macros, and matrices. If you look back at the ereturn list and return list output, you will see that regress saves examples of all three, whereas summarize saves just scalars. (regress also saves the "function" e(sample), as do all the other e-class commands; see [U] **20.6 Specifying the estimation subsample**.)

Regardless of the type of e(*name*) or r(*name*), you can just refer to e(*name*) or r(*name*). That was the rule we gave in [U] **13.6 Accessing results from Stata commands**, and that rule is sufficient for most uses. There is, however, another way to refer to saved results. Rather than referring to r(*name*) and e(*name*), you can embed the reference in macro-substitution characters ' ' to produce 'r(*name*)' and 'e(*name*)'. The result is the same as macro substitution; the saved result is evaluated, and then the evaluation is substituted:

```
. display "You can refer to " e(cmd) " or to 'e(cmd)'"
You can refer to regress or to regress
```

This means, for instance, that typing 'e(cmd)' is the same as typing regress because e(cmd) contains "regress":

```
. 'e(cmd)'
```

Source	SS	df	MS
Model	1595.40969	2	797.704846

(remaining output omitted)

```
Number of obs =      74
F( 2,    71) =   66.79
Prob > F      =  0.0000
```

In the ereturn list, e(cmd) was listed as a macro, and when you place a macro's name in single quotes, the macro's contents are substituted, so this is hardly a surprise.

What is surprising is that you can do this with scalar and even matrix saved results. e(N) is a scalar equal to 74 and may be used as such in any expression such as "display e(mss)/e(N)" or "local meanss = e(mss)/e(N)". 'e(N)' substitutes to the string "74" and may be used in any context whatsoever, such as "local val'e(N)' = e(N)" (which would create a macro named val74). The rules for referring to saved results are

1. You may refer to r(*name*) or e(*name*) without single quotes in any expression and only in an expression. (Referring to s-class s(*name*) without single quotes is not allowed.)

    1.1 If *name* does not exist, missing value (.) is returned; it is not an error to refer to a nonexistent saved result.

    1.2 If *name* is a scalar, the full double-precision value of *name* is returned.

    1.3 If *name* is a macro, it is examined to determine whether its contents can be interpreted as a number. If so, the number is returned; otherwise, the first 80 characters of *name* are returned.

    1.4 If *name* is a matrix, the full *matrix* is returned.

2. You may refer to 'r(*name*)', 'e(*name*)', or 's(*name*)'—note the presence of quotes indicating macro substitution—in any context whatsoever.

    2.1 If *name* does not exist, nothing is substituted; it is not an error to refer to a nonexistent saved result. The resulting line is the same as if you had never typed 'r(*name*)', 'e(*name*)', or 's(*name*)'.

    2.2 If *name* is a scalar, a string representation of the number accurate to no less than 12 digits of precision is substituted.

    2.3 If *name* is a macro, the full contents are substituted.

    2.4 If *name* is a matrix, the word matrix is substituted.

In general, you should refer to scalar and matrix saved results without quotes—r(*name*) and e(*name*)—and to macro saved results with quotes—'r(*name*)', 'e(*name*)', and 's(*name*)'—but it is sometimes convenient to switch. Say that returned result r(example) contains the number of periods patients are observed, and assume that r(example) was saved as a macro and not as a scalar. You could still refer to r(example) without the quotes in an expression context and obtain the expected result. It would have made more sense for you to have stored r(example) as a scalar, but really it would not matter, and the user would not even have to know how the saved result was stored.

Switching the other way is sometimes useful, too. Say that returned result r(N) is a scalar that contains the number of observations used. You now want to use some other command that has an option n(#) that specifies the number of observations used. You could not type n(r(N)) because the syntax diagram says that option n() expects its argument to be a literal number. Instead, you could type n('r(N)').

18.9   Accessing results calculated by estimation commands

Estimation results are saved in e(), and you access them in the same way you access any saved result; see [U] **18.8 Accessing results calculated by other programs** above. In summary,

1. Estimation commands—regress, logistic, etc.—save results in e().

2. Estimation commands save their name in e(cmd). For instance, regress saves "regress" and poisson saves "poisson" in e(cmd).

3. Estimation commands save the command they executed in e(cmdline). For instance, if you typed reg mpg displ, saved in e(cmdline) would be "reg mpg displ".

4. Estimation commands save the number of observations used in e(N), and they identify the estimation subsample by setting e(sample). You could type, for instance, summarize if e(sample) to obtain summary statistics on the observations used by the estimator.

5. Estimation commands save the entire coefficient vector and variance–covariance matrix of the estimators in e(b) and e(V). These are matrices, and they may be manipulated like any other matrix:

```
. matrix list e(b)
e(b)[1,3]
         weight       displ       _cons
y1   -.00656711    .00528078    40.084522
. matrix y = e(b)*e(V)*e(b)'
. matrix list y
symmetric y[1,1]
            y1
y1   6556.982
```

6. Estimation commands set _b[*name*] and _se[*name*] as convenient ways to use coefficients and their standard errors in expressions; see [U] **13.5 Accessing coefficients and standard errors**.

7. Estimation commands may set other e() scalars, macros, or matrices containing more information. This is documented in the *Saved Results* section of the particular command in the command reference.

▷ Example 2

If you are writing a command for use after regress, early in your code you should include the following:

```
if "`e(cmd)'" != "regress" {
        error 301
}
```

This is how you verify that the estimation results that are stored have been set by regress and not by some other estimation command. Error 301 is Stata's "last estimates not found" error.

◁

18.10 Saving results

If your program calculates something, it should save the results of the calculation so that other programs can access them. In this way your program not only can be used interactively but also can be used as a subroutine for other commands.

Saving results is easy:

1. On the program line, specify the option rclass, eclass, or sclass according to whether you intend to return results in r(), e(), or s().

2. Code

```
return scalar name = exp        (same syntax as scalar without the return)
return local  name ...          (same syntax as local without the return)
return matrix name matname      (moves matname to r(name))
```

to save results in r().

3. Code

ereturn <i>name</i> = <i>exp</i>   (same syntax as scalar without the ereturn)
ereturn local <i>name</i> ...   (same syntax as local without the ereturn)
ereturn matrix <i>name matname</i> (moves <i>matname</i> to e(<i>name</i>))

to save results in e(). You do not save the coefficient vector and variance matrix e(b) and e(V) in this way; instead you use ereturn post.

4. Code

sreturn local <i>name</i> ...   (same syntax as local without the sreturn)

to save results in s(). (The s-class has only macros.)

A program must be exclusively r-class, e-class, or s-class.

18.10.1   Saving results in r()

In [U] **18.8 Accessing results calculated by other programs**, we showed an example that reported the mean and standard error of the mean. A better version would save in r() the results of its calculations and would read

```
program meanse, rclass
        quietly summarize '1'
        local mean = r(mean)
        local sem  = sqrt(r(Var)/r(N))
        display "      mean = " 'mean'
        display "SE of mean = " 'sem'
        return scalar mean = 'mean'
        return scalar se = 'sem'
end
```

Running meanse now sets r(mean) and r(se):

```
. meanse mpg
      mean = 21.297297
SE of mean = .67255109

. return list

scalars:
        r(se)     =  .6725510870764975
        r(mean)   =  21.2972972972973
```

In this modification, we added option rclass to the program statement, and we added two return commands to the end of the program.

Although we placed the return statements at the end of the program, they may be placed at the point of calculation if that is more convenient. A more concise version of this program would read

```
program meanse, rclass
        quietly summarize '1'
        return scalar mean = r(mean)
        return scalar se = sqrt(r(Var)/r(N))
        display "      mean = " return(mean)
        display "SE of mean = " return(se)
end
```

The `return()` function is just like the `r()` function, except that `return()` refers to the results that this program *will* return rather than to the saved results that currently *are* returned (which here are due to `summarize`). That is, when you code the `return` command, the result is not immediately posted to `r()`. Rather, Stata holds onto the result in `return()` until your program concludes, and then it copies the contents of `return()` to `r()`. While your program is active, you may use the `return()` function to access results you have already "returned". (`return()` works just like `r()` works after your program returns, meaning that you may code '`return()`' to perform macro substitution.)

18.10.2   Saving results in e()

Saving in `e()` is in most ways similar to saving in `r()`: you add the `eclass` option to the `program` statement, and then you use `ereturn` ... just as you used `return` ... to store results. There are, however, some significant differences:

1. Unlike `r()`, estimation results are saved in `e()` the instant you issue an `ereturn scalar`, `ereturn local`, or `ereturn matrix` command. Estimation results can consume considerable memory, and Stata does not want to have multiple copies of the results floating around. That means you must be more organized and post your results at the end of your program.

2. In your code when you have your estimates and are ready to begin posting, you will first clear the previous estimates, set the coefficient vector `e(b)` and corresponding variance matrix `e(V)`, and set the estimation-sample function `e(sample)`. How you do this depends on how you obtained your estimates:

    2.1 If you obtained your estimates by using Stata's likelihood maximizer `ml`, this is automatically handled for you; skip to step 3.

    2.2 If you obtained estimates by "stealing" an existing estimator, `e(b)`, `e(V)`, and `e(sample)` already exist, and you will not want to clear them; skip to step 3.

    2.3 If you write your own code from start to finish, you use the `ereturn post` command; see [P] **ereturn**. You will code something like "`ereturn post 'b' 'V', esample('touse')`", where '`b`' is the name of the coefficient vector, '`V`' is the name of the corresponding variance matrix, and '`touse`' is the name of a variable containing 1 if the observation was used and 0 if it was ignored. `ereturn post` clears the previous estimates and moves the coefficient vector, variance matrix, and variable into `e(b)`, `e(V)`, and `e(sample)`.

    2.4 A variation on (2.3) is when you use an existing estimator to produce the estimates but do not want all the other `e()` results stored by the estimator. Then you code

    ```
    tempvar touse
    tempname b V
    matrix 'b' = e(b)
    matrix 'V' = e(V)
    qui gen byte 'touse' = e(sample)
    ereturn post 'b' 'V', esample('touse')
    ```

3. You now save anything else in `e()` that you wish by using the `ereturn scalar`, `ereturn local`, or `ereturn matrix` command.

4. Save `e(cmdline)` by coding

    ```
    ereturn local cmdline '"'0'"'
    ```

    This is not required, but it is considered good style.

5. You code `ereturn local cmd "`*cmdname*`"`. Stata does not consider estimation results complete until this command is posted, and Stata considers the results to be complete when this is posted, so you must remember to do this and to do this last. If you set `e(cmd)` too early and the user pressed *Break*, Stata would consider your estimates complete when they are not.

Say that you wish to write the estimation command with syntax

> myest *depvar var*$_1$ *var*$_2$ [if *exp*] [in *range*], *optset1 optset2*

where *optset1* affects how results are displayed and *optset2* affects the estimation results themselves. One important characteristic of estimation commands is that, when typed without arguments, they redisplay the previous estimation results. The outline is

```
program myest, eclass
        local options "optset1"
        if replay() {
                if "`e(cmd)'"!="myest" {
                        error 301               /* last estimates not found      */
                }
                syntax [, `options']
        }
        else {
                syntax varlist [if] [in] [, `options' optset2]
                marksample touse
```
Code contains either this,
```
                tempnames b V
```
commands for performing estimation
assume produces `b` *and* `V`
```
                ereturn post `b' `V', esample(`touse')
                ereturn local depvar "`depv'"
```
or this,
```
                ml model ... if `touse' ...
```
and regardless, concludes,
```
                perhaps other ereturn commands appear here
                ereturn local cmdline `"`0'"'
                ereturn local cmd "myest"
        }
                                                /* (re)display results ...       */
```
code typically reads
code to output header above coefficient table
```
        ereturn display                         /* displays coefficient table    */
```
or
```
        ml display                              /* displays header and coef. table */
end
```

Here is a list of the commonly saved `e()` results. Of course, you may create any `e()` results that you wish.

`e(N)` (scalar)
Number of observations.

`e(df_m)` (scalar)
Model degrees of freedom.

`e(df_r)` (scalar)
"Denominator" degrees of freedom if estimates are nonasymptotic.

`e(r2_p)` (scalar)
Value of the pseudo-R^2 if it is calculated. (If a "real" R^2 is calculated as it would be in linear regression, it is stored in (scalar) `e(r2)`.)

e(F) (scalar)
   Test of the model against the constant-only model, if relevant, and if results are nonasymptotic.

e(ll) (scalar)
   Log-likelihood value, if relevant.

e(ll_0) (scalar)
   Log-likelihood value for constant-only model, if relevant.

e(N_clust) (scalar)
   Number of clusters, if any.

e(chi2) (scalar)
   Test of the model against the constant-only model, if relevant, and if results are asymptotic.

e(cmd) (macro)
   Name of the estimation command.

e(cmdline) (macro)
   Command as typed.

e(depvar) (macro)
   Names of the dependent variables.

e(wtype) and e(wexp) (macros)
   If weighted estimation was performed, e(wtype) contains the weight type (fweight, pweight, etc.) and e(wexp) contains the weighting expression.

e(title) (macro)
   Title in estimation output.

e(clustvar) (macro)
   Name of the cluster variable, if any.

e(vcetype) (macro)
   Text to appear above standard errors in estimation output; typically Robust, Bootstrap, Jack-knife, or "".

e(vce) (macro)
   vcetype specified in vce().

e(chi2type) (macro)
   LR or Wald or other depending on how e(chi2) was performed.

e(crittype) (macro)
   Type of optimization criterion used, such as log likelihood or deviance.

e(properties) (macro)
   Typically contains b V.

e(predict) (macro)
   Name of the command that predict is to use; if this is blank, predict uses the default _predict.

e(b) and e(V) (matrices)
   The coefficient vector and corresponding variance matrix. Saved when you coded ereturn post.

e(sample) (function)
   This function was defined by ereturn post's esample() option if you specified it. You specified a variable containing 1 if you used an observation and 0 otherwise. ereturn post stole the variable and created e(sample) from it.

18.10.3 Saving results in s()

S is a strange class because, whereas the other classes allow scalars, macros, and matrices, s allows only macros.

S is seldom used and is for subroutines that you might write to assist in parsing the user's input prior to evaluating any user-supplied expressions.

Here is the problem s solves: say that you create a nonstandard syntax for some command so that you have to parse through it yourself. The syntax is so complicated that you want to create subroutines to take pieces of it and then return information to your main routine. Assume that your syntax contains expressions that the user might type. Now, say that one of the expressions the user types is, e.g., r(mean)/sqrt(r(Var))—perhaps the user is using results left behind by summarize.

If, in your parsing step, you call subroutines that return results in r(), you will wipe out r(mean) and r(Var) before you ever get around to seeing them, much less evaluating them. So, you must be careful to leave r() intact until your parsing is complete; you must use no r-class commands, and any subroutines you write must not touch r(). You must use s-class subroutines because s-class routines return results in s() rather than r(). S-class provides macros only because that is all you need to solve parsing problems.

To create an s-class routine, specify the sclass option on the program line and then use sreturn local to return results.

S-class results are posted to s() at the instant you issue the sreturn() command, so you must organize your results. Also, s() is never automatically cleared, so occasionally coding sreturn clear at appropriate points in your code is a good idea.

Few programs need s-class subroutines.

18.11 Ado-files

Ado-files were introduced in [U] **17 Ado-files**.

When a user types 'gobbledygook', Stata first asks itself if gobbledygook is one of its built-in commands. If so, the command is executed. Otherwise, it asks itself if gobbledygook is a defined program. If so, the program is executed. Otherwise, Stata looks in various directories for gobbledygook.ado. If there is no such file, the process ends with the "unrecognized command" error.

If Stata finds the file, it quietly issues to itself the command 'run gobbledygook.ado' (specifying the path explicitly). If that runs without error, Stata asks itself again if gobbledygook is a defined program. If not, Stata issues the "unrecognized command" error. (Here somebody wrote a bad ado-file.) If the program is defined, as it should be, Stata executes it.

Thus, you can arrange for programs you write to be loaded automatically. For instance, if you were to create hello.ado containing

```
                                             top of hello.ado
      program hello
              display "hi there"
      end
                                             end of hello.ado
```

and store the file in your current directory or your personal directory (see [U] **17.5.2 Where is my personal ado directory?**), you could type hello and be greeted by a reassuring

```
   . hello
   hi there
```

You could, at that point, think of hello as just another part of Stata.

There are two places to put your personal ado-files. One is the current directory, and that is a good choice when the ado-file is unique to a project. You will want to use it only when you are in that directory. The other place is your *personal ado-directory*, which is probably something like `C:\ado\personal` if you use Windows, `/ado/personal` if you use Unix, and `/ado/personal` if you use a Macintosh. We are guessing.

To find your personal ado-directory, enter Stata and type

```
. personal
```

❏ Technical Note

Stata looks in various directories for ado-files, defined by the c-class value `c(adopath)`, which contains

$$\text{UPDATES;BASE;SITE;.;PERSONAL;PLUS;OLDPLACE}$$

The words in capital letters are codenames for directories, and the mapping from codenames to directories can be obtained by typing the `sysdir` command. Here is what `sysdir` shows on one particular Windows computer:

```
. sysdir
    STATA:   C:\Program Files\Stata10\
  UPDATES:   C:\Program Files\Stata10\ado\updates\
     BASE:   C:\Program Files\Stata10\ado\base\
     SITE:   C:\Program Files\Stata10\ado\site\
     PLUS:   C:\ado\plus\
 PERSONAL:   C:\ado\personal\
 OLDPLACE:   C:\ado\
```

Even if you use Windows, your mapping might be different because it all depends on where you installed Stata. That is the point of the codenames. They make it possible to refer to directories according to their logical purposes rather than their physical location.

The c-class value `c(adopath)` is the search path, so in looking for an ado-file, Stata first looks in UPDATES then in BASE, and so on, until it finds the file. Actually, Stata not only looks in UPDATES but also takes the first letter of the ado-file it is looking for and looks in the lettered subdirectory. For files with the extension `.style`, Stata will look in a subdirectory named `style` rather than a lettered subdirectory. Say that Stata was looking for `gobbledygook.ado`. Stata would look up UPDATES (`C:\Program Files\Stata10\ado\updates` in our example) and, if the file were not found there, it would look in the g subdirectory of UPDATES (`C:\Program Files\Stata10\ado\updates\g`) before looking in BASE, whereupon it would follow the same rules. If Stata were looking for `gobbledygook.style`, Stata would look up UPDATES (`C:\Program Files\Stata10\ado\updates` in our example) and, if the file were not found there, it would look in the `style` subdirectory of UPDATES (`C:\Program Files\Stata10\ado\updates\style`) before looking in BASE, whereupon it would follow the same rules.

Why the extra complication? We distribute hundreds of ado-files, help files, and other file types with Stata, and some operating systems have difficulty dealing with so many files in the same directory. All operating systems experience at least a performance degradation. To prevent this, the ado-directory we ship is split 28 ways (letters *a–z*, underscore, and `style`). Thus, the Stata command `ci`, which is implemented as an ado-file, can be found in the subdirectory c of BASE.

If you write ado-files, you can structure your personal ado-directory this way, too, but there is no reason to do so until you have more than, say, 250 files in one directory.

❏

❑ Technical Note

After finding and running *gobbledygook*.ado, Stata calculates the total size of all programs that it has automatically loaded. If this exceeds adosize (see [P] **sysdir**), Stata begins discarding the oldest automatically loaded programs until the total is less than adosize. Oldest here is measured by the time last used, not the time loaded. This discarding saves memory and does not affect you, since any program that was automatically loaded could be automatically loaded again if needed.

It does, however, affect performance. Loading the program takes time, and you will again have to wait if you use one of the previously loaded-and-discarded programs. Increasing adosize reduces this possibility, but at the cost of memory. The set adosize command allows you to change this parameter; see [P] **sysdir**. The default value of adosize is 500. A value of 500 for adosize means that up to 500 K can be allocated to autoloaded programs. Experimentation has shown that this is a good number—increasing it does not improve performance much.

❑

18.11.1 Version

We recommend that the first line following program in your ado-file declare the Stata release under which you wrote the program; hello.ado would read better as

```
——————————————————————————————————— top of hello.ado ———————————
program hello
        version 10
        display "hi there"
end
——————————————————————————————————— end of hello.ado ———————————
```

We introduced the concept of version in [U] **16.1.1 Version**. In regular do-files, we recommend that the version line appear as the first line of the do-file. For ado-files, the line appears after the program because loading the ado-file is one step and executing the program is another. It is when Stata executes the program defined in the ado-file that we want to stipulate the interpretation of the commands.

The inclusion of the version line is of more importance in ado-files than in do-files because (1) ado-files have longer lives than do-files, so it is more likely that you will use an ado-file with a later release and (2) ado-files tend to use more of Stata's features, increasing the probability that any change to Stata will affect them.

18.11.2 Comments and long lines in ado-files

Comments in ado-files are handled the same way as in do-files: you enclose the text in /* comment */ brackets, or you begin the line with an asterisk (*), or you interrupt the line with //; see [U] **16.1.2 Comments and blank lines in do-files**.

Logical lines longer than physical lines are also handled as they are in do-files: either you change the delimiter to a semicolon (;) or you comment out the new line by using /// at the end of the previous physical line.

18.11.3   Debugging ado-files

Debugging ado-files is a little tricky because it is Stata and not you that controls when the ado-file is loaded.

Assume that you wanted to change `hello` to say "Hi, Mary". Assume that your editor is called `vi` and that you are in the habit of calling your editor from Stata, so you do this,

```
. !vi hello.ado
```

and make the obvious change to the program. Equivalently, you can pretend that you are using a windowed operating system and jump out of Stata—leaving it running—to modify the `hello.ado` file. Anyway, you change `hello.ado` to read

───────────────────────────────────── top of hello.ado ─────────
```
program hello
        version 10
        display "hi, Mary"
end
```
───────────────────────────────────── end of hello.ado ─────────

Back in Stata, you try it:

```
. hello
hi there
```

Stata ran the old copy of `hello`—the copy it still has in its memory. Stata wants to be fast about executing ado-files, so when it loads one, it keeps it around a while—waiting for memory to get short—before clearing it from its memory. Naturally, Stata can drop `hello` anytime because it can always reload it from disk.

You changed the copy on disk, but Stata still has the old copy loaded into memory. You type `discard` to tell Stata to forget these automatically loaded things and to force itself to get new copies of the ado-files from disk:

```
. discard
. hello
hi, Mary
```

You had to type `discard` only because you changed the ado-file while Stata was running. Had you exited Stata and returned later to use `hello`, the `discard` would not have been necessary because Stata forgets things between sessions anyway.

18.11.4   Local subroutines

An ado-file can contain more than one `program`, and if it does, the other programs defined in the ado-file are assumed to be subroutines of the main program. For example,

───────────────────────────────────── top of decoy.ado ─────────
```
program decoy
        ...
        duck ...
        ...
end
program  duck
        ...
end
```
───────────────────────────────────── end of decoy.ado ─────────

duck is considered a local subroutine of decoy. Even after decoy.ado was loaded, if you typed duck, you would be told "unrecognized command". To emphasize what *local* means, assume that you have also written an ado-file named duck.ado:

——————————————————————————————— top of duck.ado ———————

```
program duck
        . . .
end
```

——————————————————————————————— end of duck.ado ———————

Even so, when decoy called duck, it would be the program duck defined in decoy.ado that was called. To further emphasize what *local* means, assume that decoy.ado contains

——————————————————————————————— top of decoy.ado ———————

```
program decoy
        . . .
        manic . . .
        . . .
        duck . . .
        . . .
end
program duck
        . . .
end
```

——————————————————————————————— end of decoy.ado ———————

and that manic.ado contained

——————————————————————————————— top of manic.ado ———————

```
program manic
        . . .
        duck . . .
        . . .
end
```

——————————————————————————————— end of manic.ado ———————

Here is what would happen when you executed decoy:

1. decoy in decoy.ado would begin execution. decoy calls manic.
2. manic in manic.ado would begin execution. manic calls duck.
3. duck in duck.ado (yes) would begin execution. duck would do whatever and return.
4. manic regains control and eventually returns.
5. decoy is back in control. decoy calls duck.
6. duck in decoy.ado would execute, complete, and return.
7. decoy would regain control and return.

When manic called duck, it was the global ado-file duck.ado that was executed, yet when decoy called duck, it was the local program duck that was executed.

Stata does not find this confusing and neither should you.

18.11.5   Development of a sample ado-command

Below we demonstrate how to create a new Stata command. We will program an influence measure for use with linear regression. It is an interesting statistic in its own right, but even if you are not interested in linear regression and influence measures, the focus here is on programming, not on the particular statistic chosen.

Belsley, Kuh, and Welsch (1980, 24) present a measure of influence in linear regression defined as

$$\frac{\text{Var}\left(\widehat{y}_i^{(i)}\right)}{\text{Var}(\widehat{y}_i)}$$

which is the ratio of the variance of the ith fitted value based on regression estimates obtained by omitting the ith observation to the variance of the ith fitted value estimated from the full dataset. This ratio is estimated using

$$\text{FVARATIO}_i \equiv \frac{n-k}{n-(k+1)} \left\{ 1 - \frac{d_i^2}{1-h_{ii}} \right\} (1-h_{ii})^{-1}$$

where n is the sample size; k is the number of estimated coefficients; $d_i^2 = e_i^2/\mathbf{e}'\mathbf{e}$ and e_i is the ith residual; and h_{ii} is the ith diagonal element of the hat matrix. The ingredients of this formula are all available through Stata, so, after estimating the regression parameters, we can easily calculate FVARATIO_i. For instance, we might type

```
. regress mpg weight displ
. predict hii if e(sample), hat
. predict ei if e(sample), resid
. quietly count if e(sample)
. scalar nreg = r(N)
. gen eTe = sum(ei*ei)
. gen di2 = (ei*ei)/eTe[_N]
. gen FVi = (nreg - 3) / (nreg - 4) * (1 - di2/(1-hii)) / (1-hii)
```

The number 3 in the formula for `FVi` represents k, the number of estimated parameters (which is an intercept plus coefficients on `weight` and `displ`), and the number 4 represents $k+1$.

❏ Technical Note

Do you understand why this works? `predict` can create h_{ii} and e_i, but the trick is in getting $\mathbf{e}'\mathbf{e}$—the sum of the squared e_is. Stata's `sum()` function creates a running sum. The first observation of `eTe` thus contains e_1^2; the second, $e_1^2 + e_2^2$; the third, $e_1^2 + e_2^2 + e_3^2$; and so on. The last observation, then, contains $\sum_{i=1}^{N} e_i^2$, which is $\mathbf{e}'\mathbf{e}$. (We specified `if e(sample)` on our `predict` commands to restrict calculations to the estimation subsample, so `hii` and `eii` might have missing values, but that does not matter because `sum()` treats missing values as contributing zero to the sum.) We use Stata's explicit subscripting feature and then refer to `eTe[_N]`, the last observation. (See [U] **13.3 Functions** and [U] **13.7 Explicit subscripting**.) After that, we plug into the formula to obtain the result.

❏

Assuming that we often wanted this influence measure, it would be easier and less prone to error if we canned this calculation in a program. Our first draft of the program reflects exactly what we would have typed interactively:

```
                                                    ——— top of fvaratio.ado, version 1 ———
program fvaratio
        version 10
        predict hii if e(sample), hat
        predict ei if e(sample), resid
        quietly count if e(sample)
        scalar nreg = r(N)
        gen eTe = sum(ei*ei)
        gen di2 = (ei*ei)/eTe[_N]
        gen FVi = (nreg - 3) / (nreg - 4) * (1 - di2/(1-hii)) / (1-hii)
        drop hii ei eTe di2
end
                                                    ——— end of fvaratio.ado, version 1 ———
```

All we have done is to enter what we would have typed into a file, bracketing it with `program fvaratio` and `end`. Since our command is to be called `fvaratio`, the file must be named `fvaratio.ado` and must be stored in either the current directory or our personal ado-directory (see [U] **17.5.2 Where is my personal ado directory?**).

Now when we type `fvaratio`, Stata will be able to find it, load it, and execute it. In addition to copying the interactive lines into a program, we added the line 'drop hii ...' to eliminate the working variables we had to create along the way.

So, now we can interactively type

```
. regress mpg weight displ
. fvaratio
```

and add the new variable `FVi` to our data.

Our program is not general. It is suitable for use after fitting a regression model on two, and only two, independent variables because we coded a 3 in the formula for k. Stata statistical commands such as `regress` store information about the problem and answer in `e()`. Looking in *Saved Results* in [R] **regress**, we find that `e(df_m)` contains the model degrees of freedom, which is $k-1$, assuming that the model has an intercept. Also, the sample size of the dataset used in the regression is stored in `e(N)`, eliminating our need to count the observations and define a scalar containing this count. Thus, the second draft of our program reads

```
                                                    ——— top of fvaratio.ado, version 2 ———
program fvaratio
        version 10
        predict hii if e(sample), hat
        predict ei if e(sample), resid
        gen eTe = sum(ei*ei)
        gen di2 = (ei*ei)/eTe[_N]
        gen FVi = (e(N)-(e(df_m)+1)) / (e(N)-(e(df_m)+2)) *      /// changed this
                  (1 - di2/(1-hii)) / (1-hii)                    // version
        drop hii ei eTe di2
end
                                                    ——— end of fvaratio.ado, version 2 ———
```

In the formula for `FVi`, we substituted `(e(df_m)+1)` for the literal number 3, `(e(df_m)+2)` for the literal number 4, and `e(N)` for the sample size.

Back to the substance of our problem, `regress` also saves the residual sum of squares in `e(rss)`, so calculating `eTe` is not really necessary:

```
                                                    top of fvaratio.ado, version 3
program fvaratio
        version 10
        predict hii if e(sample), hat
        predict ei if e(sample), resid
        gen di2 = (ei*ei)/e(rss)                        // changed this version
        gen FVi = (e(N)-(e(df_m)+1)) / (e(N)-(e(df_m)+2)) *    ///
                (1 - di2/(1-hii)) / (1-hii)
        drop hii ei di2
end
                                                    end of fvaratio.ado, version 3
```

Our program is now shorter and faster, and it is completely general. This program is probably good enough for most users; if you were implementing this solely for your own occasional use, you could stop right here. The program does, however, have the following deficiencies:

1. When we use it with data with missing values, the answer is correct, but we see messages about the number of missing values generated. (These messages appear when the program is generating the working variables.)

2. We cannot control the name of the variable being produced—it is always called FVi. Moreover, when FVi already exists (say, from a previous regression), we get an error message that FVi already exists. We then have to drop the old FVi and type fvaratio again.

3. If we have created any variables named hii, ei, or di2, we also get an error that the variable already exists, and the program refuses to run.

Fixing these problems is not difficult. The fix for problem 1 is easy; we embed the entire program in a quietly block:

```
                                                    top of fvaratio.ado, version 4
program fvaratio
        version 10
        quietly {                                       // new this version
                predict hii if e(sample), hat
                predict ei if e(sample), resid
                gen di2 = (ei*ei)/e(rss)
                gen FVi = (e(N)-(e(df_m)+1)) / (e(N)-(e(df_m)+2)) *    ///
                        (1 - di2/(1-hii)) / (1-hii)
                drop hii ei di2
        }                                               // new this version
end
                                                    end of fvaratio.ado, version 4
```

The output for the commands between the quietly { and } is now suppressed—the result is the same as if we had put quietly in front of each command.

Solving problem 2—that the resulting variable is always called FVi—requires use of the syntax command. Let's put that off and deal with problem 3—that the working variables have nice names like hii, ei, and di2, and so prevent users from using those names in their data.

One solution would be to change the nice names to unlikely names. We could change hii to MyHiiVaR, which would not guarantee the prevention of a conflict but would certainly make it unlikely. It would also make our program difficult to read, an important consideration should we want to change it in the future. There is a better solution. Stata's tempvar command (see [U] **18.7.1 Temporary variables**) places names into local macros that are guaranteed to be unique:

```
                                          top of fvaratio.ado, version 5
program fvaratio
        version 10
        tempvar hii ei di2                              // new this version
        quietly {
                predict `hii' if e(sample), hat    // changed, as are other lines
                predict `ei' if e(sample), resid
                gen `di2' = (`ei'*`ei')/e(rss)
                gen FVi = (e(N)-(e(df_m)+1)) / (e(N)-(e(df_m)+2)) *      ///
                        (1 - `di2'/(1-`hii')) / (1-`hii')
        }
end
                                          end of fvaratio.ado, version 5
```

At the top of our program, we declare the temporary variables. (We can do it outside or inside the quietly—it makes no difference—and we do not have to do it at the top or even all at once; we could declare them as we need them, but at the top is prettiest.) When we refer to a temporary variable, we do not refer directly to it (such as by typing hii); we refer to it indirectly by typing open and close single quotes around the name (`hii`). And at the end of our program, we no longer bother to drop the temporary variables—temporary variables are dropped automatically by Stata when a program concludes.

❑ Technical Note

Why do we type single quotes around the names? tempvar creates local macros containing the real temporary variable names. hii in our program is now a local macro, and `hii` refers to the contents of the local macro, which is the variable's actual name.

❑

We now have an excellent program—its only fault is that we cannot specify the name of the new variable to be created. Here is the solution to that problem:

```
                                          top of fvaratio.ado, version 6
program fvaratio
        version 10
        syntax newvarname                              // new this version
        tempvar hii ei di2
        quietly {
                predict `hii' if e(sample), hat
                predict `ei' if e(sample), resid
                gen `di2' = (`ei'*`ei')/e(rss)
                gen `typlist' `varlist' = ///              changed this version
                        (e(N)-(e(df_m)+1)) / (e(N)-(e(df_m)+2)) *      ///
                        (1 - `di2'/(1-`hii')) / (1-`hii')
        }
end
                                          end of fvaratio.ado, version 6
```

It took a change to one line and the addition of another to obtain the solution. This magic all happens because of syntax (see [U] **18.4.4 Parsing standard Stata syntax** above).

'syntax newvarname' specifies that one new variable name must be specified (had we typed 'syntax [newvarname]', the new varname would have been optional; had we typed 'syntax newvarlist', the user would have been required to specify at least one new variable and allowed to specify more). In any case, syntax compares what the user types to what is allowed. If what the user types does not match what we have declared, syntax will issue the appropriate error message and stop our program. If it does match, our program will continue, and what the user typed will be broken out and stored in local macros for us. For a newvarname, the new name typed by the user is placed in the local macro varlist, and the type of the variable (float, double, ...) is placed in typlist (even if the user did not specify a storage type, in which case the type is the current default storage type).

This is now an excellent program. There are, however, two more improvements we could make. First, we have demonstrated that, by the use of 'syntax newvarname', we can allow the user to define not only the name of the created variable but also the storage type. However, when it comes to the creation of intermediate variables, such as 'hii' and 'di2', it is good programming practice to keep as much precision as possible. We want our final answer to be precise as possible, regardless of how we ultimately decide to store it. Any calculation that uses a previously generated variable would benefit if the previously generated variable were stored in double precision. Below we modify our program appropriately:

```
─────────────────────────────────── top of fvaratio.ado, version 7 ───────────
program fvaratio
        version 10
        syntax newvarname
        tempvar hii ei di2
        quietly {
                predict double 'hii' if e(sample), hat        // changed, as are
                predict double''ei' if e(sample), resid       // other lines
                gen double 'di2' = ('ei'*'ei')/e(rss)
                gen 'typlist' 'varlist' = ///
                        (e(N)-(e(df_m)+1)) / (e(N)-(e(df_m)+2)) *        ///
                        (1 - 'di2'/(1-'hii')) / (1-'hii')
        }
end
─────────────────────────────────── end of fvaratio.ado, version 7 ───────────
```

As for the second improvement we could make, fvaratio is intended to be used sometime after regress. How do we know the user is not misusing our program and executing it after, say, logistic? e(cmd) will tell us the name of the last estimation command; see [U] **18.9 Accessing results calculated by estimation commands** and [U] **18.10.3 Saving results in s()** above. We should change our program to read

```
─────────────────────────────────── top of fvaratio.ado, version 9 ───────────
program fvaratio
        version 10
        if "'e(cmd)'"!="regress" {                           // new this version
                error 301
        }
        syntax newvarname
        tempvar hii ei di2
        quietly {
                predict double 'hii' if e(sample), hat
                predict double 'ei' if e(sample), resid
                gen double 'di2' = ('ei'*'ei')/e(rss)
                gen 'typlist' 'varlist' = ///
                        (e(N)-(e(df_m)+1)) / (e(N)-(e(df_m)+2)) *        ///
                        (1 - 'di2'/(1-'hii')) / (1-'hii')
        }
end
─────────────────────────────────── end of fvaratio.ado, version 8 ───────────
```

The error command issues one of Stata's prerecorded error messages and stops our program. Error 301 is "last estimates not found"; see [P] **error**. (Try typing error 301 at the command line.)

In any case, this is a perfect program.

❏ Technical Note

You do not have to go to all the trouble we did to program the FVARATIO measure of influence or any other statistic that appeals to you. Whereas version 1 was not really an acceptable solution—it was too specialized—version 2 was acceptable. Version 3 was better, and version 4 better yet, but the improvements were of less and less importance.

Putting aside the details of Stata's language, you should understand that final versions of programs do not just happen—they are the results of drafts that have been refined. How much refinement depends on how often and who will be using the program. In this sense, the "official" ado-files that come with Stata are poor examples. They have been subject to substantial refinement because they will be used by strangers with no knowledge of how the code works. When writing programs for yourself, you may want to stop refining at an earlier draft.

❏

18.11.6   Writing online help

When you write an ado-file, you should also write a help file to go with it. This file is a standard ASCII text file, named *command*.sthlp, that you place in the same directory as your ado-file *command*.ado. This way, when users type help followed by the name of your new command (or pulls down **Help**), they will see something better than "help for . . . not found".

You can obtain examples of help files by examining the .sthlp files in the official ado-directory; type "sysdir" and look in the lettered subdirectories of the directory defined as BASE:

```
. sysdir
    STATA:  C:\Program Files\Stata10\
  UPDATES:  C:\Program Files\Stata10\ado\updates\
     BASE:  C:\Program Files\Stata10\ado\base\
     SITE:  C:\Program Files\Stata10\ado\site\
     PLUS:  C:\ado\plus\
 PERSONAL:  C:\ado\personal\
 OLDPLACE:  C:\ado\
```

Here you would find examples of .sthlp files in the a, b, ... subdirectories of C:\Program Files\Stata10\ado\base.

Help files are physically written on the disk in ASCII text format, but their contents are SMCL—Stata Markup and Control Language. For the most part, you can ignore that. If the file contains a line that reads

```
Also see help for the finishup command
```

it will display in just that way. However, SMCL contains many special directives, so that if the line in the file were to read

```
Also see {hi:help} for the {help finishup} command
```

what would be displayed would be

Also see **help** for the **finishup** command

and moreover, **finishup** would appear as a hypertext link, meaning that if users clicked on it, they would see help on finishup.

You can read about the details of SMCL in [P] **smcl**. The following is a SMCL help file:

── top of examplehelpfile.sthlp ────────────

```
{smcl}
{* *! version 1.0.5  18may2007}{...}
{cmd:help whatever}
{hline}

{title:Title}

{phang}
{bf:whatever -- Calculate whatever statistic}


{title:Syntax}

{p 8 17 2}
{cmdab:wh:atever}
[{varlist}]
{ifin}
{weight}
[{cmd:,}
{it:options}]

{synoptset 20 tabbed}{...}
{synopthdr}
{synoptline}
{syntab:Main}
{synopt:{opt d:etail}}display additional statistics{p_end}
{synopt:{opt mean:only}}suppress the display; calculate only the mean; programmer's
                        option{p_end}
{synopt:{opt f:ormat}}use variable's display format{p_end}
{synopt:{opt sep:arator(#)}}draw separator line after every {it:#} variables; default
                        is {cmd:separator(5)}{p_end}
{synopt:{opth g:enerate(newvar)}}create variable name {it:newvar}{p_end}
{synoptline}
{p2colreset}{...}
{p 4 6 2}
{cmd:by} is allowed; see {manhelp by D}.{p_end}
{p 4 4 2}
{cmd:fweight}s are allowed; see {help weight}.


{title:Description}

{pstd}
{cmd:whatever} calculates the whatever statistic for the variables in
{varlist} when the data are not stratified.


{title:Options}

{dlgtab:Main}

{phang}
{opt detail} displays detailed output of the calculation.

{phang}
{opt meanonly} restricts the calculation to be based on only the
means.  The default is to use a trimmed mean.

{phang}
{opt format} requests that the summary statistics be displayed using the display formats
associated with the variables, rather than the default {cmd:g} display format; see {bf:[U]
12.5 Formats: controlling how data are displayed}.
```

{phang}
{opt separator(#)} specifies how often to insert separation lines
into the output. The default is {cmd:separator(5)}, meaning that a
line is drawn after every 5 variables. {cmd:separator(10)} would
draw a line after every 10 variables. {cmd:separator(0)} suppresses
the separation line.
{phang}
{opth generate(newvar)} creates {it:newvar} containing the whatever values.

{title:Remarks}
{pstd}
For detailed information on the whatever statistic, see {bf:[R] intro}.

{title:Examples}
{phang}{cmd:. whatever mpg weight}
{phang}{cmd:. whatever mpg weight, meanonly}

{title:Also see}
{psee}
Manual: {hi:[U] 18.11.6 Writing online help}
{psee}
Online: {manhelp help R}
{p_end}

——————————————————————————— end of examplehelpfile.sthlp ———————

If you were to select **Help > Stata Command**, and type examplehelpfile and click **OK**, or if you
were to type help examplehelpfile, this is what you would see:

---

help whatever

Title

    whatever -- Calculate whatever statistic

Syntax

        whatever [*varlist*] [*if*] [*in*] [*weight*] [, *options*]

options	description
Main	
detail	display additional statistics
meanonly	suppress the display; calculate only the mean; programmer's option
format	use variable's display format
separator(#)	draw separator line after every # variables; default is **separator(5)**
generate(*newvar*)	create variable name *newvar*

by is allowed; see **[D] by**.
fweights are allowed; see **weight**.

Description

    whatever calculates the whatever statistic for the variables in *varlist* when
    the data are not stratified.

Options

┌─── Main ───┐

detail displays detailed output of the calculation.

meanonly restricts the calculation to be based on only the means. The default is to use a trimmed mean.

format requests that the summary statistics be displayed using the display formats associated with the variables, rather than the default **g** display format; see **[U] 12.5 Formats: controlling how data are displayed**.

separator(#) specifies how often to insert separation lines into the output. The default is **separator(5)**, meaning that a line is drawn after every 5 variables. **separator(10)** would draw a line after every 10 variables. **separator(0)** suppresses the separation line.

generate(_newvar_**)** creates _newvar_ containing the whatever values.

Remarks

For detailed information on the whatever statistic, see **[R] intro**.

Examples

. **whatever mpg weight**

. **whatever mpg weight, meanonly**

Also see

Manual: **[U] 18.11.6 Writing online help**

Online: **[R] help**

Users will find it easier to understand your programs if you document them the same way that we document ours. We offer the following guidelines:

1. The first line must be

   {smcl}

   This notifies Stata that the help file is in SMCL format.

2. The second line should be

   {* *! version #.#.# _date_}{...}

   The * indicates a comment and the {...} will suppress the blank line. Whenever you edit the help file, update the version number and the date found in the comment line.

3. The third line should contain

   {cmd:help _yourcmd_}
   {hline}

   assuming the help file is named _yourcmd_.sthlp.

4. Include one blank line, and then place the title.

   {title:Title}

   {phang}
   {bf:_yourcmd_ -- _Your title_}

5. Include two blank lines, and place the Syntax title, syntax diagram, and options table:

> {title:Syntax}
>
> {p 8 17 2}
> *syntax line*
>
> {p 8 17 2}
> *second syntax line, if necessary*
>
> {synoptset 20 tabbed}{...}
> {synopthdr}
> {synoptline}
> {syntab:*tab*}
> {synopt:{*option*}}*brief description of option*{p_end}
> {synoptline}
> {p2colreset}{...}
>
> {p 4 6 2}
> *clarifying text, if required*

6. Include two blank lines, and place the Description title and text:

> {title:Description}
>
> {pstd}
> *description text*

Briefly describe what the command does. Do not burden the user with details yet. Assume that the user is at the point of asking whether this is what he or she is looking for.

7. If your command allows options, include two blank lines, and place the Options title and descriptions:

> {title:Options}
>
> {phang}
> {cmd:*optionname*} *option description*
>
> {pmore}
> *continued option description, if necessary*
>
> {phang}
> {cmd:*optionname*} *second option description*

Options should be included in the order in which they appear in the option table. Option paragraphs are reverse indented, with the option name on the far left, where it is easily spotted. If an option requires more than one paragraph, subsequent paragraphs are set using {pmore}. One blank line separates one option from another.

8. Optionally include two blank lines, and place the Remarks title and text:

> {title:Remarks}
>
> {pstd}
> *text*

Include whatever long discussion you feel necessary. Stata's official online help files often omit this because the discussions appear in the manual. Stata's official help files for features added between releases (obtained from the *Stata Journal*, the Stata web site, etc.), however, include this section because the appropriate *Stata Journal* may not be as accessible as the manuals.

9. Optionally include two blank lines, and place the Examples title and text:

    {title:Examples}

    {phang}
    {cmd:. *first example*}

    {phang}
    {cmd:. *second example*}

Nothing communicates better than providing something beyond theoretical discussion. Examples rarely need much explanation.

10. Optionally include two blank lines, and place the Author title and text:

    {title:Author}

    {pstd}
    Name, affiliation, etc.

Exercise caution. If you include a telephone number, expect your phone to ring. An email address may be more appropriate.

11. Optionally include two blank lines, and place the References title and text:

    {title:References}

    {pstd}
    Author. year.
    Title. Location: Publisher.

12. Include two blank lines, and place the Also see title and text:

    {title:Also see}

    {psee}
    Manual: {hi:*Manual reference, if any*}

    {psee}
    Online: {manhelp *name*}, {manhelp *name*}, ...

We also warn that it is easy to use too much {hi:highlighting}. Use it sparingly. In text, use {cmd:...} to show what would be shown in typewriter typeface it the documentation were printed in this manual.

❑ Technical Note

Sometimes it is more convenient to describe two or more related commands in the same .sthlp file. Thus, xyz.sthlp might document both the xyz and abc commands. To arrange that typing help abc displays xyz.sthlp, create the file abc.sthlp, containing

```
                                                              top of abc.sthlp
    .h xyz
                                                              end of abc.sthlp
```

When a .sthlp file contains one line of the form '.h *refname*', Stata interprets that as an instruction to display help for *refname*.

❑

❏ Technical Note

If you write a collection of programs, you need to somehow index the programs so that users (and you) can find the command they want. We do that with our `contents.sthlp` entry. You should create a similar kind of entry. We suggest that you call your private entry `user.sthlp` in your personal ado-directory; see [U] **17.5.2 Where is my personal ado directory?**. This way, to review what you have added, you can type `help user`.

We suggest that Unix users at large sites also add `site.sthlp` to the SITE directory (typically `/usr/local/ado`, but type `sysdir` to be sure). Then you can type `help site` for a list of the commands available sitewide.

❏

18.12   A compendium of useful commands for programmers

You can use any Stata command in your programs and ado-files. Also, some commands are intended solely for use by Stata programmers. You should see the section under the *Programming* heading in the subject table of contents at the beginning of the *Quick Reference and Index*.

18.13   References

Belsley, D. A., E. Kuh, and R. E. Welsch. 1980. *Regression Diagnostics: Identifying Influential Data and Sources of Collinearity*. New York: Wiley.

Gould, W. 2001. pr0001: Statistical software certification. *Stata Journal* 1: 29–50.

19 Immediate commands

Contents

19.1 Overview

An *immediate* command is a command that obtains data not from the data stored in memory but from numbers typed as arguments. Immediate commands, in effect, turn Stata into a glorified hand calculator.

There are many instances when you may not have the data, but you do know something about the data, and what you know is adequate to perform statistical tests. For instance, you do not have to have individual-level data to obtain the standard error of the mean, and thereby a confidence interval, if you know the mean, standard deviation, and number of observations. In other instances, you may actually have the data, and you could enter the data and perform the test, but it would be easier if you could just ask for the statistic based on a summary. For instance, you flip a coin 10 times, and it comes up heads twice. You could enter a 10-observation dataset with two ones (standing for heads) and eight zeros (meaning tails).

Immediate commands are meant to solve those problems. Immediate commands have the following properties:

1. They never disturb the data in memory. You can perform an immediate calculation as an aside without changing your data.

2. The syntax for these commands is the same, the command name followed by numbers, which are the summary statistics from which the statistic is calculated. The numbers are almost always summary statistics, and the order in which they are specified is in some sense "natural".

3. Immediate commands all end in the letter *i*, although the converse is not true. Usually, if there is an immediate command, there is a nonimmediate form also, that is, a form that works on the data in memory. For every statistical command in Stata, we have included an immediate form if it is reasonable to assume that you might know the requisite summary statistics without having the underlying data and if typing those statistics is not absurdly burdensome.

4. Immediate commands are documented along with their nonimmediate counterparts. Thus, if you want to obtain a confidence interval, whether it be from summary data with an immediate command or using the data in memory, use the table of contents or index to discover that [R] **ci** discusses confidence intervals. There, you learn that `ci` calculates confidence intervals by using the data in memory and that `cii` does the same with the data specified immediately following the command.

19.1.1   Examples

▷ Example 1

Let's take the example of confidence intervals. Professional papers often publish the mean, standard deviation, and number of observations for variables used in the analysis. Those statistics are sufficient for calculating a confidence interval. If we know that the mean mileage rating of cars in some sample is 24, that the standard deviation is 6, and that there are 97 cars in the sample, we can calculate

```
. cii 97 24 6
```

Variable	Obs	Mean	Std. Err.	[95% Conf. Interval]
	97	24	.6092077	22.79073 25.20927

We learn that the mean's standard error is 0.61 and its 95% confidence interval is $[22.8, 25.2]$. To obtain this, we typed cii (the immediate form of the ci command) followed by the number of observations, the mean, and the standard deviation. We knew the order in which to specify the numbers because we had read [R] **ci**.

We could use the immediate form of the ttest command to test the hypothesis that the true mean is 22:

```
. ttesti 97 24 6 22
```

One-sample t test

	Obs	Mean	Std. Err.	Std. Dev.	[95% Conf. Interval]
x	97	24	.6092077	6	22.79073 25.20927

```
         mean = mean(x)                                     t =    3.2830
Ho: mean = 22                              degrees of freedom =        96

    Ha: mean < 22            Ha: mean != 22             Ha: mean > 22
 Pr(T < t) = 0.9993      Pr(|T| > |t|) = 0.0014       Pr(T > t) = 0.0007
```

The first three numbers were as we specified in the cii command. ttesti requires a fourth number, which is the constant against which the mean is being tested; see [R] **ttest**.

◁

▷ Example 2

We mentioned flipping a coin 10 times and having it come up heads twice. The 99% confidence interval can also be obtained from ci:

```
. cii 10 2, level(99)
```

Variable	Obs	Mean	Std. Err.	— Binomial Exact — [99% Conf. Interval]
	10	.2	.1264911	.0108505 .6482012

In the previous example, we specified cii with three numbers following it; in this example, we specify 2. Immediate commands often determine what to do by the number of arguments following the command. With two arguments, ci assumes that we are specifying the number of trials and successes from a binomial experiment; see [R] **ci**.

The immediate form of the `bitest` command performs exact hypothesis testing:

```
. bitesti 10 2 .5
          N    Observed k    Expected k    Assumed p    Observed p
         10            2             5      0.50000       0.20000
   Pr(k >= 2)              = 0.989258    (one-sided test)
   Pr(k <= 2)              = 0.054688    (one-sided test)
   Pr(k <= 2 or k >= 8) = 0.109375      (two-sided test)
```

For a full explanation of this output, see [R] **bitest**.

◁

▷ Example 3

Stata's `tabulate` command makes tables and calculates various measures of association. The immediate form, `tabi`, does the same, but we specify the contents of the table following the command:

```
. tabi 5 10 \ 2 14
              |        col
          row |     1        2 |     Total
     ---------+------------------+---------
            1 |     5       10 |        15
            2 |     2       14 |        16
     ---------+------------------+---------
        Total |     7       24 |        31

             Fisher's exact =              0.220
     1-sided Fisher's exact =              0.170
```

The `tabi` command is slightly different from most immediate commands because it uses '\' to indicate where one row ends and another begins.

◁

(Continued on next page)

19.1.2   A list of the immediate commands

Command	Reference	Description
bitesti	[R] **bitest**	Binomial probability test
cci csi iri mcci	[ST] **epitab**	Tables for epidemiologists
cii	[R] **ci**	Confidence intervals for means, proportions, and counts
prtesti	[R] **prtest**	One- and two-sample tests of proportions
sampsi	[R] **sampsi**	Sample size and power determination
sdtesti	[R] **sdtest**	Variance comparison tests
symmi	[R] **symmetry**	Symmetry and marginal homogeneity tests
tabi	[R] **tabulate twoway**	Two-way tables of frequencies
ttesti	[R] **ttest**	Mean comparison tests
twoway lpolyci twoway pci twoway pcarrowi twoway scatteri	[G] **graph twoway lpolyci** [G] **graph twoway pci** [G] **graph twoway pcarrowi** [G] **graph twoway scatteri**	Local polynomial smooth plots with CIs Paired-coordinate plot with spikes or lines Paired-coordinate plot with arrows Twoway scatterplot

19.2   The display command

display is not really an immediate command, but it can be used as a hand calculator.

```
. display 2+5
7
. display sqrt(2+sqrt(3^2-4*2*-2))/(2*3)
.44095855
```

See [R] **display**.

20 Estimation and postestimation commands

Contents

20.1   All estimation commands work the same way

All Stata commands that fit statistical models—commands such as `regress`, `logit`, `sureg`, and so on—work the same way. Most single-equation estimation commands have the syntax

$$command\ varlist\ \big[\,if\,\big]\ \big[\,in\,\big]\ \big[\,weight\,\big]\ \big[\,,\ options\,\big]$$

and most multiple-equation estimation commands have the syntax

$$command\ (varlist)\ (varlist)\ \dots\ (varlist)\ \big[\,if\,\big]\ \big[\,in\,\big]\ \big[\,weight\,\big]\ \big[\,,\ options\,\big]$$

Adopt a loose definition of single and multiple equation in interpreting this. For instance, heckman is a two-equation system, mathematically speaking, yet we categorize it, syntactically, with single-equation commands because most researchers think of it as a linear regression with an adjustment for the censoring. The important thing is that most estimation commands have one or the other of these two syntaxes.

In single-equation commands, the first variable in the *varlist* is the dependent variable, and the remaining variables are the independent variables, with some exceptions. For instance, anova allows you to specify terms in addition to variables for the independent variables.

Prefix commands may be specified in front of an estimation command to modify what it does. The syntax is

$$prefix: command \ldots$$

where the prefix commands are

prefix command	description	manual entry
rolling	time-series rolling estimation	[TS] **rolling**
*nestreg	nested model statistics	[R] **nestreg**
statsby	collect results across subsets of data	[D] **statsby**
*stepwise	stepwise estimation	[R] **stepwise**
*svy	estimation for complex survey data	[SVY] **svy**
xi	interaction expansion	[R] **xi**

    *Available for some but not all estimation commands

Two other prefix commands—bootstrap and jackknife—also work with estimation commands— see [R] **bootstrap** and [R] **jackknife**—but usually it is easier to specify the estimation-command option vce(bootstrap) or vce(jackknife).

Also, all estimation commands—whether single or multiple equation—share the following features:

1. You can use the standard features of Stata's syntax—if *exp* and in *range*—to specify the estimation subsample; you do not have to make a special dataset.

2. You can retype the estimation command without arguments to redisplay the most recent estimation results. For instance, after fitting a model with regress, you can see the estimates again by typing regress by itself. You do not have to do this immediately—any number of commands can occur between the estimation and the replaying, and, in fact, you can even replay the last estimates after the data have changed or you have dropped the data altogether. Stata never forgets (unless you type discard; see [P] **discard**).

3. You can specify option level() at the time of estimation, or when you redisplay results if that makes sense, to specify the width of the confidence intervals for the coefficients. The default is level(95), meaning 95% confidence intervals. You can reset the default with set level; see [R] **level**.

4. You can use the postestimation command mfx to display model results in terms of marginal effects (dy/dx or even $df(y)/dx$), which can be displayed as either derivatives or elasticities; see [R] **mfx**.

5. You can use the postestimation command adjust to obtain tables of adjusted means; see [R] **adjust**.

6. You can use the postestimation command `estat` to obtain common statistics associated with the model. What statistics are available are documented in the postestimation section following the documentation of the estimation command, for instance, in [R] **regress postestimation** following [R] **regress**.

   You can always use the postestimation command `estat vce` to obtain the variance–covariance matrix of the estimators (VCE), presented as either a correlation matrix or a covariance matrix. (You can also obtain the estimated coefficients and covariance matrix as vectors and matrices and manipulate them with Stata's matrix capabilities; see [U] **14.5 Accessing matrices created by Stata commands**.)

7. You can use the postestimation command `predict` to obtain predictions, residuals, influence statistics, and the like, either for the data on which you just estimated or for some other data. You can use postestimation command `predictnl` to obtain point estimates, standard errors, etc., for customized predictions.

8. You can refer to the values of coefficients and standard errors in expressions (such as with `generate`) by using standard notation; see [U] **13.5 Accessing coefficients and standard errors**. You can refer in expressions to the values of other estimation-related statistics by using `e(`*resultname*`)`. For instance, all commands define `e(N)` recording the number of observations in the estimation subsample. After estimation, type `ereturn list` to see a list of all that is available. See the *Saved Results* section in the estimation command's documentation for their definition.

   An especially useful `e()` result is `e(sample)`: it returns 1 if an observation was used in the estimation and 0 otherwise, so you can add `if e(sample)` to the end of other commands to restrict them to the estimation subsample. You could type, for instance, `summarize if e(sample)`.

9. You can use the postestimation command `test` to perform tests on the estimated parameters (Wald tests of linear hypotheses), `testnl` to perform Wald tests of nonlinear hypotheses, and (`lrtest` to perform likelihood-ratio tests. You can use the postestimation command `lincom` to obtain point estimates and confidence intervals for linear combinations of the estimated parameters and the postestimation command `nlcom` to obtain nonlinear combinations.

10. You can use the postestimation command `estimates` to store estimation results by name for later retrieval or for displaying/comparing multiple models by using `estimates`, or for saving them in a file; see [R] **estimates**.

11. You can use the postestimation command `_estimates` to hold estimates, perform other estimation commands, and then restore the prior estimates. This is of particular interest to programmers.

12. You can use the postestimation command `suest` to obtain the joint parameter vector and variance–covariance matrix for coefficients from two different models by using seemingly unrelated estimation. This is especially useful for testing the equality, say, of coefficients across models; see [R] **suest**.

13. You can use the postestimation command `hausman` to perform Hausman model-specification tests by using `hausman`; see [R] **hausman**.

14. With some exceptions, you can specify option `vce(robust)` at the time of estimation to obtain the Huber/White/robust alternate estimate of variance, or you can specify option `vce(cluster `*clustvar*`)` to relax the assumption of independence of the observations.

   Most estimation commands also allow a `vce(`*vcetype*`)` option to specify other alternative variance estimators—which ones are allowed are documented with the estimator—and usually `vce(opg)`, `vce(bootstrap)`, and `vce(jackknife)` are available.

20.2 Standard syntax

You can combine Stata's if *exp* and in *range* with any estimation command. Estimation commands also allow by *varlist*:, where it would be sensible.

▷ Example 1

We have data on 74 automobiles that record the mileage rating (mpg), weight (weight), and whether the car is domestic or foreign produced (foreign). We can fit a linear regression model of mpg on weight and weightsq, using just the foreign-made automobiles, by typing

```
. use http://www.stata-press.com/data/r10/auto2
(1978 Automobile Data)
. regress mpg weight weightsq if foreign
```

Source	SS	df	MS		Number of obs =	22
					F(2, 19) =	8.31
Model	428.256889	2	214.128444		Prob > F =	0.0026
Residual	489.606747	19	25.7687762		R-squared =	0.4666
					Adj R-squared =	0.4104
Total	917.863636	21	43.7077922		Root MSE =	5.0763

mpg	Coef.	Std. Err.	t	P>|t|	[95% Conf. Interval]	
weight	-.0132182	.0275711	-0.48	0.637	-.0709252	.0444888
weightsq	5.50e-07	5.41e-06	0.10	0.920	-.0000108	.0000119
_cons	52.33775	34.1539	1.53	0.142	-19.14719	123.8227

We can run separate regressions for the domestic and foreign-produced automobiles with the by *varlist*: prefix:

```
. by foreign: regress mpg weight weightsq
```

```
-> foreign = Domestic
```

Source	SS	df	MS		Number of obs =	52
					F(2, 49) =	91.64
Model	905.395466	2	452.697733		Prob > F =	0.0000
Residual	242.046842	49	4.93973146		R-squared =	0.7891
					Adj R-squared =	0.7804
Total	1147.44231	51	22.4988688		Root MSE =	2.2226

mpg	Coef.	Std. Err.	t	P>|t|	[95% Conf. Interval]	
weight	-.0131718	.0032307	-4.08	0.000	-.0196642	-.0066794
weightsq	1.11e-06	4.95e-07	2.25	0.029	1.19e-07	2.11e-06
_cons	50.74551	5.162014	9.83	0.000	40.37205	61.11896

```
-> foreign = Foreign
```

Source	SS	df	MS		Number of obs =	22
					F(2, 19) =	8.31
Model	428.256889	2	214.128444		Prob > F =	0.0026
Residual	489.606747	19	25.7687762		R-squared =	0.4666
					Adj R-squared =	0.4104
Total	917.863636	21	43.7077922		Root MSE =	5.0763

mpg	Coef.	Std. Err.	t	P>|t|	[95% Conf. Interval]	
weight	-.0132182	.0275711	-0.48	0.637	-.0709252	.0444888
weightsq	5.50e-07	5.41e-06	0.10	0.920	-.0000108	.0000119
_cons	52.33775	34.1539	1.53	0.142	-19.14719	123.8227

Although all estimation commands allow if *exp* and in *range*, only some allow the by *varlist*: prefix. For by(), the duration of Stata's memory is limited: it remembers the *last* set of estimates only. This means that, if we were to use any of the other features described below, they would use the last regression estimated, which right now is mpg on weight and weightsq for the Foreign subsample.

◁

20.3 Replaying prior results

When you type an estimation command without arguments, it redisplays prior results.

▷ Example 2

To perform a regression of mpg on the variables weight and displacement, we could type

(Continued on next page)

```
. regress mpg weight displacement
```

Source	SS	df	MS
Model	1595.40969	2	797.704846
Residual	848.049768	71	11.9443629
Total	2443.45946	73	33.4720474

```
Number of obs =      74
F(  2,    71) =   66.79
Prob > F      =  0.0000
R-squared     =  0.6529
Adj R-squared =  0.6432
Root MSE      =  3.4561
```

mpg	Coef.	Std. Err.	t	P>|t|	[95% Conf. Interval]	
weight	-.0065671	.0011662	-5.63	0.000	-.0088925	-.0042417
displacement	.0052808	.0098696	0.54	0.594	-.0143986	.0249602
_cons	40.08452	2.02011	19.84	0.000	36.05654	44.11251

We now go on to do other things, summarizing data, listing observations, performing hypothesis tests, or anything else. If we decide that we want to see the last set of estimates again, we type the estimation command without arguments.

```
. regress
```

Source	SS	df	MS
Model	1595.40969	2	797.704846
Residual	848.049768	71	11.9443629
Total	2443.45946	73	33.4720474

```
Number of obs =      74
F(  2,    71) =   66.79
Prob > F      =  0.0000
R-squared     =  0.6529
Adj R-squared =  0.6432
Root MSE      =  3.4561
```

mpg	Coef.	Std. Err.	t	P>|t|	[95% Conf. Interval]	
weight	-.0065671	.0011662	-5.63	0.000	-.0088925	-.0042417
displacement	.0052808	.0098696	0.54	0.594	-.0143986	.0249602
_cons	40.08452	2.02011	19.84	0.000	36.05654	44.11251

This feature works with *every* estimation command, so we could just as well have done it with, say, stcox or logit.   ◁

20.4   Cataloging estimation results

Stata keeps only the results of the most recently fitted model in active memory. You can use Stata's estimates command, however, to temporarily store estimation results for displaying, comparing, cross-model testing, etc., during the same session. You can also save estimation results to disk, but that will be the subject of the next section. You may temporarily store up to 300 sets of estimation results.

▷ Example 3

Continuing with our automobile data, we fit four models and estimates store them. We fit the models quietly to keep the output to a minimum.

```
. quietly regress mpg weight displ
. estimates store r_base
. quietly regress mpg weight displ foreign
. estimates store r_alt
```

```
. quietly qreg mpg weight displ
. estimates store q_base
. quietly qreg mpg weight displ foreign
. estimates store q_alt
```

We saved the four models under the names r_base, r_alt, q_base, and q_alt but, if we forget, we can ask to see a directory of what is stored:

```
. estimates dir
```

model	command	depvar	npar	title
r_base	regress	mpg	3	
r_alt	regress	mpg	4	
q_base	qreg	mpg	3	
q_alt	qreg	mpg	4	

We can ask that any of the previous models be replayed:

```
. estimates replay r_base
```

Model **r_base**

Source	SS	df	MS		
Model	1595.40969	2	797.704846		
Residual	848.049768	71	11.9443629		
Total	2443.45946	73	33.4720474		

Number of obs =   74
F(2,   71) =   66.79
Prob > F   = 0.0000
R-squared   = 0.6529
Adj R-squared = 0.6432
Root MSE   = 3.4561

mpg	Coef.	Std. Err.	t	P>|t|	[95% Conf. Interval]
weight	-.0065671	.0011662	-5.63	0.000	-.0088925 -.0042417
displacement	.0052808	.0098696	0.54	0.594	-.0143986 .0249602
_cons	40.08452	2.02011	19.84	0.000	36.05654 44.11251

Or we can ask to see all the models in a combined table:

```
. estimates table _all
```

Variable	r_base	r_alt	q_base	q_alt
weight	-.00656711	-.00677449	-.00581173	-.00595056
displacement	.00528078	.00192865	.00428411	.00018552
foreign		-1.6006312		-2.1326004
_cons	40.084522	41.847949	37.559865	39.213348

estimates displayed just the coefficients, but we could ask for other statistics. Finally, we can also select one of the stored estimates to be made active, meaning things are just as if we had just fitted the model:

(Continued on next page)

```
. estimates restore r_alt
(results r_alt are active now)

. regress
```

Source	SS	df	MS
Model	1619.71935	3	539.906448
Residual	823.740114	70	11.7677159
Total	2443.45946	73	33.4720474

```
Number of obs =      74
F(  3,     70) =   45.88
Prob > F      =  0.0000
R-squared     =  0.6629
Adj R-squared =  0.6484
Root MSE      =  3.4304
```

mpg	Coef.	Std. Err.	t	P>|t|	[95% Conf. Interval]	
weight	-.0067745	.0011665	-5.81	0.000	-.0091011	-.0044479
displacement	.0019286	.0100701	0.19	0.849	-.0181556	.0220129
foreign	-1.600631	1.113648	-1.44	0.155	-3.821732	.6204699
_cons	41.84795	2.350704	17.80	0.000	37.15962	46.53628

◁

You can do a lot more with estimates; see [R] **estimates**. In particular, estimates makes it easy to perform cross-model tests, such as the Hausman specification test.

20.5   Saving estimation results

estimates can also save estimation results into a file.

```
. estimates save alt
    file alt.ster saved
```

That saved the active estimation results, the ones we just estimated or, in our case, the ones we just restored. Later, even in another Stata session, we could reload our estimates:

```
. estimates use alt

. regress
```

Source	SS	df	MS
Model	1619.71935	3	539.906448
Residual	823.740114	70	11.7677159
Total	2443.45946	73	33.4720474

```
Number of obs =      74
F(  3,     70) =   45.88
Prob > F      =  0.0000
R-squared     =  0.6629
Adj R-squared =  0.6484
Root MSE      =  3.4304
```

mpg	Coef.	Std. Err.	t	P>|t|	[95% Conf. Interval]	
weight	-.0067745	.0011665	-5.81	0.000	-.0091011	-.0044479
displacement	.0019286	.0100701	0.19	0.849	-.0181556	.0220129
foreign	-1.600631	1.113648	-1.44	0.155	-3.821732	.6204699
_cons	41.84795	2.350704	17.80	0.000	37.15962	46.53628

There is one important difference between storing results in memory and saving them in a file: e(sample) is lost. We have not discussed e(sample) yet, but it allows us to identify the observations among those currently in memory that were used in the estimation. For instance, after estimation, we could type

```
. summarize mpg weight displ foreign if e(sample)
    (output omitted )
```

and see the summary statistics of the relevant data. We could do that after an `estimates restore`, too. But we cannot do it after an `estimates use`. Part of the reason is that we might not even have the relevant data in memory. Even if we do, however, here is what will happen:

```
. summarize mpg weight displ foreign if e(sample)
     Variable |       Obs        Mean    Std. Dev.       Min        Max
--------------+--------------------------------------------------------
          mpg |         0
       weight |         0
 displacement |         0
      foreign |         0
```

Stata will just assume that none of the data in memory played a role in obtaining the estimation results.

There is more worth knowing. You could, for instance, type `estimates describe` to see the command line that produced the estimates. See [R] **estimates**.

20.6   Specifying the estimation subsample

You specify the estimation subsample—the sample to be used in estimation—by specifying the `if` *exp* and/or `in` *range* modifiers with the estimation command.

Once an estimation command has been run or previous estimates restored, Stata remembers the estimation subsample, and you can use the modifier `if e(sample)` on the end of other Stata commands. The term *estimation subsample* refers to the set of observations used to produce the active estimation results. That might turn out to be all the observations (as it was in the above example) or some of the observations:

```
. generate excellent = rep78==5 if rep78 < .
(5 missing values generated)
. regress mpg weight excellent if foreign
       Source |       SS           df       MS            Number of obs =      21
--------------+----------------------------           F(  2,     18) =   10.21
        Model |  423.317154         2  211.658577      Prob > F       =  0.0011
     Residual |   372.96856        18  20.7204756      R-squared      =  0.5316
--------------+----------------------------           Adj R-squared  =  0.4796
        Total |  796.285714        20  39.8142857      Root MSE       =   4.552

          mpg |      Coef.   Std. Err.      t    P>|t|     [95% Conf. Interval]
--------------+----------------------------------------------------------------
       weight |  -.0131402   .0029684    -4.43   0.000    -.0193765   -.0069038
    excellent |   5.052676    2.13492     2.37   0.029     .5673764    9.537977
        _cons |   52.86088   6.540147     8.08   0.000     39.12054    66.60122
```

```
. summarize mpg weight excellent if e(sample)
     Variable |       Obs        Mean    Std. Dev.       Min        Max
--------------+--------------------------------------------------------
          mpg |        21    25.28571    6.309856        17         41
       weight |        21    2263.333    364.7099      1760       3170
    excellent |        21    .4285714    .5070926         0          1
```

Twenty-one observations were used in the above regression and we subsequently obtained the means for those same 21 observations by typing `summarize ... if e(sample)`. There are two reasons observations were dropped: we specified `if foreign` when we ran the regression, and there were observations for which `excellent` was missing. The reason does not matter; `e(sample)` is true if the observation was used and false otherwise.

You can use if e(sample) on the end of any Stata command that allows an if *exp*.

Here Stata has a shorthand command that produces the same results as summarize ... if e(sample):

```
. estat summarize, label
Estimation sample regress                                    Number of obs =      21
```

Variable	Mean	Std. Dev.	Min	Max	Label
mpg	25.28571	6.309856	17	41	Mileage (mpg)
weight	2263.333	364.7099	1760	3170	Weight (lbs.)
excellent	.4285714	.5070926	0	1	

See [R] **estat**.

20.7   Specifying the width of confidence intervals

You can specify the width of the confidence intervals for the coefficients using the level() option at estimation or when you play back the results.

▷ Example 4

To obtain narrower, 90% confidence intervals when we fit the model, we type

```
. regress mpg weight displ, level(90)
```

Source	SS	df	MS		
Model	1595.40969	2	797.704846		
Residual	848.049768	71	11.9443629		
Total	2443.45946	73	33.4720474		

```
Number of obs =      74
F(  2,    71) =   66.79
Prob > F      = 0.0000
R-squared     = 0.6529
Adj R-squared = 0.6432
Root MSE      = 3.4561
```

mpg	Coef.	Std. Err.	t	P>|t|	[90% Conf. Interval]
weight	-.0065671	.0011662	-5.63	0.000	-.0085108 -.0046234
displacement	.0052808	.0098696	0.54	0.594	-.0111679 .0217294
_cons	40.08452	2.02011	19.84	0.000	36.71781 43.45124

If we subsequently typed regress, without arguments, 95% confidence intervals would be reported. If we initially fitted the model with 95% confidence intervals, we could later type regress, level(90) to redisplay results with 90% confidence intervals.

Also, we could type set level 90 to make 90% intervals our default; see [R] **level**.

Stata allows noninteger confidence intervals between 10.00 and 99.99, with a maximum of two digits following the decimal point. For instance, we could type

```
. regress mpg weight displ, level(92.5)
```

Source	SS	df	MS		
Model	1595.40969	2	797.704846		
Residual	848.049768	71	11.9443629		
Total	2443.45946	73	33.4720474		

```
Number of obs =      74
F(  2,    71) =   66.79
Prob > F      = 0.0000
R-squared     = 0.6529
Adj R-squared = 0.6432
Root MSE      = 3.4561
```

mpg	Coef.	Std. Err.	t	P>|t|	[92.5% Conf. Interval]	
weight	-.0065671	.0011662	-5.63	0.000	-.0086745	-.0044597
displacement	.0052808	.0098696	0.54	0.594	-.0125535	.023115
_cons	40.08452	2.02011	19.84	0.000	36.43419	43.73485

◁

20.8   Obtaining the variance–covariance matrix

Typing estat vce displays the variance–covariance matrix of the estimators in active memory.

▷ Example 5

In example 2, we typed regress mpg weight displacement. The full variance–covariance matrix of the estimators can be displayed at any time after estimation:

```
. estat vce
Covariance matrix of coefficients of regress model
      e(V) |      weight  displace~t        _cons

     weight | 1.360e-06
displacement | -.0000103    .00009741
      _cons | -.00207455   .01188356    4.0808455
```

Typing estat vce with the corr option presents this matrix as a correlation matrix:

```
. estat vce, corr
Correlation matrix of coefficients of regress model
      e(V) |      weight   displa~t        _cons

     weight |     1.0000
displacement |    -0.8949     1.0000
      _cons |    -0.8806     0.5960       1.0000
```

See [R] estat.

Also, Stata's matrix commands understand that e(V) refers to the matrix:

```
. matrix list e(V)
symmetric e(V)[3,3]
                  weight  displacement        _cons
     weight    1.360e-06
displacement   -.0000103     .00009741
      _cons   -.00207455     .01188356    4.0808455
. mat Vinv = invsym(e(V))
. mat list Vinv
symmetric Vinv[3,3]
                  weight  displacement        _cons
     weight     60175851
displacement   4081161.2      292709.46
      _cons    18706.732      1222.3339    6.1953911
```

See [U] **14.5 Accessing matrices created by Stata commands**.

◁

20.9   Obtaining predicted values

Our discussion below, although cast in terms of predicted values, applies equally to the other statistics generated by predict; see [R] **predict**.

When Stata fits a model, whether it is regression or anything else, it internally saves the results, including the estimated coefficients and the variable names. The predict command allows you to use that information.

▷ Example 6

Let's perform a linear regression of mpg on weight and weightsq:

```
. regress mpg weight weightsq

      Source |       SS       df       MS              Number of obs =      74
-------------+------------------------------           F(  2,    71) =   72.80
       Model | 1642.52197     2  821.260986            Prob > F      =  0.0000
    Residual | 800.937487    71  11.2808097            R-squared     =  0.6722
-------------+------------------------------           Adj R-squared =  0.6630
       Total | 2443.45946    73  33.4720474            Root MSE      =  3.3587

------------------------------------------------------------------------------
         mpg |      Coef.   Std. Err.      t    P>|t|     [95% Conf. Interval]
-------------+----------------------------------------------------------------
      weight |  -.0141581   .0038835    -3.65   0.001    -.0219016   -.0064145
    weightsq |   1.32e-06   6.26e-07     2.12   0.038     7.67e-08    2.57e-06
       _cons |   51.18308   5.767884     8.87   0.000     39.68225    62.68392
------------------------------------------------------------------------------
```

After the regression, predict is defined to be

$$-.0141581\text{weight} + 1.32 \cdot 10^{-6}\text{weightsq} + 51.18308$$

(Actually, it is more precise because the coefficients are internally stored at much higher precision than shown in the output.) Thus we can create a new variable—call it fitted—equal to the prediction by typing predict fitted and then use scatter to display the fitted and actual values separately for domestic and foreign automobiles:

```
. predict fitted
(option xb assumed; fitted values)
. scatter mpg fitted weight, by(foreign, total) c(. l) m(o i) sort
```

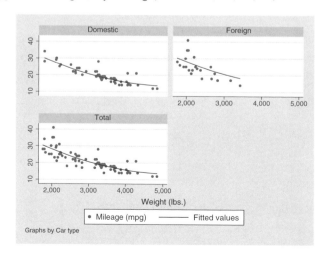

predict can calculate much more than just predicted values. For predict after linear regression, predict can calculate residuals, standardized residuals, studentized residuals, influence statistics, etc. In any case, we specify what is to be calculated via an option, so if we wanted the residuals stored in new variable r, we would type

```
. predict r, resid
```

The options that may be specified following predict vary according to the estimation command previously used; the predict options are documented along with the estimation command. For instance, to discover all the things predict can do following regress, see [R] **regress**.

◁

20.9.1   Using predict

The use of predict is not limited to linear regression. predict can be used after any estimation command.

▷ Example 7

You fit a logistic regression model of whether a car is manufactured outside the United States on the basis of its weight and mileage rating using either the logistic or the logit command; see [R] **logistic** and [R] **logit**. We will use logit.

```
. use http://www.stata-press.com/data/r10/auto

. logit foreign weight mpg

Iteration 0:   log likelihood =  -45.03321
Iteration 1:   log likelihood = -29.898968
Iteration 2:   log likelihood = -27.495771
Iteration 3:   log likelihood = -27.184006
Iteration 4:   log likelihood = -27.175166
Iteration 5:   log likelihood = -27.175156
```

```
Logistic regression                             Number of obs   =         74
                                                LR chi2(2)      =      35.72
                                                Prob > chi2     =     0.0000
Log likelihood = -27.175156                     Pseudo R2       =     0.3966
```

foreign	Coef.	Std. Err.	z	P>|z|	[95% Conf. Interval]	
weight	-.0039067	.0010116	-3.86	0.000	-.0058894	-.001924
mpg	-.1685869	.0919174	-1.83	0.067	-.3487418	.011568
_cons	13.70837	4.518707	3.03	0.002	4.851864	22.56487

After logit, predict without options calculates the probability of a positive outcome (we learned that by looking at [R] **logit**). To obtain the predicted probabilities that each car is manufactured outside the United States, we type

```
. predict probhat
(option pr assumed; Pr(foreign))

. summarize probhat
```

Variable	Obs	Mean	Std. Dev.	Min	Max
probhat	74	.2972973	.3052979	.000729	.8980594

```
. list make mpg weight foreign probhat in 1/5
```

	make	mpg	weight	foreign	probhat
1.	AMC Concord	22	2,930	Domestic	.1904363
2.	AMC Pacer	17	3,350	Domestic	.0957767
3.	AMC Spirit	22	2,640	Domestic	.4220815
4.	Buick Century	20	3,250	Domestic	.0862625
5.	Buick Electra	15	4,080	Domestic	.0084948

◁

20.9.2   Making in-sample predictions

predict does not retrieve a vector of prerecorded values—it calculates the predictions on the basis of the recorded coefficients and the data currently in memory. In the above examples, when we typed things like

```
. predict probhat
```

predict filled in the prediction everywhere that it could be calculated.

Sometimes we have more data in memory than were used by the estimation command, either because we explicitly ignored some of the observations by specifying an if *exp* with the estimation command or because there are missing values. In such cases, if we want to restrict the calculation to the estimation subsample, we would do that in the usual way by adding if e(sample) to the end of the command:

```
. predict probhat if e(sample)
```

20.9.3   Making out-of-sample predictions

Because predict makes its calculations on the basis of the recorded coefficients and the data in memory, predict can do more than calculate predicted values for the data on which the estimation took place—it can make out-of-sample predictions, as well.

If you fit your model on a subset of the observations, you could then predict the outcome for all the observations:

```
. logit foreign weight mpg if rep78 > 3
. predict pall
```

If you do not specify if e(sample) at the end of the predict command, predict calculates the predictions for all observations possible.

In fact, because predict works from the active estimation results, you can use predict with *any* dataset that contains the necessary variables.

▷ Example 8

Continuing with our previous logit example, assume that we have a second dataset containing the mpg and weight of a different sample of cars. We have just fitted your model and now continue:

```
. use otherdat, clear
(Different cars)
```

```
. predict probhat                    Stata remembers the previous model
(option pr assumed; Pr(foreign))
. summarize probhat foreign
```

Variable	Obs	Mean	Std. Dev.	Min	Max
probhat	12	.2505068	.3187104	.0084948	.8920776
foreign	12	.1666667	.3892495	0	1

◁

▷ Example 9

There are many ways to obtain out-of-sample predictions. Above, we estimated on one dataset and then used another. If our first dataset had contained both sets of cars, marked, say, by the variable difcars being 0 if from the first sample and 1 if from the second, we could type

```
. logit foreign weight mpg if difcars==0
same output as above appears

. predict probhat
(option pr assumed; Pr(foreign))

. summarize probhat foreign if difcars==1
same output as directly above appears
```

If we just had a few additional cars, we could even input them after estimation. Assume that our data once again contain only the first sample of cars, and assume that we are interested in an additional sample of only two rather than 12 cars; we could type

```
. use http://www.stata-press.com/data/r10/auto

. keep make mpg weight foreign

. logit foreign weight mpg
same output as above appears

. input
                    make      mpg    weight   foreign
75. "Merc. Zephyr" 20 2830 0           we type in our new data
76. "VW Dasher" 23 2160 1
77. end

. predict probhat                    obtain all the predictions
(option pr assumed; Pr(foreign))

. list in -2/1
```

	make	mpg	weight	foreign	probhat
75.	Merc. Zephyr	20	2830	Domestic	.3275397
76.	VW Dasher	23	2160	Foreign	.8009743

◁

20.9.4 Obtaining standard errors, tests, and confidence intervals for predictions

When you use predict, you create, for each observation in the prediction sample, a statistic that is a function of the data and the estimated model parameters. You also could have generated your own customized predictions by using generate. In either case, to get standard errors, Wald tests, and confidence intervals for your predictions, use predictnl. For example, if we wanted the standard errors for our predicted probabilities, we could type

```
. drop probhat
. predictnl probhat = predict(), se(phat_se)
. list in 1/5
```

	make	mpg	weight	foreign	probhat	phat_se
1.	AMC Concord	22	2,930	Domestic	.1904363	.0658386
2.	AMC Pacer	17	3,350	Domestic	.0957767	.0536296
3.	AMC Spirit	22	2,640	Domestic	.4220815	.0892845
4.	Buick Century	20	3,250	Domestic	.0862625	.0461927
5.	Buick Electra	15	4,080	Domestic	.0084948	.0093079

Comparing this output to our previous listing of the first five predicted probabilities, you will notice that the output is identical, except that we now have an additional variable, phat_se, which contains the estimated standard error for each predicted probability.

We first had to drop probhat since predictnl will regenerate it. Note also the use of predict() within predictnl—it specified that we wanted to generate a point estimate (and standard error) for the default prediction after logit; see [R] **predictnl** for more details.

20.10   Accessing estimated coefficients

You can access coefficients and standard errors after estimation by referring to _b[*name*] and _se[*name*]; see [U] **13.5 Accessing coefficients and standard errors**.

▷ Example 10

Let us return to linear regression. We are doing a study of earnings of men and women at a particular company. In addition to each person's earnings, we have information on their educational attainment and tenure with the company. We type the following:

```
. generate femten = female*tenure
. generate femed = female*ed
. regress lnearn ed tenure female femed femten
  (output omitted )
```

We now wish to predict everyone's income as if they were male and then compare these as-if earnings with the actual earnings:

```
. generate asif = _b[_cons] + _b[ed]*ed + _b[tenure]*tenure
```

◁

▷ Example 11

We are analyzing the mileage of automobiles and are using a slightly more sophisticated model than any we have used so far. As we have previously, we assume that mpg is a function of weight and weightsq, but we also add the interaction of foreign multiplied by weight (called fweight), the car's gear ratio (gear_ratio), and foreign multiplied by gear_ratio (fgear_ratio).

```
. use http://www.stata-press.com/data/r10/auto2, clear
(1978 Automobile Data)
. generate fweight = foreign*weight
```

```
. generate fgear_ratio = foreign*gear_ratio
. regress mpg weight weightsq fweight gear_ratio fgear_ratio
```

Source	SS	df	MS		
Model	1737.05293	5	347.410585		
Residual	706.406534	68	10.3883314		
Total	2443.45946	73	33.4720474		

```
Number of obs =      74
F(  5,    68) =   33.44
Prob > F      =  0.0000
R-squared     =  0.7109
Adj R-squared =  0.6896
Root MSE      =  3.2231
```

mpg	Coef.	Std. Err.	t	P>|t|	[95% Conf. Interval]
weight	-.0118517	.0045136	-2.63	0.011	-.0208584 -.002845
weightsq	9.81e-07	7.04e-07	1.39	0.168	-4.25e-07 2.39e-06
fweight	-.0032241	.0015577	-2.07	0.042	-.0063326 -.0001157
gear_ratio	1.159741	1.553418	0.75	0.458	-1.940057 4.259539
fgear_ratio	1.597462	1.205313	1.33	0.189	-.8077036 4.002627
_cons	44.61644	8.387943	5.32	0.000	27.87856 61.35432

If you are not experienced in both regression technology and automobile technology, you may find it difficult to interpret this regression. Putting aside issues of statistical significance, we find that mileage decreases with a car's weight but increases with the square of weight; decreases even more rapidly with weight for foreign cars; increases with higher gear ratio; and increases even more rapidly with higher gear ratio in foreign cars.

Thus, do foreign cars yield better or worse gas mileage? Results are mixed. As the foreign cars' weight increases, they do more poorly in relation to domestic cars, but they do better at higher gear ratios. One way to compare the results is to predict what mileage foreign cars would have *if they were manufactured domestically*. The regression provides all the information necessary for making that calculation; mileage for domestic cars is estimated to be

$$-.012\,\texttt{weight} + 9.81 \cdot 10^{-7}\,\texttt{weightsq} + 1.160\,\texttt{gear\_ratio} + 44.6$$

We can use that equation to predict the mileage of foreign cars and then compare it with the true outcome. The _b[] function simplifies reference to the estimated coefficients. We can type

```
. gen asif=_b[weight]*weight + _b[weightsq]*weightsq +
        _b[gear_ratio]*gear_ratio + _b[_cons]
```

_b[weight] refers to the estimated coefficient on weight, _b[weightsq] to the estimated coefficient on weightsq, and so on.

We might now ask how the actual mileage of a Honda compares with the asif prediction:

```
. list make asif mpg if strpos(make,"Honda")
```

	make	asif	mpg
61.	Honda Accord	26.52597	25
62.	Honda Civic	30.62202	28

Notice the way we constructed our if clause to select Hondas. strpos() is the string function that returns the location in the first string where the second string is found or, if the second string does not occur in the first, zero. Thus, any recorded make that contains the string "Honda" anywhere in it would be listed; see [D] **functions**.

We find that both Honda models yield slightly lower gas mileage than the asif domestic car–based prediction. (We do not endorse this model as a complete model of the determinants of mileage, nor do we single out the Honda for any special scorn. In fact, please note that the observed values are within the root mean squared error of the average prediction.)

We might wish to compare the overall average mpg and the asif prediction over all foreign cars in the data:

```
. summarize mpg asif if foreign

    Variable │      Obs       Mean    Std. Dev.       Min        Max
─────────────┼──────────────────────────────────────────────────────
         mpg │       22   24.77273   6.611187         14         41
        asif │       22   26.67124   3.142912   19.70466   30.62202
```

We find that, on average, foreign cars yield slightly lower mileage than our asif prediction. This might lead us to ask if any foreign cars do better than the asif prediction:

```
. list make asif mpg if foreign & mpg>asif, sep(0)

     ┌──────────────────────────────────┐
     │ make                asif     mpg  │
     ├──────────────────────────────────┤
 55. │ BMW 320i        24.31697      25  │
 57. │ Datsun 210      28.96818      35  │
 63. │ Mazda GLC       29.32015      30  │
 66. │ Subaru          28.85993      35  │
 68. │ Toyota Corolla  27.01144      31  │
 71. │ VW Diesel       28.90355      41  │
     └──────────────────────────────────┘
```

We find six such automobiles.

◁

20.11   Performing hypothesis tests on the coefficients

20.11.1   Linear tests

After estimation, test is used to perform tests of linear hypotheses on the basis of the variance–covariance matrix of the estimators (Wald tests).

▷ Example 12

(test has many syntaxes and features, so do not use this example as an excuse for not reading [R] test.) Using the automobile data, we perform the following regression:

```
. use http://www.stata-press.com/data/r10/auto, clear
(1978 Automobile Data)
. generate weightsq=weight^2
```

```
. regress mpg weight weightsq foreign
```

Source	SS	df	MS
Model	1689.15372	3	563.05124
Residual	754.30574	70	10.7757963
Total	2443.45946	73	33.4720474

```
Number of obs =      74
F(  3,    70) =   52.25
Prob > F      =  0.0000
R-squared     =  0.6913
Adj R-squared =  0.6781
Root MSE      =  3.2827
```

mpg	Coef.	Std. Err.	t	P>|t|	[95% Conf. Interval]	
weight	-.0165729	.0039692	-4.18	0.000	-.0244892	-.0086567
weightsq	1.59e-06	6.25e-07	2.55	0.013	3.45e-07	2.84e-06
foreign	-2.2035	1.059246	-2.08	0.041	-4.3161	-.0909002
_cons	56.53884	6.197383	9.12	0.000	44.17855	68.89913

We can use the `test` command to calculate the joint significance of `weight` and `weightsq`:

```
. test weight weightsq
 ( 1)  weight = 0
 ( 2)  weightsq = 0
       F(  2,    70) =   60.83
            Prob > F =    0.0000
```

We are not limited to testing whether the coefficients are zero. We can test whether the coefficient on `foreign` is -2 by typing

```
. test foreign = -2
 ( 1)  foreign = -2
       F(  1,    70) =    0.04
            Prob > F =    0.8482
```

We can even test more complicated hypotheses since `test` can perform basic algebra. Here is an absurd hypothesis:

```
. test 2*(weight+weightsq)=-3*(foreign-(weight-weightsq))
 ( 1)  - weight + 5.0 weightsq + 3.0 foreign = 0
       F(  1,    70) =    4.31
            Prob > F =    0.0416
```

`test` simplified the algebra of our hypothesis and then presented the test results. We discover that the hypothesis may be absurd but we cannot reject it at the 1% or even 4% level. We can also use `test`'s `accumulate` option to combine this test with another test:

```
. test foreign+weight=0, accum
 ( 1)  - weight + 5.0 weightsq + 3.0 foreign = 0
 ( 2)  weight + foreign = 0
       F(  2,    70) =    9.12
            Prob > F =    0.0003
```

There are limitations. `test` can test only linear hypotheses. If we attempt to test a nonlinear hypothesis, `test` will tell us that it is not possible:

```
. test weight/foreign=0
not possible with test
r(131);
```

Testing nonlinear hypotheses is discussed in [U] **20.11.4 Nonlinear Wald tests** below.

20.11.2 Using test

test bases its results on the estimated variance–covariance matrix of the estimators (i.e., performs a Wald test), so it can be used after any estimation command. For maximum likelihood estimation, you will have to decide whether you want to perform tests on the basis of the information matrix instead of constraining the equation, reestimating it, and then calculating the likelihood-ratio test (see [U] **20.11.3 Likelihood-ratio tests**). Since test bases its results on the information matrix, its results have the same standing as the asymptotic Z statistic presented in the coefficient table.

▷ Example 13

Let's examine the repair records of the cars in our automobile data as rated by *Consumer Reports*:

```
. tabulate rep78 foreign
```

Repair Record 1978	Car type Domestic	Foreign	Total
1	2	0	2
2	8	0	8
3	27	3	30
4	9	9	18
5	2	9	11
Total	48	21	69

The values are coded 1–5, corresponding to well below average to well above average. We will fit this variable by using a maximum-likelihood ordered logit model (the nolog option suppresses the iteration log, saving us some paper):

```
. ologit rep78 price foreign weight weightsq displ, nolog
```

Ordered logistic regression

				Number of obs	=	69
				LR chi2(5)	=	33.12
				Prob > chi2	=	0.0000
Log likelihood = -77.133082				Pseudo R2	=	0.1767

rep78	Coef.	Std. Err.	z	P>|z|	[95% Conf. Interval]	
price	-.000034	.0001188	-0.29	0.775	-.0002669	.000199
foreign	2.685648	.9320398	2.88	0.004	.8588833	4.512412
weight	-.0037447	.0025609	-1.46	0.144	-.0087639	.0012745
weightsq	7.87e-07	4.50e-07	1.75	0.080	-9.43e-08	1.67e-06
displacement	-.0108919	.0076805	-1.42	0.156	-.0259455	.0041617
/cut1	-9.417196	4.298201			-17.84151	-.9928766
/cut2	-7.581864	4.23409			-15.88053	.7168002
/cut3	-4.82209	4.147679			-12.95139	3.307212
/cut4	-2.79344	4.156219			-10.93948	5.352599

We now wonder whether all our variables other than foreign are jointly significant. We test the hypothesis just as we would after linear regression:

```
. test weight weightsq displ price

 ( 1)  [rep78]weight = 0
 ( 2)  [rep78]weightsq = 0
 ( 3)  [rep78]displacement = 0
 ( 4)  [rep78]price = 0

         chi2(  4) =     3.63
       Prob > chi2 =    0.4590
```

To compare this with the results performed by a likelihood-ratio test, see [U] **20.11.3 Likelihood-ratio tests**. Here results differ little.

◁

20.11.3   Likelihood-ratio tests

After maximum likelihood estimation, you can obtain likelihood-ratio tests by fitting both the unconstrained and constrained models, storing the results using estimates store, and then running lrtest. See [R] **lrtest** for the full details.

▷ Example 14

In [U] **20.11.2 Using test** above, we fitted an ordered logit on rep78 and then tested the significance of all the explanatory variables except foreign.

To obtain the likelihood-ratio test, sometime after fitting the full model, we type estimates store *full_model_name*, where *full_model_name* is just a label that we assign to these results.

```
. ologit rep78 price foreign weight weightsq displ
 (output omitted )
. estimates store myfullmodel
```

This command saves the current model results with the name myfullmodel.

Next we fit the constrained model. After that, typing 'lrtest myfullmodel .' compares the current model with the model we saved:

```
. ologit rep78 foreign

Iteration 0:   log likelihood = -93.692061
Iteration 1:   log likelihood = -79.696089
Iteration 2:   log likelihood = -79.044933
Iteration 3:   log likelihood = -79.029267
Iteration 4:   log likelihood = -79.029243

Ordered logistic regression                   Number of obs   =          69
                                              LR chi2(1)      =       29.33
                                              Prob > chi2     =      0.0000
Log likelihood = -79.029243                   Pseudo R2       =      0.1565
```

rep78	Coef.	Std. Err.	z	P>|z|	[95% Conf. Interval]	
foreign	2.98155	.6203637	4.81	0.000	1.76566	4.197441
/cut1	-3.158382	.7224269			-4.574313	-1.742452
/cut2	-1.362642	.3557343			-2.059868	-.6654154
/cut3	1.232161	.3431227			.5596533	1.90467
/cut4	3.246209	.5556646			2.157127	4.335292

```
. lrtest myfullmodel .
Likelihood-ratio test                                     LR chi2(4)   =       3.79
(Assumption: . nested in myfullmodel)                     Prob > chi2 =      0.4348
```

When we tested the same constraint with `test` (which performed a Wald test), we obtained a χ^2 of 3.63 and a significance level of 0.4590. We used . (the dot) to specify the results in active memory, although we also could have stored them with `estimates store` and referred to them by name instead. Also, the order in which you specify the two models to `lrtest` doesn't matter; `lrtest` is smart enough to know the full model from the constrained model.

◁

Two other postestimation commands work in the same way as `lrtest`, meaning that they accept names of stored estimation results as their input: `hausman` for performing Hausman specification tests and `suest` for seemingly unrelated estimation. We do not cover these commands here; see [R] **hausman** and [R] **suest** for more details.

20.11.4 Nonlinear Wald tests

`testnl` can be used to test nonlinear hypotheses about the parameters of the active estimation results. `testnl`, like `test`, bases its results on the variance–covariance matrix of the estimators (i.e., performs a Wald test), so it can be used after any estimation command; see [R] **testnl**.

▷ Example 15

We fit the model

```
. regress price mpg weight foreign
(output omitted )
```

and then type

```
. testnl (38*_b[mpg]^2 = _b[foreign]) (_b[mpg]/_b[weight]=4)
 (1)   38*_b[mpg]^2 = _b[foreign]
 (2)   _b[mpg]/_b[weight]=4
           F(2, 70) =        0.02
           Prob > F =      0.9806
```

We performed this test on linear regression estimates, but tests of this type could be performed after any estimation command.

◁

20.12   Obtaining linear combinations of coefficients

lincom computes point estimates, standard errors, t or z statistics, p-values, and confidence intervals for a linear combination of coefficients after any estimation command. Results can optionally be displayed as odds ratios, incidence-rate ratios, or relative-risk ratios.

▷ Example 16

We fit a linear regression:

```
. use http://www.stata-press.com/data/r10/regress, clear
. regress y x1 x2 x3
```

Source	SS	df	MS		Number of obs =	148
					F(3, 144) =	96.12
Model	3259.3561	3	1086.45203		Prob > F =	0.0000
Residual	1627.56282	144	11.3025196		R-squared =	0.6670
					Adj R-squared =	0.6600
Total	4886.91892	147	33.2443464		Root MSE =	3.3619

y	Coef.	Std. Err.	t	P>|t|	[95% Conf. Interval]	
x1	1.457113	1.07461	1.36	0.177	-.6669339	3.581161
x2	2.221682	.8610358	2.58	0.011	.5197797	3.923583
x3	-.006139	.0005543	-11.08	0.000	-.0072345	-.0050435
_cons	36.10135	4.382693	8.24	0.000	27.43863	44.76407

Suppose that we want to see the difference of the coefficients of x2 and x1. We type

```
. lincom x2 - x1
 ( 1) - x1 + x2 = 0
```

y	Coef.	Std. Err.	t	P>|t|	[95% Conf. Interval]	
(1)	.7645682	.9950282	0.77	0.444	-1.20218	2.731316

◁

lincom is handy for computing the odds ratio of one covariate group relative to another.

▷ Example 17

We estimate the parameters of a logistic model of low birthweight:

```
. use http://www.stata-press.com/data/r10/lbw3
(Hosmer & Lemeshow data)
```

(*Continued on next page*)

```
. logit low age lwd black other smoke ptd ht ui

Iteration 0:   log likelihood =    -117.336
Iteration 1:   log likelihood = -99.431174
Iteration 2:   log likelihood = -98.785718
Iteration 3:   log likelihood =    -98.778
Iteration 4:   log likelihood = -98.777998
```

```
Logistic regression                          Number of obs   =         189
                                             LR chi2(8)      =       37.12
                                             Prob > chi2     =      0.0000
Log likelihood = -98.777998                  Pseudo R2       =      0.1582
```

low	Coef.	Std. Err.	z	P>|z|	[95% Conf. Interval]	
age	-.0464796	.0373888	-1.24	0.214	-.1197603	.0268011
lwd	.8420615	.4055338	2.08	0.038	.0472299	1.636893
black	1.073456	.5150752	2.08	0.037	.0639273	2.082985
other	.815367	.4452979	1.83	0.067	-.0574008	1.688135
smoke	.8071996	.404446	2.00	0.046	.0145001	1.599899
ptd	1.281678	.4621157	2.77	0.006	.3759478	2.187408
ht	1.435227	.6482699	2.21	0.027	.1646415	2.705813
ui	.6576256	.4666192	1.41	0.159	-.2569313	1.572182
_cons	-1.216781	.9556797	-1.27	0.203	-3.089878	.656317

If we want to obtain the odds ratio for black smokers relative to white nonsmokers (the reference group), we type

```
. lincom black + smoke, or

 ( 1)  black + smoke = 0
```

low	Odds Ratio	Std. Err.	z	P>|z|	[95% Conf. Interval]	
(1)	6.557805	4.744692	2.60	0.009	1.588176	27.07811

`lincom` computed $\exp(\beta_{\text{black}} + \beta_{\text{smoke}}) = 6.56$.

◁

20.13   Obtaining nonlinear combinations of coefficients

`lincom` is limited to estimating linear combinations of coefficients, e.g., `black + smoke`, or exponentiated linear combinations, as in the above. For general nonlinear combinations, use `nlcom`.

▷ Example 18

Continuing our previous example: suppose that we wanted the ratio of the coefficients (and standard errors, Wald test, confidence interval, etc.) of `black` and `other`:

```
. nlcom _b[black]/_b[other]

      _nl_1:  _b[black]/_b[other]
```

low	Coef.	Std. Err.	z	P>|z|	[95% Conf. Interval]	
_nl_1	1.316531	.7359262	1.79	0.074	-.1258574	2.75892

The Wald test given is that of the null hypothesis that the nonlinear combination is zero versus the two-sided alternative—this is probably not informative for a ratio. If we would instead like to test whether this ratio is one, we can rerun nlcom, this time subtracting one from our ratio estimate.

```
. nlcom _b[black]/_b[other] - 1
      _nl_1:  _b[black]/_b[other] - 1
```

low	Coef.	Std. Err.	z	P>|z|	[95% Conf. Interval]
_nl_1	.3165314	.7359262	0.43	0.667	-1.125857 1.75892

We can interpret this as not much evidence that the "ratio minus 1" is different from zero, meaning that we cannot reject the null hypothesis that the ratio equals one.

When using nlcom, we needed to refer to the model coefficients by their "proper" names, e.g., _b[black], and not by the shorthand black, such as when using lincom. If we had typed

```
. nlcom black/other
```

Stata would have reported an error. Consider this a limitation of Stata.

◁

20.14   Obtaining marginal effects

Stata's mfx command computes the marginal effects of the independent variables on predicted values.

▷ Example 19

Consider the logistic regression model that we previously fitted on the automobile data:

```
. use http://www.stata-press.com/data/r10/auto
. logit foreign weight mpg
Iteration 0:    log likelihood =  -45.03321
Iteration 1:    log likelihood = -29.898968
Iteration 2:    log likelihood = -27.495771
Iteration 3:    log likelihood = -27.184006
Iteration 4:    log likelihood = -27.175166
Iteration 5:    log likelihood = -27.175156
```

```
Logistic regression                          Number of obs   =         74
                                             LR chi2(2)      =      35.72
                                             Prob > chi2     =     0.0000
Log likelihood = -27.175156                  Pseudo R2       =     0.3966
```

foreign	Coef.	Std. Err.	z	P>|z|	[95% Conf. Interval]
weight	-.0039067	.0010116	-3.86	0.000	-.0058894 -.001924
mpg	-.1685869	.0919174	-1.83	0.067	-.3487418 .011568
_cons	13.70837	4.518707	3.03	0.002	4.851864 22.56487

Typing mfx compute gives the marginal effects for the default prediction, which, for logit, is the predicted probability that the automobile is manufactured outside the United States.

```
. mfx compute
```

Marginal effects after logit
     y = Pr(foreign) (predict)
       = .15733364

variable	dy/dx	Std. Err.	z	P>|z|	[95% C.I.]	X
weight	-.0005179	.00014	-3.73	0.000	-.00079 -.000246	3019.46
mpg	-.0223512	.0127	-1.76	0.079	-.04725 .002548	21.2973

Given the above output, we see that both `weight` and `mpg` have a negative effect on the predicted probability. For example, increased weight (from the mean weight of 3,019.46 lbs.) decreases the likelihood that the automobile is `foreign` when controlling for gas mileage.

◁

`mfx` can also calculate elasticities, calculate at covariate values other than the covariate means (the default), and calculate marginal effects for predictions other than the default prediction; see [R] **mfx** for details.

20.15   Obtaining robust variance estimates

Estimates of variance refer to estimated standard errors or, more completely, the estimated variance–covariance matrix of the estimators of which the standard errors are a subset, being the square root of the diagonal elements. Call this matrix the variance. All estimation commands produce an estimate of variance and, using that, produce confidence intervals and significance tests.

In addition to the conventional estimator of variance, there is another estimator that has been called by various names because it has been derived independently in different ways by different authors. Two popular names associated with the calculation are Huber and White, but it is also known as the sandwich estimator of variance (because of how the calculation formula physically appears) and the robust estimator of variance (because of claims made about it). Also, this estimator also has an independent and long tradition in the survey literature.

The conventional estimator of variance is derived by starting with a model. Let's start with the regression model

$$y_i = \mathbf{x}_i\boldsymbol{\beta} + \epsilon_i, \qquad \epsilon_i \sim N(0, \sigma^2)$$

although it is not important for the discussion that we are using regression. Under the model-based approach, we assume that the model is true and thereby derive an estimator for β and its variance.

The estimator of the standard error of $\widehat{\beta}$ we develop is based on the assumption that the model is true in every detail. y_i is not exactly equal to $\mathbf{x}_i\boldsymbol{\beta}$ (so that we would only need to solve an equation to obtain precisely that value of β) because the observed y_i has noise ϵ_i added to it, the noise is Gaussian, and it has constant variance. That noise leads to the uncertainty about β, and it is from the characteristics of that noise that we are able to calculate a sampling distribution for $\widehat{\beta}$.

The key thought here is that the standard error of $\widehat{\beta}$ arises because of ϵ and is valid only because the model is absolutely, without question, true; we just do not happen to know the particular values of β and σ^2 that make the model true. The implication is that, in an infinite-sized sample, the estimator $\widehat{\beta}$ for β would converge to the true value of β and that its variance would go to 0.

Now, here is another interpretation of the estimation problem: We are going to fit the model

$$y_i = \mathbf{x}_i \mathbf{b} + e_i$$

and, to obtain estimates of $\mathbf{b}$, we are going to use the calculation formula

$$\widehat{\mathbf{b}} = (\mathbf{X}'\mathbf{X})^{-1}\mathbf{X}'\mathbf{y}$$

We have made no claims that the model is true or any claims about e_i or its distribution. We shifted our notation from β and ϵ_i to $\mathbf{b}$ and e_i to emphasize this. All we have stated are the physical actions we intend to carry out on the data. Interestingly, it is possible to calculate a standard error for $\widehat{\mathbf{b}}$ here. At least, it is possible if you will agree with us on what the standard error measures are.

We are going to define the standard error as measuring the standard error of the calculated $\widehat{\mathbf{b}}$ if we were to repeat the data collection followed by estimation over and over again.

This is a different concept of the standard error from the conventional, model-based ideas, but it is related. Both measure uncertainty about $\mathbf{b}$ (or β). The regression model–based derivation states from where the variation arises and so can make grander statements about the applicability of the measured standard error. The weaker second interpretation makes fewer assumptions and so produces a standard error suitable for one purpose.

There is a subtle difference in interpretation of these identically calculated point estimates. $\widehat{\beta}$ is the estimate of β under the assumption that the model is true. $\widehat{\mathbf{b}}$ is the estimate of $\mathbf{b}$, which is merely what the estimator would converge to if we collected more and more data.

Is the estimate of $\mathbf{b}$ unbiased? If we mean, "Does $\mathbf{b} = \beta$?" that depends on whether the model is true. $\widehat{\mathbf{b}}$ is, however, an unbiased estimate of $\mathbf{b}$, which admittedly is not saying much.

What if $\mathbf{x}$ and e are correlated? Don't we have a problem then? We may have an interpretation problem—$\mathbf{b}$ may not measure what we want to measure, namely, β—but we measure $\widehat{\mathbf{b}}$ to be such and such and expect, if the experiment and estimation were repeated, that we would observe results in the range we have reported.

So, we have two different understandings of what the parameters mean and how the variance in their estimators arises. However, both interpretations must confront the issue of how to make valid statistical inference about the coefficient estimates when the data do not come from either a simple random sample or the distribution of $(\mathbf{x}_i, \epsilon_i)$ is not independent and identically distributed (i.i.d.). In essence, we need an estimator of the standard errors that is robust to this deviation from the standard case.

Hence, the name *the robust estimate of variance*; its associated authors are Huber (1967) and White (1980, 1982) (who developed it independently), although many others have extended its development, including Gail, Tan, and Piantadosi (1988); Kent (1982); Royall (1986); and Lin and Wei (1989). In the survey literature, this same estimator has been developed; see Kish and Frankel (1974), Fuller (1975), and Binder (1983).

Many of Stata's estimation commands can produce this alternative estimate of variance, and, if they can, they have a vce(robust) option. Without vce(robust), we get one measure of variance:

```
. use http://www.stata-press.com/data/r10/auto7
(1978 Automobile Data)

. regress mpg weight foreign
```

Source	SS	df	MS
Model	1619.2877	2	809.643849
Residual	824.171761	71	11.608053
Total	2443.45946	73	33.4720474

```
                                    Number of obs =      74
                                    F(  2,    71) =   69.75
                                    Prob > F      =  0.0000
                                    R-squared     =  0.6627
                                    Adj R-squared =  0.6532
                                    Root MSE      =  3.4071
```

mpg	Coef.	Std. Err.	t	P>|t|	[95% Conf. Interval]	
weight	-.0065879	.0006371	-10.34	0.000	-.0078583	-.0053175
foreign	-1.650029	1.075994	-1.53	0.130	-3.7955	.4954422
_cons	41.6797	2.165547	19.25	0.000	37.36172	45.99768

With vce(robust), we get another:

```
. regress mpg weight foreign, vce(robust)
Linear regression
```

```
                                    Number of obs =      74
                                    F(  2,    71) =   73.81
                                    Prob > F      =  0.0000
                                    R-squared     =  0.6627
                                    Root MSE      =  3.4071
```

mpg	Coef.	Robust Std. Err.	t	P>|t|	[95% Conf. Interval]	
weight	-.0065879	.0005462	-12.06	0.000	-.007677	-.0054988
foreign	-1.650029	1.132566	-1.46	0.150	-3.908301	.6082424
_cons	41.6797	1.797553	23.19	0.000	38.09548	45.26392

Either way, the point estimates are the same. (See [R] **regress** for an example where specifying vce(robust) produces strikingly different standard errors.)

How do we interpret these results? Let's consider the model-based interpretation. Suppose that

$$y_i = \mathbf{x}_i \boldsymbol{\beta} + \epsilon_i,$$

where $(\mathbf{x}_i, \epsilon_i)$ are independently and identically distributed (i.i.d.) with variance σ^2. For the model-based interpretation, we also must assume that $\mathbf{x}_i$ and ϵ_i are uncorrelated. With these assumptions and a few technical regularity conditions, our first regression gives us consistent parameter estimates and standard errors that we can use for valid statistical inference about the coefficients. Now suppose that we weaken our assumptions so that $(\mathbf{x}_i, \epsilon_i)$ are independently and—but not necessarily—identically distributed. Our parameter estimates are still consistent, but the standard errors from the first regression can no longer be used to make valid inference. We need estimates of the standard errors that are robust to the fact that the error term is not identically distributed. The standard errors in our second regression are just what we need. We can use them to make valid statistical inference about our coefficients, even though our data are not identically distributed.

Now consider a non–model-based interpretation. If our data come from a survey design that ensures that $(\mathbf{x}_i, e_i)$ are i.i.d., then we can use the nonrobust standard errors for valid statistical inference about the population parameters $\mathbf{b}$. For this interpretation, we do not need to assume that $\mathbf{x}_i$ and e_i are uncorrelated. If they are uncorrelated, the population parameters $\mathbf{b}$ and the model parameters $\boldsymbol{\beta}$ are the same. However, if they are correlated, then the population parameters $\mathbf{b}$ that we are estimating are not the same as the model-based β. So, what we are estimating is different, but we still need standard errors that allow us to make valid statistical inference. So, if the process that we used to collect the data caused $(\mathbf{x}_i, e_i)$ to be independently but not identically distributed, then we need to use the robust standard errors to make valid statistical inference about the population parameters $\mathbf{b}$.

The robust estimator of variance has one feature that the conventional estimator does not have: the ability to relax the assumption of independence of the observations. That is, if you specify the vce(cluster *clustvar*) option, it can produce "correct" standard errors (in the measurement sense), even if the observations are correlated.

For the automobile data, it is difficult to believe that the models of the various manufacturers are truly independent. Manufacturers, after all, use common technology, engines, and drive trains across their model lines. The VW Dasher in the above regression has a measured residual of -2.80. Having been told that, do you really believe that the residual for the VW Rabbit is as likely to be above 0 as below? (The residual is -2.32.) Similarly, the measured residual for the Chevrolet Malibu is 1.27. Does that provide information about the expected value of the residual of the Chevrolet Monte Carlo (which turns out to be 1.53)?

We need to be careful about picking examples from data; we have not told you about the Datsun 210 and 510 (residuals $+8.28$ and -1.01) or the Cadillac Eldorado and Seville (residuals -1.99 and $+7.58$), but you should, at least, question the assumption of independence. It may be believable that the measured mpg given the weight of one manufacturer's vehicles is independent of other manufacturers' vehicles, but it is at least questionable whether a manufacturer's vehicles are independent of one another.

In commands with the vce(robust) option, another option—vce(cluster *clustvar*)—relaxes the independence assumption and requires only that the observations be independent across the clusters:

```
. regress mpg weight foreign, vce(cluster manufacturer)

Linear regression                              Number of obs =       74
                                               F(  2,    22) =    90.93
                                               Prob > F      =   0.0000
                                               R-squared     =   0.6627
                                               Root MSE      =   3.4071

                 (Std. Err. adjusted for 23 clusters in manufacturer)
```

mpg	Coef.	Robust Std. Err.	t	P>|t|	[95% Conf. Interval]	
weight	-.0065879	.0005339	-12.34	0.000	-.0076952	-.0054806
foreign	-1.650029	1.039033	-1.59	0.127	-3.804852	.5047939
_cons	41.6797	1.844559	22.60	0.000	37.85432	45.50508

It turns out that, in these data, whether or not we specify vce(cluster *clustvar*) makes little difference. The VW and Chevrolet examples above were not representative; had they been, the confidence intervals would have widened. (In the above, manuf is a variable that takes on values such as "Chev." or "VW", recording the manufacturer of the vehicle. This variable was created from variable make, which contains values such as "Chev. Malibu" or "VW Rabbit", by extracting the first word.)

As a demonstration of how well clustering can work, in [R] **regress** we fitted a random-effects model with regress, vce(robust) and then compared the results with ordinary least squares and the GLS random-effects estimator. Here we will simply summarize the results.

We start with a dataset on 4,711 women aged 14–46 years. Subjects appear an average of 6.056 times in the data; there are a total of 28,534 observations. The model we use is log wage on age,

age-squared, and job tenure. The focus of the example is the estimated coefficient on tenure. We obtain the following results:

Estimator	point estimate	confidence interval
(inappropriate) least squares	.039	[.038, .041]
robust clusterered	.039	[.036, .042]
GLS random effects	.026	[.025, .027]

Notice how well the robust clustered estimate does compared with the GLS random-effects model. We then run a Hausman specification test, obtaining $\chi^2(3) = 336.62$, which casts grave doubt on the assumptions justifying the use of the GLS estimator and hence on the GLS results. At this point, we will simply quote our comments:

> Meanwhile, our robust regression results still stand, as long as we are careful about the interpretation. The correct interpretation is that, if the data collection were repeated (on women sampled the same way as in the original sample), and if we were to refit the model, 95% of the time we would expect the estimated coefficient on tenure to be in the range $[.036, .042]$.

> Even with robust regression, we must be careful about going beyond that statement. Here the Hausman test is probably picking up something that differs within and between person, which would cast doubt on our robust regression model in terms of interpreting $[.036, .042]$ to contain the rate of return for keeping a job, economywide, for all women, without exception.

The formula for the robust estimator of variance is

$$\widehat{\mathcal{V}} = \widehat{\mathbf{V}} \left(\sum_{j=1}^{N} \mathbf{u}_j' \mathbf{u}_j \right) \widehat{\mathbf{V}}$$

where $\widehat{\mathbf{V}} = (-\partial^2 \ln L / \partial \beta^2)^{-1}$ (the conventional estimator of variance) and $\mathbf{u}_j$ (a row vector) is the contribution from the jth observation to $\partial \ln L / \partial \beta$.

In the example above, observations are assumed to be independent. Assume for a moment that the observations denoted by j are not independent but that they can be divided into M groups G_1, $G_2, \ldots, G_M$ that are independent. The robust estimator of variance is

$$\widehat{\mathcal{V}} = \widehat{\mathbf{V}} \left(\sum_{k=1}^{M} \mathbf{u}_k^{(G)'} \mathbf{u}_k^{(G)} \right) \widehat{\mathbf{V}}$$

where $\mathbf{u}_k^{(G)}$ is the contribution of the kth group to $\partial \ln L / \partial \beta$. That is, application of the robust variance formula merely involves using a different decomposition of $\partial \ln L / \partial \beta$, namely, $\mathbf{u}_k^{(G)}$, $k = 1, \ldots, M$ rather than $\mathbf{u}_j$, $j = 1, \ldots, N$. Moreover, if the log-likelihood function is additive in the observations denoted by j

$$\ln L = \sum_{j=1}^{N} \ln L_j$$

then $\mathbf{u}_j = \partial \ln L_j / \partial \beta$, so

$$\mathbf{u}_k^{(G)} = \sum_{j \in G_k} \mathbf{u}_j$$

That is what the vce(cluster *clustvar*) option does. (This point was first made in writing by Rogers [1993], although he considered the point an obvious generalization of Huber [1967] and the calculation—implemented by Rogers—had appeared in Stata a year earlier.)

❑ Technical Note

What is written above is asymptotically correct but ignores a finite-sample adjustment to $\widehat{\mathcal{V}}$. For maximum likelihood estimators, when you specify vce(robust) but not vce(cluster *clustvar*), a better estimate of variance is $\widehat{\mathcal{V}}^* = \{N/(N-1)\}\widehat{\mathcal{V}}$. When you also specify the vce(cluster *clustvar*) option, this becomes $\widehat{\mathcal{V}}^* = \{M/(M-1)\}\widehat{\mathcal{V}}$.

For linear regression, the finite-sample adjustment is $N/(N-k)$ without vce(cluster *clustvar*)—where k is the number of regressors—and $\{M/(M-1)\}\{(N-1)/(N-k)\}$ with vce(cluster *clustvar*). Also, two data-dependent modifications to the calculation for $\widehat{\mathcal{V}}^*$, suggested by MacKinnon and White (1985), are also provided by regress; see [R] **regress**.

❑

20.16   Obtaining scores

Many of the estimation commands that provide the vce(robust) option also provide the ability to generate equation-level score variables via the predict command. With the score option, predict returns an important ingredient into the robust variance calculation that is sometimes useful in its own right. As explained in [U] **20.15 Obtaining robust variance estimates** above, ignoring the finite-sample corrections, the robust estimate of variance is

$$\widehat{\mathcal{V}} = \widehat{\mathbf{V}}\left(\sum_{j=1}^{N}\mathbf{u}_j'\mathbf{u}_j\right)\widehat{\mathbf{V}}$$

where $\widehat{\mathbf{V}} = (-\partial^2\ln L/\partial\boldsymbol{\beta}^2)^{-1}$ is the conventional estimator of variance. Let us consider likelihood functions that are additive in the observations

$$\ln L = \sum_{j=1}^{N}\ln L_j$$

then $\mathbf{u}_j = \partial\ln L_j/\partial\boldsymbol{\beta}$. In general, function L_j is a function of $\mathbf{x}_j$ and $\boldsymbol{\beta}$, $L_j(\boldsymbol{\beta};\mathbf{x}_j)$. For many likelihood functions, however, it is only the linear form $\mathbf{x}_j\boldsymbol{\beta}$ that enters the function. In those cases,

$$\frac{\partial\ln L_j(\mathbf{x}_j\boldsymbol{\beta})}{\partial\boldsymbol{\beta}} = \frac{\partial\ln L_j(\mathbf{x}_j\boldsymbol{\beta})}{\partial(\mathbf{x}_j\boldsymbol{\beta})}\frac{\partial(\mathbf{x}_j\boldsymbol{\beta})}{\partial\boldsymbol{\beta}} = \frac{\partial\ln L_j(\mathbf{x}_j\boldsymbol{\beta})}{\partial(\mathbf{x}_j\boldsymbol{\beta})}\mathbf{x}_j$$

By writing $u_j = \partial\ln L_j(\mathbf{x}_j\boldsymbol{\beta})/\partial(\mathbf{x}_j\boldsymbol{\beta})$, this becomes simply $u_j\mathbf{x}_j$. Thus, the formula for the robust estimate of variance can be rewritten as

$$\widehat{\mathcal{V}} = \widehat{\mathbf{V}}\left(\sum_{j=1}^{N}u_j^2\mathbf{x}_j'\mathbf{x}_j\right)\widehat{\mathbf{V}}$$

We refer to u_j as the equation-level score (in the singular), and it is u_j that is returned when you use predict with the score option. u_j is like a residual in that

1. $\sum_j u_j = 0$ and

2. correlation of u_j and $\mathbf{x}_j$, calculated over $j = 1, \dots, N$, is 0.

In fact, for linear regression, u_j is the residual, normalized,

$$\frac{\partial \ln L_j}{\partial (\mathbf{x}_j \beta)} = \frac{\partial}{\partial (\mathbf{x}_j \beta)} \ln f \left\{ (y_j - \mathbf{x}_j \beta)/\sigma \right\}$$
$$= (y_j - \mathbf{x}_j \beta)/\sigma$$

where $f()$ is the standard normal density.

▷ Example 20

probit provides both the vce(robust) option and predict, score. Equation-level scores play an important role in calculating the robust estimate of variance, but we can use predict, score regardless of whether we specify vce(robust):

```
. use http://www.stata-press.com/data/r10/auto

. probit foreign mpg weight
Iteration 0:   log likelihood =  -45.03321
Iteration 1:   log likelihood = -29.244141
Iteration 2:   log likelihood = -27.041557
Iteration 3:   log likelihood =  -26.84658
Iteration 4:   log likelihood = -26.844189
Iteration 5:   log likelihood = -26.844189
```

Probit regression					Number of obs	=	74
					LR chi2(2)	=	36.38
					Prob > chi2	=	0.0000
Log likelihood = -26.844189					Pseudo R2	=	0.4039

foreign	Coef.	Std. Err.	z	P>|z|	[95% Conf. Interval]	
mpg	-.1039503	.0515689	-2.02	0.044	-.2050235	-.0028772
weight	-.0023355	.0005661	-4.13	0.000	-.003445	-.0012261
_cons	8.275464	2.554142	3.24	0.001	3.269438	13.28149

```
. predict double u, scores

. summarize u
```

Variable	Obs	Mean	Std. Dev.	Min	Max
u	74	-3.87e-16	.5988325	-1.655439	1.660787

```
. correlate u mpg weight
(obs=74)
```

	u	mpg	weight
u	1.0000		
mpg	-0.0000	1.0000	
weight	-0.0000	-0.8072	1.0000

```
. list make foreign mpg weight u if abs(u)>1.65
```

	make	foreign	mpg	weight	u
24.	Ford Fiesta	Domestic	28	1,800	-1.6554395
64.	Peugeot 604	Foreign	14	3,420	1.6607871

The light, high-mileage Ford Fiesta is surprisingly domestic, whereas the heavy, low-mileage Peugeot 604 is surprisingly foreign.

◁

❏ Technical Note

For some estimation commands, one score is not enough. Consider a likelihood that can be written as $L_j(\mathbf{x}_j\boldsymbol{\beta}_1, \mathbf{z}_j\boldsymbol{\beta}_2)$, a function of two linear forms (or linear equations). Then $\partial\ln L_j/\partial\boldsymbol{\beta}$ can be written $(\partial\ln L_j/\partial\boldsymbol{\beta}_1, \partial\ln L_j/\partial\boldsymbol{\beta}_2)$. Each of the components can in turn be written as $[\partial\ln L_j/\partial(\boldsymbol{\beta}_1\mathbf{x})]\mathbf{x} = u_1\mathbf{x}$ and $[\partial\ln L_j/\partial(\boldsymbol{\beta}_2\mathbf{z})]\mathbf{z} = u_2\mathbf{z}$. There are then two equation-level scores, u_1 and u_2, and, in general, there could be more.

Stata's streg, distribution(weibull) command is an example of this: it estimates $\boldsymbol{\beta}$ and a shape parameter, $\ln p$, the latter of which can be thought of as a degenerate linear form $(\ln p)\mathbf{z}$ with $\mathbf{z} = 1$. predict, scores after this command requires that you specify two new variable names, or you can specify *stub**, which will generate two variables, *stub*1 and *stub*2; the first will be defined containing u_1—the score associated with $\boldsymbol{\beta}$—and the second will be defined containing u_2—the score associated with $\ln p$.

❏

❏ Technical Note

Using Stata's matrix commands—see [P] **matrix**—we can make the robust variance calculation for ourselves and then compare it with that made by Stata.

```
. quietly probit foreign mpg weight, score(u)

. matrix accum S =  mpg weight [iweight=u^2*74/73]
(obs=26.53642547)

. matrix rV = e(V)*S*e(V)

. matrix list rV

symmetric rV[3,3]
                  mpg       weight        _cons
    mpg      .00352299
 weight      .00002216    2.434e-07
  _cons    -.14090346    -.00117031    6.4474172

. quietly probit foreign mpg weight, vce(robust)

. matrix list e(V)

symmetric e(V)[3,3]
                  mpg       weight        _cons
    mpg      .00352299
 weight      .00002216    2.434e-07
  _cons    -.14090346    -.00117031    6.4474172
```

The results are the same.

There is an important lesson here for programmers. Given the scores, conventional variance estimates can be easily transformed to robust estimates. If we were writing a new estimation command, it would not be difficult to include a vce(robust) option.

It is, in fact, easy if we ignore clustering. With clustering, it is more work since the calculation involves forming sums within clusters. For programmers interested in implementing robust variance calculations, Stata provides an _robust command to ease the task. This is documented in [P] **_robust**.

To use _robust, you first produce conventional results (a vector of coefficients and covariance matrix) along with a variable containing the scores u_j (or variables if the likelihood function has more than one stub). You then call _robust, and it will transform your conventional variance estimate into the robust estimate. _robust will handle the work associated with clustering and the details of the finite-sample adjustment, and it will even label your output so that the word *Robust* appears above the standard error when the results are displayed.

Of course, this is all even easier if you write your commands with Stata's ml maximum likelihood optimization, in which case you merely pass the vce(robust) option on to ml. ml will then call _robust itself and do all the work for you.

❏

❏ Technical Note

For some estimation commands, predict, score computes parameter-level scores $\partial L_j/\partial\beta$ instead of equation-level scores $\partial L_j/\partial\mathbf{x}_j\beta$. Those commands are

    asclogit

    asmprobit

    asroprobit

    nlogit

These models share the characteristic that there are multiple observations per independent event.

In making the robust variance calculation, what are really needed are parameter-level scores $\partial L_j/\partial\beta$, and so you may be asking yourself why predict, score does not always produce parameter-level scores. In the usual case we can obtain them from equation-level scores via the chain rule, and fewer variables are required if we adopt this approach. In the cases above, however, the likelihood is calculated at the group level and is not split into contributions from the individual observations. Thus the chain rule cannot be used and we must use the parameter level scores directly.

_robust can be tricked into using them if each parameter appears to be in its own equation as a constant. This requires resetting the row and column stripes on the covariance matrix before _robust is called. The equation names for each row and column must each be unique and the variable names must all be _cons.

❏

20.17   Weighted estimation

[U] **11.1.6 weight** introduced the syntax for weights. Stata provides four kinds of weights: fweights, or frequency weights; pweights, or sampling weights; aweights, or analytic weights; and iweights, or importance weights. The syntax for using each is the same. Type

    . regress y x1 x2

and you obtain unweighted estimates; type

    . regress y x1 x2 [pweight=pop]

and you obtain (in this example) pweighted estimation.

The sections below explain how each type of weight is used in estimation.

20.17.1 Frequency weights

Frequency weights—fweights—are integers and are nothing more than replication counts. The weight is statistically uninteresting, but from a data processing perspective it is important. Consider the following data

y	x1	x2
22	1	0
22	1	0
22	1	1
23	0	1
23	0	1
23	0	1

and the estimation command

    . regress y x1 x2

Equivalent is the following, more compressed data

y	x1	x2	pop
22	1	0	2
22	1	1	1
23	0	1	3

and the corresponding estimation command

    . regress y x1 x2 [fweight=pop]

When you specify frequency weights, you are treating each observation as one or more real observations.

❑ Technical Note

You might occasionally run across a command that does not allow weights at all, especially among user-written commands. expand (see [D] **expand**) can be used with such commands to obtain frequency-weighted results. The expand command duplicates observations so that the data become self-weighting. Suppose that you want to run the command usercmd, which does something or other, and you would like to type usercmd y x1 x2 [fw=pop]. Unfortunately, usercmd does not allow weights. Instead, you type

    . expand pop
    . usercmd y x1 x2

to obtain your result. Moreover, there is an important principle here: the results of running any command with frequency weights should be the same as running the command on the unweighted, expanded data. Unweighted, duplicated data and frequency-weighted data are merely two ways of recording identical information.

❑

(Continued on next page)

20.17.2   Analytic weights

Analytic weights—*analytic* is a term we made up—statistically arise in one particular problem: linear regression on data that are themselves observed means. That is, think of the model

$$y_i = \mathbf{x}_i\boldsymbol{\beta} + \epsilon_i, \qquad \epsilon_i \sim N(0, \sigma^2)$$

and now think about fitting this model on data $(\overline{y}_j, \overline{\mathbf{x}}_j)$ that are themselves observed averages. For instance, a piece of the underlying data for $(y_i, \mathbf{x}_i)$ might be $(3, 1)$, $(4, 2)$, and $(2, 2)$, but you do not know that. Instead, you have one observation $\{(3 + 4 + 2)/3, (1 + 2 + 2)/3\} = (3, 1.67)$ and know only that the $(3, 1.67)$ arose as the average of 3 underlying observations. All your data are like that.

`regress` with `aweights` is the solution to that problem:

```
. regress y x [aweight=pop]
```

There is a history of misusing such weights. A researcher does not have cell-mean data but instead has a probability-weighted random sample. Long before Stata existed, some researchers were using aweights to produce estimates from such samples. We will come back to this point in [U] **20.17.3 Sampling weights** below.

Anyway, the statistical problem that aweights resolve can be written as

$$y_i = \mathbf{x}_i\boldsymbol{\beta} + \epsilon_i, \qquad \epsilon_i \sim N(0, \sigma^2/w_i)$$

where the w_i are the analytic weights. The details of the solution are to make linear regression calculations using the weights as if they were fweights but to normalize them to sum to N before doing that.

Most commands that allow aweights handle them in this manner. That is, if you specify aweights, they are

1. normalized to sum to N and then

2. inserted in the calculation formulas in the same way as fweights.

20.17.3   Sampling weights

Sampling weights—probability weights or pweights—refer to probability-weighted random samples. Actually, what you specify in [pweight=...] is a variable recording the number of subjects in the full population that the sampled observation in your data represents. That is, an observation that had probability $1/3$ of being included in your sample has pweight 3.

Some researchers have used aweights with this kind of data. If they do, they are probably making a mistake. Consider the regression model

$$y_i = \mathbf{x}_i\boldsymbol{\beta} + \epsilon_i, \qquad \epsilon_i \sim N(0, \sigma^2)$$

Begin by considering the exact nature of the problem of fitting this model on cell-mean data—for which aweights are the solution: heteroskedasticity arising from the grouping. The error term ϵ_i is homoskedastic (meaning that it has constant variance σ^2). Say that the first observation in the data is the mean of 3 underlying observations. Then

$$y_1 = \mathbf{x}_1\boldsymbol{\beta} + \epsilon_1, \qquad \epsilon_i \sim N(0, \sigma^2)$$
$$y_2 = \mathbf{x}_2\boldsymbol{\beta} + \epsilon_2, \qquad \epsilon_i \sim N(0, \sigma^2)$$
$$y_3 = \mathbf{x}_3\boldsymbol{\beta} + \epsilon_3, \qquad \epsilon_i \sim N(0, \sigma^2)$$

and taking the mean

$$(y_1 + y_2 + y_3)/3 = \{(\mathbf{x}_1 + \mathbf{x}_2 + \mathbf{x}_3)/3\}\beta + (\epsilon_1 + \epsilon_2 + \epsilon_3)/3$$

For another observation in the data—which may be the result of summing of a different number of observations—the variance will be different. Hence, the model for the data is

$$\overline{y}_j = \overline{x}_j\beta + \overline{\epsilon}_j, \qquad \overline{\epsilon}_j \sim N(0, \sigma^2/N_j)$$

This makes intuitive sense. Consider 2 observations, one recording means over two subjects and the other means over 100,000 subjects. You would expect the variance of the residual to be less in the 100,000-subject observation; i.e., there is more information in the 100,000-subject observation than in the two-subject observation.

Now instead say that you are fitting the same model, $y_i = \mathbf{x}_i\beta + \epsilon_i$, $\epsilon_i \sim N(0, \sigma^2)$, on probability-weighted data. Each observation in your data is one subject, but the different subjects have different chances of being included in your sample. Therefore, for each subject in your data

$$y_i = \mathbf{x}_i\beta + \epsilon_i, \qquad \epsilon_i \sim N(0, \sigma^2)$$

That is, there is no heteroskedasticity problem. The use of the aweighted estimator cannot be justified on these grounds.

As a matter of fact, from the argument just given, you do not need to adjust for the weights at all, although the argument does not justify not making an adjustment. If you do not adjust, you are holding tightly to the assumed truth of your model. Two issues arise when considering adjustment for sampling weights:

1. the efficiency of the point estimate $\widehat{\beta}$ of β and

2. the reported standard errors (and, more generally, the variance matrix of $\widehat{\beta}$).

Efficiency argues in favor of adjustment, and that, by the way, is why many researchers have used aweights with pweighted data. The adjustment implied by pweights to the point estimates is the same as the adjustment implied by aweights.

With regard to the second issue, the use of aweights produces incorrect results because it interprets larger weights as designating more accurately measured points. For pweights, however, the point is no more accurately measured—it is still just one observation with one residual ϵ_j and variance σ^2. In [U] **20.15 Obtaining robust variance estimates** above, we introduced another estimator of variance that measures the variation that would be observed if the data collection followed by the estimation were repeated. Those same formulas provide the solution to pweights, and they have the added advantage that they are not conditioned on the model's being true. If we have any hopes of measuring the variation that would be observed were the data collection followed by estimation repeated, we must include the probability of the observations being sampled in the calculation.

In Stata, when you type

```
. regress y x1 x2 [pw=pop]
```

the results are the same as if you had typed

```
. regress y x1 x2 [pw=pop], vce(robust)
```

That is, specifying pweights implies the vce(robust) option and, hence, the robust variance calculation (but weighted). In this example, we use regress simply for illustration. The same is true of probit and all of Stata's estimation commands. Estimation commands that do not have a vce(robust) option (there are a few) do not allow pweights.

pweights are adequate for handling random samples where the probability of being sampled varies. pweights may be all you need. If, however, the observations are not sampled independently but are sampled in groups—called clusters in the jargon—you should specify the estimator's vce(cluster *clustvar*) option as well:

```
. regress y x1 x2 [pw=pop], vce(cluster block)
```

There are two ways of thinking about this:

1. The robust estimator answers the question of which variation would be observed were the data collection followed by the estimation repeated; if that question is to be answered, the estimator must account for the clustered nature of how observations are selected. If observations 1 and 2 are in the same cluster, then you cannot select observation 1 without selecting observation 2 (and, by extension, you cannot select observations like 1 without selecting observations like 2).

2. If you prefer, you can think about potential correlations. Observations in the same cluster may not really be independent—that is an empirical question to be answered by the data. For instance, if the clusters are neighborhoods, it would not be surprising that the individual neighbors are similar in their income, their tastes, and their attitudes, and even more similar than two randomly drawn persons from the area at large with similar characteristics, such as age and sex.

Either way of thinking leads to the same (robust) estimator of variance.

Sampling weights usually arise from complex sampling designs, which often involve not only unequal probability sampling and cluster sampling but also stratified sampling. There is a family of commands in Stata designed to work with the features of complex survey data, and those are the commands that begin with svy. To fit a linear regression model with stratification, for example, you would use the svy:regress command.

Non-svy commands that allow pweights and clustering give essentially identical results to the svy commands. If the sampling design is simple enough that it can be accommodated by the non-svy command, that is a fine way to perform the analysis. The svy commands differ in that they have more features, and they do all the little details correctly for real survey data. See [SVY] **survey** for a brief discussion of some of the issues involved in the analysis of survey data and a list of all the differences between the svy and non-svy commands.

Not all model estimation commands in Stata allow pweights. This is often because they are computationally or statistically difficult to implement.

20.17.4 Importance weights

Stata's iweights—importance weights—are the emergency exit. These weights are for those who want to take control and create special effects. For example, programmers have used regress with iweights to compute iteratively reweighted least-squares solutions for various problems.

iweights are treated much like aweights, except that they are not normalized. Stata's iweight rule is that

1. the weights are not normalized and

2. they are generally inserted into calculation formulas in the same way as fweights. There are exceptions; see the *Methods and Formulas* for the particular command.

iweights are used mostly by programmers who are often on the way to implementing one of the other kinds of weights.

20.18   A list of postestimation commands

The following commands can be used after estimation:

[R]	**adjust**	adjusted predictions of $\mathbf{x}\beta$, probabilities, or $\exp(\mathbf{x}\beta)$
[R]	**estat**	AIC, BIC, VCE, and estimation sample summary
[R]	**estimates**	cataloging estimation results
[R]	**hausman**	Hausman specification test
[R]	**lincom**	point estimates, standard errors, testing, and inference for linear combinations of coefficients
[R]	**linktest**	specification link test for single-equation models
[R]	**lrtest**	likelihood-ratio test
[R]	**mfx**	marginal effects or elasticities
[R]	**nlcom**	point estimates, standard errors, testing, and inference for generalized predictions
[R]	**predict**	predictions, residuals, influence statistics, and other diagnostic measures
[R]	**predictnl**	point estimates, standard errors, testing, and inference for generalized predictions
[R]	**suest**	seemingly unrelated estimation
[R]	**test**	Wald tests for simple and composite linear hypotheses
[R]	**testnl**	Wald tests of nonlinear hypotheses

Also see [U] **13.5 Accessing coefficients and standard errors** for accessing coefficients and standard errors.

Halbert Lynn White Jr. (1950–) was born in Kansas City. After receiving economics degrees at Princeton and MIT, he has taught and researched economics at the University of Rochester and, since 1980, at the University of California, San Diego. He is also active as a consultant. White's research interests include artificial neural networks, finance, econometrics, and forecasting, and his hobbies include music and antiques.

Peter Jost Huber (1934–) was born in Wohlen (Aargau, Switzerland). He gained mathematics degrees from ETH Zürich, including a Ph.D. thesis on homotopy theory, and then studied statistics at Berkeley on postdoctoral fellowships. This visit yielded a celebrated 1964 paper on robust estimation, and Huber's later monographs on robust statistics were crucial in directing that field. Thereafter his career took him back and forth across the Atlantic, with periods at Cornell, ETH Zürich, Harvard, MIT, and Bayreuth. His work has touched several other major parts of statistics, theoretical and applied, including regression, exploratory multivariate analysis, large datasets, and statistical computing. Huber also has a major long-standing interest in Babylonian astronomy.

20.19 References

Binder, D. A. 1983. On the variances of asymptotically normal estimators from complex surveys. *International Statistical Review* 51: 279–292.

Deaton, A. 1997. *The Analysis of Household Surveys: A Microeconometric Approach to Development Policy.* Baltimore, MD: Johns Hopkins University Press.

Fuller, W. A. 1975. Regression analysis for sample survey. *Sankhyā, Series C* 37: 117-132.

Gail, M. H., W. Y. Tan, and S. Piantadosi. 1988. Tests for no treatment effect in randomized clinical trials. *Biometrika* 75: 57–64.

Hampel, F. R. 1992. Introduction to Huber (1964) Robust estimation of a location parameter. In *Breakthroughs in Statistics. Volume II: Methodology and Distribution,* ed. S. Kotz and N. L. Johnson, 479–491. New York: Springer.

Huber, P. J. 1967. The behavior of maximum likelihood estimates under non-standard conditions. In *Proceedings of the Fifth Berkeley Symposium on Mathematical Statistics and Probability.* Berkeley: University of California Press, 1, 221–233.

Kent, J. T. 1982. Robust properties of likelihood ratio tests. *Biometrika* 69: 19–27.

Kish, L., and M. R. Frankel. 1974. Inference from complex samples. *Journal of the Royal Statistical Society, Series B* 36: 1–37.

Lin, D. Y., and L. J. Wei. 1989. The robust inference for the Cox proportional hazards model. *Journal of the American Statistical Association* 84: 1074–1078.

Long, J. S., and J. Freese. 2000a. sg145: Scalar measures of fit for regression models. *Stata Technical Bulletin* 56: 34–40. Reprinted in *Stata Technical Bulletin Reprints,* vol. 10, pp. 197–205.

——. 2000b. sg152: Listing and interpreting transformed coefficients from certain regression models. *Stata Technical Bulletin* 57: 27–34. Reprinted in *Stata Technical Bulletin Reprints,* vol. 10, pp. 231–240.

MacKinnon, J. G., and H. White. 1985. Some heteroskedasticity consistent covariance matrix estimators with improved finite sample properties. *Journal of Econometrics* 29: 305–325.

Rogers, W. H. 1993. sg17: Regression standard errors in clustered samples. *Stata Technical Bulletin* 13: 19–23. Reprinted in *Stata Technical Bulletin Reprints,* vol. 3, 88–94.

Royall, R. M. 1986. Model robust confidence intervals using maximum likelihood estimators. *International Statistical Review* 54: 221–226.

Weesie, J. 2000. sg127: Summary statistics for estimation sample. *Stata Technical Bulletin* 53: 32–35. Reprinted in *Stata Technical Bulletin Reprints,* vol. 9, pp. 275–277.

White, H. 1980. A heteroskedasticity-consistent covariance matrix estimator and a direct test for heteroskedasticity. *Econometrica* 48: 817–830.

——. 1982. Maximum likelihood estimation of misspecified models. *Econometrica* 50: 1–25.

Advice

21 Inputting data

Contents

21.1   Overview

To input data into Stata, you can use

[D] **edit** and [D] **input**	to enter data from the keyboard
[D] **insheet**	to read tab- or comma-separated data
[D] **infile (free format)**	to read unformatted data
[D] **infile (fixed format)** or [D] **infix (fixed format)**	to read formatted data
[D] **xmlsave** (where `xmluse` is documented)	to use datasets in XML and Excel format
[D] **odbc**	to read from an ODBC source
[D] **fdasave** (where `fdause` is documented)	to read datasets in SAS XPORT format
[TS] **haver**	to read data in Haver Analytics' format
[U] **21.4 Transfer programs**	to transfer data

Since dataset formats differ, you should familiarize yourself with each method.

[D] **infile (fixed format)** and [D] **infix (fixed format)** are two different commands that do the same thing. Read about both, and then use whichever appeals to you.

Alternatively, `edit` and `input` both allow you to enter data from the keyboard. `edit` opens a Viewer, and `input` allows you to type as the command line.

After you have read this chapter, also see [D] **infile** for more examples of the different commands to input data.

21.2   Determining which input method to use

Below are several rules that, when applied sequentially, will direct you to the appropriate method for entering your data. After the rules is a description of each command, as well as a reference to the corresponding entry in the *Reference* manuals.

1. If you have a few data and simply wish to type the data directly into Stata at the keyboard, see [D] **input**—there are many examples, and doing so should be easy. Also see [D] **edit**.

2. If your dataset is in binary format or the internal format of some software package, you have several options:

   a. If the data are in a spreadsheet, copy and paste the data into Stata's Data Editor; see [D] **edit** for details.

   b. If the data are in an Excel spreadsheet, use Excel to export the data as XML, and then use Stata's `xmluse` command to read them; see [D] **xmlsave**.

   c. If the data are in SAS XPORT format, use `fdause` to read the data; see [D] **fdasave**.

   d. If the data in Haver Analytics' `.dat` format (Haver Analytics provides economics and financial databases), and you are using Stata for Windows, use `haver` to read the data; see [TS] **haver**.

   e. Translate the data into ASCII (also known as character) format by using the other software. For instance, in Excel you can save spreadsheets as tab-delimited or comma-separated text. Then, see [D] **insheet**.

   f. Other software packages are available that will convert non–Stata format data files into Stata-format files; see [U] **21.4 Transfer programs**.

   g. If the data are located in an ODBC source, which typically includes databases and spreadsheets, you can use the `odbc load` command to import the data; see [D] **odbc**. Currently `odbc` is available for Windows, Macintosh, and Linux versions of Stata.

3. If the dataset has one observation per line and the data are tab- or comma separated, use `insheet`; see [D] **insheet**. This is the easiest way to read data.

4. If the dataset is formatted and that formatting information is required to interpret the data, you can use `infile` with a dictionary or `infix`; see [D] **infile (fixed format)** or [D] **infix (fixed format)**.

5. If there are no string variables, you can use `infile` without a dictionary: see [D] **infile (free format)**.

6. If all the string variables in the data are enclosed in (single or double) quotes, you can use `infile` without a dictionary; see [D] **infile (free format)**.

7. If the undelimited string variables have no blanks, you can use `infile` without a dictionary; see [D] **infile (free format)**.

8. If you make it to here, see [D] **infile (fixed format)** or [D] **infix (fixed format)**.

21.2.1   Entering data interactively

If you have a few data, you can type the data directly into Stata; see [D] **input** or [D] **edit**. Otherwise, we assume that your data are stored on disk.

21.2.2   If the dataset is in binary format

Stata can read ASCII datasets, which is technical jargon for datasets composed of characters—datasets that can be typed on your screen or printed on your printer. The alternative, binary datasets, cannot be read by Stata. Binary datasets are popular, and almost every software package has its own binary format. Stata .dta datasets are an example of a binary format that Stata can read. The Excel format is a binary format that Stata cannot read.

If your dataset is in binary format or in the internal format of another software package, you must translate it into an XML format that Stata can understand (see [D] **xmlsave**), translate it into ASCII, or use some other program for conversion to Stata format. If this dataset is located in a database or an ODBC source, see [U] **21.5 ODBC sources**. If the dataset is in SAS XPORT format, you can read it by using Stata's `fdause` command; see [D] **fdasave**. If the dataset is in Haver Analytics' .dat format, you can read it by using Stata's `haver` command; see [TS] **haver**.

Detecting whether data are stored in binary format can be tricky. For instance, many Windows users wish to read data that have been entered into a word processor—let's assume Word. Unwittingly, they have stored the dataset as a Word document. The dataset looks like ASCII to them: When they look at it in Word, they see readable characters. The dataset seems to even pass the printing test in that Word can print it. Nevertheless, the dataset is not ASCII; it is stored in an internal Word format, and the data cannot really pass the printing test since only Word can print it. To read the dataset, Windows users must use it in Word and then store it as an MS-DOS text file, MS-DOS text being the term Word decided to use to mean ASCII.

So, how do you know whether your dataset is binary? Here's a simple test: regardless of the operating system you use, start Stata and type `type` followed by the name of the file:

```
. type myfile.raw
output will appear
```

You do not have to print the entire file; press *Break* when you have seen enough.

Do you see things that look like hieroglyphics? If so, the dataset is binary. See [U] **21.4 Transfer programs** below.

If it looks like data, however, the file is (probably) ASCII.

Let us assume that you have an ASCII dataset that you wish to read. The data's format will determine the command you need to use. The different formats are discussed in the following sections.

21.2.3   If the data are simple

The easiest way to read data is with `insheet`; see [D] **insheet**.

`insheet` is smart: it looks at the dataset, determines what it contains, and then reads it. That is, `insheet` is smart given certain restrictions, such as that the dataset has one observation per line and that the values are tab- or comma separated. `insheet` can read this

———————————————————————————— top of data1.raw ————————
```
M,Joe Smith,288,14
M,K Marx,238,12
F,Farber,211,7
```
———————————————————————————— end of data1.raw ————————

or this (which has variable names on the first line)

```
───────────────────────────────────── top of data2.raw ─────────────
sex, name, dept, division
M,Joe Smith,288,14
M,K Marx,238,12
F,Farber,211,7
────────────────────────────────────── end of data2.raw ────────────
```

or this (which has one tab character separating the values):

```
───────────────────────────────────── top of data3.raw ─────────────
M        Joe Smith        288      14
M        K Marx   238     12
F        Farber   211     7
────────────────────────────────────── end of data3.raw ────────────
```

This looks odd because of how tabs work; data3.raw could similarly have a variable header, but insheet cannot read

```
───────────────────────────────────── top of data4.raw ─────────────
M        Joe Smith        288      14
M        K Marx           238      12
F        Farber           211      7
────────────────────────────────────── end of data4.raw ────────────
```

which has spaces rather than tabs.

There is a way to tell data3.raw from data4.raw: Ask Stata to type the data and show the tabs by typing

```
. type data3.raw, showtabs
M<T>Joe Smith<T>288<T>14
M<T>K Marx<T>238<T>12
F<T>Farber<T>211<T>7

. type data4.raw, showtabs
M        Joe Smith        288      14
M        K Marx           238      12
F        Farber           211      7
```

21.2.4   If the dataset is formatted and the formatting is significant

If the dataset is formatted and formatting information is required to interpret the data, see [D] **infile (fixed format)** or [D] **infix (fixed format)**.

Using infix or infile with a data dictionary is something new users want to avoid if at all possible.

The purpose of this section is only to take you to the most complicated of all cases if there is no alternative. Otherwise, you should wait and see if it is necessary. Do not misinterpret this section and say, "Ah, my dataset is formatted, so at last I have a solution."

Just because a dataset is formatted does not mean that you have to exploit the formatting information. The following dataset is formatted

```
─────────────────────────────────────────────── top of data5.raw ──────────
   1    27.39     12
   2     1.00      4
   3   100.10    100
─────────────────────────────────────────────── end of data5.raw ──────────
```

in that the numbers line up in neat columns, but you do not need to know the information to read it. Alternatively, consider the same data run together:

```
─────────────────────────────────────────────── top of data6.raw ──────────
   1 27.39 12
   2  1.00  4
   3100.10100
─────────────────────────────────────────────── end of data6.raw ──────────
```

This dataset is formatted, too, and you must know the formatting information to make sense of "3100.10100". You must know that variable 2 starts in column 4 and is six characters long to extract the 100.10. It is datasets like `data6.raw` that you should be looking for at this stage—datasets that make sense only if you know the starting and ending columns of data elements. To read data such as `data6.raw`, you must use either `infix` or `infile` with a data dictionary.

Reading unformatted data is easier. If you need the formatting information to interpret the data, then you must communicate that information to Stata, which means that you will have to type it. This is the hardest kind of data to read, but Stata can do it. See [D] **infile (fixed format)** or [D] **infix (fixed format)**.

Looking back at `data4.raw`,

```
─────────────────────────────────────────────── top of data4.raw ──────────
   M       Joe Smith      288     14
   M       K Marx         238     12
   F       Farber         211      7
─────────────────────────────────────────────── end of data4.raw ──────────
```

you may be uncertain whether you have to read it with a data dictionary. If you are uncertain, do not jump yet.

Finally, here is an obvious example of unformatted data:

```
─────────────────────────────────────────────── top of data7.raw ──────────
   1 27.39              12
   2 1 4
   3 100.1 100
─────────────────────────────────────────────── end of data7.raw ──────────
```

Here blanks separate one data element from the next and, in one case, many blanks, although there is no special meaning attached to more than one blank.

The following sections discuss datasets that are unformatted or formatted in a way that do not require a data dictionary.

21.2.5 If there are no string variables

If there are no string variables, see [D] **infile (free format)**.

Although the dataset `data7.raw` is unformatted, it can still be read using `infile` without a dictionary. This is not the case with `data4.raw` because this dataset contains undelimited string variables with embedded blanks.

❏ Technical Note

Some Stata users prefer to read data with a data dictionary, even when we suggest differently, as above. They like the convenience of the data dictionary—they can sit in front of an editor and carefully compose the list of variables and attach variable labels rather than having to type the variable list (correctly) on the Stata command line. However, they can create a do-file containing the `infile` statement and thus have all the advantages of a data dictionary without some of the (extremely technical) disadvantages of data dictionaries.

Nevertheless, we do tend to agree with such users—we, too, prefer data dictionaries. Our recommendations, however, are designed to work in all cases. If the dataset is unformatted and contains no string variables, it can always be read without a data dictionary, whereas only sometimes can it be read with a data dictionary.

The distinction is that `infile` without a data dictionary performs stream I/O, whereas with a data dictionary it performs record I/O. The difference is intentional—it guarantees that you will be able to read your data into Stata somehow. Some datasets require stream I/O, others require record I/O, and still others can be read either way. Recommendations 1–5 identify datasets that either require stream I/O or can be read either way.

❏

We are now left with datasets that contain at least one string variable.

21.2.6 If all the string variables are enclosed in quotes

If all the string variables in the data are enclosed in (single or double) quotes, see [D] **infile (free format)**.

See [U] **23 Dealing with strings** for a formal definition of strings, but as a quick guide, a string variable is a variable that takes on values like "bob" or "joe", as opposed to numeric variables that take on values like 1, 27.5, and –17.393. Undelimited strings—strings not enclosed in quotes—can be difficult to read.

Here is an example including delimited string variables:

```
──────────────────────────────────────── top of data8.raw ────────────
    "M" "Joe Smith" 288 14
    "M" "K Marx" 238 12
    "F" "Farber" 211 7
                                          ──────── end of data8.raw ────────────
```

or

```
──────────────────────────────────── top of data8.raw, alternative format ────────
    "M" "Joe Smith" 288   14
    "M" "K Marx"     238  12
    "F" "Farber"     211   7
                                        ──── end of data8.raw, alternative format ────────
```

Both of these are merely variations on `data4.raw` except that the strings are enclosed in quotes. Here `infile` without a dictionary can be used to read the data.

Here is another version of `data4.raw` without delimiters or even formatting:

```
──────────────────────────────────────── top of data9.raw ────────────
    M Joe Smith 288 14
    M K Marx 238 12
    F Farber 211 7
                                          ──────── end of data9.raw ────────────
```

What makes these data difficult? Blanks sometimes separate values and sometimes are nothing more than a blank within a string. For instance, you cannot tell whether Farber has first initial F with missing sex or is instead female with a missing first initial.

Fortunately, such data rarely happen. Either the strings are delimited, as we showed in data8.raw, or the data are in columns, as in data4.raw.

21.2.7   If the undelimited strings have no blanks

There is a case in which uncolumnized, undelimited strings cause no confusion—when they contain no blanks. For instance, if our data contained only last names,

```
——————————————————————————————————————— top of data10.raw ————————
Smith 288 14
Marx 238 12
Farber 211 7
——————————————————————————————————————— end of data10.raw ————————
```

Stata could read it without a data dictionary. Caution: the last names must contain no blanks—no Van Owen's or von Beethoven's.

If the undelimited string variables have no blanks, see [D] **infile (free format)**.

21.2.8   If you make it to here

If you make it to here, see [D] **infile (fixed format)** or [D] **infix (fixed format)**.

Remember data4.raw?

```
——————————————————————————————————————— top of data4.raw ————————
M        Joe Smith     288     14
M        K Marx        238     12
F        Farber        211     7
——————————————————————————————————————— end of data4.raw ————————
```

It can be read using either infile with a dictionary or infix.

21.3   If you run out of memory

You can increase the amount of memory allocated to Stata; see [U] **6 Setting the size of memory**.

You can also try to conserve memory.

When you read the data, did you specify variable types? Stata can store integers more compactly than floats and small integers more compactly than large integers; see [U] **12 Data**.

If that is not sufficient, you will have to resort to reading the data in pieces. Both infile and infix allow you to specify an in *range* modifier, and, here the range is interpreted as the observation range to read. Thus, infile ... in 1/100 would read observations 1–100 of your data and stop.

infile ... in 101/200 would read observations 101–200. The end of the range may be specified as larger than the actual number of observations in the data. If the dataset contained only 150 observations, infile ... in 101/200 would read observations 101–150.

Another way of reading the data in pieces is to specify the `if` *exp* modifier. Say that your data contained an equal number of males and females, coded as the variable `sex` (which you will read) being 0 or 1, respectively. You could type `infile ... if sex==0` to read the males. `infile` will read an observation, determine if `sex` is zero, and if not, throw the observation away. You could read just the females by typing `infile ... if sex==1`.

If the dataset is really big, perhaps you need only a random sample of the data—you never intended to analyze the entire dataset. Since `infile` and `infix` allow `if` *exp*, you could type `infile ... if uniform()<.1`. `uniform()` is the uniformly distributed random-number generator; see [D] **functions**. This method would read an approximate 10% sample of the data. If you are serious about using random samples, do not forget to set the seed before using `uniform()`; see [D] **generate**.

The final approach is to read all the observations but only some of the variables. When reading data without a data dictionary, you can specify `_skip` for variables, indicating that the variable is to be skipped. When reading with a data dictionary or using `infix`, you can specify the actual columns to read, skipping any columns you wish to ignore.

21.4 Transfer programs

To import data from, say, Excel, you can save the data as a text file and then read it into Stata according to the rules above, read it via an ODBC source, or purchase a program to translate the dataset from Excel's format to Stata's format.

One such program is Stat/Transfer, which is available for Windows, Macintosh OS X, and Unix. It reads and writes data in a variety of formats, including Microsoft Access, dBASE, Epi Info, Excel, FoxPro, GAUSS, JMP, Lotus 1-2-3, MATLAB, MineSet, Minitab, ODBC, Paradox, Quattro Pro, S-Plus, SAS, SPSS, Statistica, SYSTAT, and, of course, Stata.

Stat/Transfer, available from...	is manufactured by...
StataCorp	Circle Systems
4905 Lakeway Drive	1001 Fourth Avenue Plaza, Suite 3200
College Station, Texas 77845	Seattle, Washington 98154
Telephone: 979-696-4600	Telephone: 206-682-3783
Fax: 979-696-4601	Fax: 206-328-4788
Email: stata@stata.com	sales@circlesys.com

There are other transfer programs available, too. Our web site, http://www.stata.com, lists programs available from other sources.

21.5 ODBC sources

If your dataset is located in a network database or shared spreadsheet, you may be able to import your data via ODBC. ODBC (Open Database Connectivity) is a standard for exchanging data between programs. Stata supports the ODBC standard for importing data via the `odbc` command and can read from any ODBC source on your computer.

This process requires a data source, such as a database located on a network. To use the `odbc` command to import data from a database requires that the database first be set up as an ODBC source on the same machine that is running Stata. The database itself does not have to be on the same machine, just the definition of that database as the ODBC source. On a Windows machine, an ODBC source is added via a Control Panel called "Data Sources." Also, typing `odbc list` from Stata displays all the ODBC sources that are provided by the computer.

If the database is functioning and the appropriate data source has been set up on the same machine as Stata, one call using `odbc load` is all that is needed to import data. For a more thorough description of this process, see [D] **odbc**.

21.6   Reference

Swagel, P. 1994. os14: A program to format raw data files. *Stata Technical Bulletin* 20: 10–12. Reprinted in *Stata Technical Bulletin Reprints*, vol. 4, pp. 80–82.

22 Combining datasets

You have two datasets that you wish to combine. Below, we will draw a dataset as a box where, in the box, the variables go across and the observations go down.

See [D] **append** if you want to combine datasets vertically:

append adds observations to the existing variables. That is an oversimplification because append does not require that the datasets have the same variables. append is appropriate, for instance, when you have data on hospital patients and then receive data on more patients.

See [D] **merge** if you want to combine datasets horizontally:

merge adds variables to the existing observations. That is an oversimplification because merge does not require that the datasets have the same observations. merge is appropriate, for instance, when you have data on survey respondents and then receive data on part 2 of the questionnaire.

See [D] **joinby** when you want to combine datasets horizontally but form all pairwise combinations within group:

joinby is similar to merge but forms all combinations of the observations where it makes sense. joinby would be appropriate, for instance, where A contained data on parents and B contained data on their children. joinby *familyid* would form a dataset of each parent joined with each of his or her children.

Also see [D] **cross** for a less frequently used command that forms every pairwise combination of two datasets.

23 Dealing with strings

Please read [U] **12 Data** before reading this entry.

23.1   Description

The word *string* is shorthand for a string of characters. "Male" and "Female", "yes" and "no", and "R. Smith" and "P. Jones" are examples of strings. The alternative to strings is numbers—0, 1, 2, 5.7, and so on. Variables containing strings—called *string variables*—occur in data for a variety of reasons. Four of these reasons are listed below.

A variable might contain strings because it is an identifying variable. Employee names in a payroll file, patient names in a hospital file, and city names in a city data file are all examples of this. This is a proper use of string variables.

A variable might contain strings because it records categorical information. "Male" and "Female" and "yes" and "no" are examples of such use, but this is not an appropriate use of string variables. It is not appropriate because the same information could be coded numerically, and, if it were, (1) it would take less memory to store the data and (2) the data would be more useful. We will explain how to convert categorical strings to categorical numbers below.

Also, a variable might contain strings because of a mistake. For example, the variable contains things like 1, 5, 8.2, but because of an error in reading the data, the data were mistakenly put into a string variable. We will explain how to fix such mistakes.

Finally, a variable might contain strings because the data simply could not be coerced into being stored numerically. "15 Jan 1992", "1/15/92", and "1A73" are examples of such use. We will explain how to deal with such complexities.

23.2   Categorical string variables

A variable might contain strings because it records categorical information.

Suppose that you have read in a dataset that contains a variable called sex, recorded as "male" and "female", yet when you attempt to run an ANOVA, the following message is displayed:

```
. use http://www.stata-press.com/data/r10/hbp2

. anova hbp sex
no observations
r(2000);
```

There are no observations because anova, along with most of Stata's "analytic" commands, cannot deal with string variables. Commands want to see numbers, and when they do not, they treat the variable as if it contained numeric missing values. Despite this limitation, it is possible to obtain tables:

```
. encode sex, gen(gender)
. anova hbp gender
```

	Number of obs =		1128	R-squared	=	0.0123
	Root MSE	= .214223		Adj R-squared =		0.0114

Source	Partial SS	df	MS	F	Prob > F
Model	.644485682	1	.644485682	14.04	0.0002
gender	.644485682	1	.644485682	14.04	0.0002
Residual	51.6737767	1126	.045891454		
Total	52.3182624	1127	.046422593		

The magic here is to convert the string variable sex into a numeric variable called gender with an associated value label, a trick accomplished by encode; see [U] **12.6.3 Value labels** and [D] **encode**.

23.3 Mistaken string variables

A variable might contain strings because of a mistake.

Suppose that you have numeric data in a variable called x, but because of a mistake, x was made a string variable when you read the data. When you list the variable, it looks fine:

```
. list x
```

	x
1.	2
2.	2.5
3.	17

(output omitted)

Yet, when you attempt to obtain summary statistics on x,

```
. summarize x
```

Variable	Obs	Mean	Std. Dev.	Min	Max
x	0				

If this happens to you, type describe to confirm that x is stored as a string:

```
. describe
Contains data
  obs:            10
  vars:            3
  size:          160 (99.9% of memory free)
```

variable name	storage type	display format	value label	variable label
x	str4	%9s		
y	float	%9.0g		
z	float	%9.0g		

```
Sorted by:
```

x is stored as a str4.

The problem is that `summarize` does not know how to calculate the mean of string variables—how to calculate the mean of "Joe" plus "Bill" plus "Roger"—even when the string variable contains what could be numbers. By using the `destring` command ([D] **destring**), the variable mistakenly stored as a `str4` can be converted to a numeric variable.

```
. destring x, replace
. summarize x
```

Variable	Obs	Mean	Std. Dev.	Min	Max
newx	10	1.76	.8071899	.7	3

An alternative to using the `destring` command is to use `generate` with the `real()` function; see [D] **functions**.

23.4 Complex strings

A variable might contain strings because the data simply could not be coerced into being stored numerically.

A complex string is a string that contains more than one piece of information. The most common example of a complex string is a date: "15 Jan 1992" contains three pieces of information—a day, a month, and a year. If your complex strings are dates or times, see [U] **24 Dealing with dates and times**.

Although Stata has functions for dealing with dates, you will have to deal with other complex strings yourself. Assume that you have data that include part numbers:

```
. list partno
```

	partno
1.	5A2713
2.	2B1311
3.	8D2712

(*output omitted*)

The first digit of the part number is a division number, and the character that follows identifies the plant at which the part was manufactured. The next three digits represent the major part number and the last digit is a modifier indicating the color. This complex variable can be decomposed using the `substr()` and `real()` functions described in [D] **functions**:

```
. gen byte div = real(substr(partno,1,1))
. gen str1 plant = substr(partno,2,1)
. gen int part = real(substr(partno,3,3))
. gen byte color = real(substr(partno,6,1))
```

We use the `substr()` function to extract pieces of the string and use the `real()` function, when appropriate, to translate the piece into a number.

For an extended discussion of numeric and string data types and how to convert from one kind to another, see Cox (2002).

23.5 Reference

Cox, N. J. 2002. Speaking Stata: On numbers and strings. *Stata Journal* 2: 314–329.

24 Dealing with dates and times

Contents

24.1 Overview

Full documentation on Stata's date and time capabilities—including documentation on relevant functions and display formats—can be found in [D] **dates and times**.

Stata can work with dates such as 21nov2006, with times such as 13:42:02.213, and with dates and times such as 21nov2006 13:42:02.213. You can write these dates and times however you wish, such as 11/21/2006, November 21, 2006, and 1:42 p.m.

Stata stores dates, times, and dates and times as integers such as $-4,102$, 0, 82, 4,227, and 1,479,735,745,213. It works like this:

1. You begin with the date/time variables in your data however they are recorded, such as 21nov2006 or 11/21/2006 or November 21, 2006, or 13:42:02.213 or 1:42 p.m. The original values are usually best stored in string variables.

2. Using functions we will describe below, you translate the original into the integers that Stata understands and store those values in a new variable.

3. You specify the appropriate display format for the new variable so that, rather than displaying as the integer values that they are, they display in a way you can read them such as 21nov2006 or 11/21/2006 or November 21, 2006, or 13:42:02.213 or 1:42 p.m.

The numeric encoding that Stata uses is centered around the first millisecond of 01jan1960, i.e., 01jan1960 00:00:00.000. That date/time is assigned integer value 0.

Integer value 1 is the millisecond after that: 01jan1960 00:00:00.001.

Integer value -1 is the millisecond before that: 31dec1959 23:59:59.999.

By that logic, 21nov2006 13:42:02.213 is integer value 1,479,735,722,213 or, at least, it is if we ignore the leap seconds that have been inserted to keep clocks in alignment with astronomical observation. If we account for leap seconds, 21nov2006 13:42:02.213 would be 23 seconds later, namely, 1,479,735,745,213. Stata can work either way.

Obtaining the number of milliseconds associated with a date/time is easy because Stata provides functions that translate things like 21nov2006 13:42:02.213 (written however you wish) to 1,479,735,722,213 or 1,479,735,745,213.

Just remember, Stata records data/time values as the number of milliseconds since the first millisecond of 01jan1960.

Stata records pure time values (clock times independent of date) the same way. Rather than thinking of the numeric value as the number of milliseconds since 01jan1960, however, think of it as the number of milliseconds since the beginning of the day. For instance, at 2 p.m. every day, the airplane takes off from Houston for London. The numeric value associated with 2 p.m. is 50,400,000 because there are that many milliseconds between the beginning of the day (00:00:00.000) and 2 p.m.

The advantage of thinking this way is that you can add dates and times. What is the date/time value for when the plane takes off on 21nov2006? Well, 21nov2006 00:00:00.000 is 1,479,686,400,000 (ignoring leap seconds), and 1,479,686,400,000 + 50,400,000 is 1,479,736,800,000.

Subtracting date/time values is useful, too. How many hours are there between 21jan1952 7:23 a.m. and 21nov2006 3:14 p.m.? Answer: $(1,479,741,240,000 - (-250,706,220,000))/3,600,000 = 480,679.85$ hours.

Variables that record the number of milliseconds since 01jan1960 and ignore leap seconds are called %tc variables.

Variables that record the number of milliseconds since 01jan1960 and account for leap seconds are called %tC variables.

Stata has seven other kinds of %t variables.

In many applications, calendar dates by themselves are sufficient. The applicant was hired on 15jan2006, for instance. You could use a %tc variable to record that value, assigning some arbitrary time that you would ignore, but it is better and easier to use a %td variable. In %td variables, 0 still corresponds to 01jan1960, but a unit change now represents an entire day rather than a millisecond. The value 1 represents 02jan1960. The value −1 represents 31dec1959. When you subtract %td variables, you obtain the number of days between dates.

In a financial application, you might use %tq variables. In %tq, 0 represents the first quarter of 1960, 1 represents the second quarter, and −1 represents the last quarter of 1959. When you subtract %tq variables, you obtain the number of quarters between dates.

Stata understands nine %t formats:

format	base	units	comment
%tc	01jan1960	milliseconds	ignores leap seconds
%tC	01jan1960	milliseconds	accounts for leap seconds
%td	01jan1960	days	calendar date format
%tw	1960-w1	weeks	52nd week may have more than 7 days
%tm	jan1960	months	calendar month format
%tq	1960-q1	quarters	financial quarter
%th	1960-h1	half-years	1 half-year = 2 quarters
%ty	0 A.D	year	1960 means year 1960
%tg	?	?	whatever you wish

All formats except %ty and %tg are based on the beginning of January 1960. The value 0 means the first millisecond, day, week, month, quarter, or half-year of 1960, depending on format. The value 1 is the millisecond, day, week, month, quarter, or half-year after that. The value −1 is the millisecond, day, week, month, quarter, or half-year before that.

Stata's %ty format records years as numeric values and it codes them the natural way: rather than 0 meaning 1960, 1960 means 1960, and so 2006 also means 2006.

Stata's %tg format is for use when none of the other formats fit your date/time data. The value 0, 1, and −1 mean whatever you want them to mean.

24.2   Inputting dates and times

Dates and time variables are best read as strings. You then use one of the string-to-numeric conversion functions to convert the string to an appropriate %t value:

format	string-to-numeric conversion function
%tc	clock(*string*, *mask*)
%tC	Clock(*string*, *mask*)
%td	date(*string*, *mask*)
%tw	weekly(*string*, *mask*)
%tm	monthly(*string*, *mask*)
%tq	quarterly(*string*, *mask*)
%th	halfyearly(*string*, *mask*)
%ty	yearly(*string*, *mask*)
%tg	no function necessary; read as numeric

The full documentation of these functions can be found in *String-to-numeric translation functions* in [D] **dates and times**.

In the above table, *string* is the string variable to be translated, and *mask* specifies the order in which the components of the date and/or time appear in *string*. For instance, the *mask* in %td function date() is made up of the letters M, D, and Y.

date(*string*, "DMY") specifies *string* contains dates in the order of day, month, year. With that specification, date() can translate 21nov2006, 21 November 2006, 21-11-2006, 21112006, and other strings that contain dates in the order day, month, year.

date(*string*, "MDY") specifies *string* contains dates in the order of month, day, year. With that specification, date() can translate November 21, 2006, 11/21/2006, 11212006, and other strings that contain dates in the order month, day, year.

You can specify a two-digit prefix in front of Y to handle two-digit years. date(*string*, "MD19Y") specifies *string* contains dates in the order of month, day, and year, and that if the year contains only two digits, it is to be prefixed with 19. With that specification, date() could not only translate November 21, 2006, 11/21/2006, and 11212006, but also Feb. 15 '98, 2/15/98, and 21598. (There is another way to deal with two-digit years so that 98 becomes 1998 while 06 becomes 2006; it involves specifying an optional third argument.)

Let's consider some %td data. We have the following raw-data file:

```
──────────────────────────────── top of bdays.raw ────────────
   Bill   21 Jan 1952   22
   May    11 Jul 1948   18
   Sam    12 Nov 1960   25
   Kay     9 Aug 1975   16
──────────────────────────────── end of bdays.raw ────────────
```

We could read these data by typing

```
. infix str name 1-5  str bday 7-17  x 20-21 using bdays
(4 observations read)
```

We read the date not as three separate variables but as one variable. Variable bday contains the entire date:

```
. list
```

	name	bday	x
1.	Bill	21 Jan 1952	22
2.	May	11 Jul 1948	18
3.	Sam	12 Nov 1960	25
4.	Kay	9 Aug 1975	16

The data look fine, but if we set about using them, we would quickly discover there is not much we could do with variable bday. Variable bday looks like a date, but it is just a string. We need to turn bday into a %t variable that Stata understands:

```
. gen birthday = date(bday, "DMY")
. list
```

	name	bday	x	birthday
1.	Bill	21 Jan 1952	22	-2902
2.	May	11 Jul 1948	18	-4191
3.	Sam	12 Nov 1960	25	316
4.	Kay	9 Aug 1975	16	5699

New variable birthday is a %td variable. The problem now is that, whereas the new variable is perfectly understandable to Stata, it is not understandable to us. Naturally enough, a %td variable needs a %td format:

```
. format birthday %td
. list
```

	name	bday	x	birthday
1.	Bill	21 Jan 1952	22	21jan1952
2.	May	11 Jul 1948	18	11jul1948
3.	Sam	12 Nov 1960	25	12nov1960
4.	Kay	9 Aug 1975	16	09aug1975

Using our new %td variable, we can create a variable recording how old each of these subjects was on 01jan2000:

```
. gen age2000 = (td(1jan2000)-birthday)/365.25
. list
```

	name	bday	x	birthday	age2000
1.	Bill	21 Jan 1952	22	21jan1952	47.94524
2.	May	11 Jul 1948	18	11jul1948	51.47433
3.	Sam	12 Nov 1960	25	12nov1960	39.13484
4.	Kay	9 Aug 1975	16	09aug1975	24.39699

td() is a function that makes it easy to type %td dates. There are also functions tc(), tC(), tw(), tm(), tq(), and th() for the other %t formats; see [D] **dates and times**.

Let's consider one more example. We have the following data:

```
. use http://www.stata-press.com/data/r10/datexmpl2
. list
```

	id	timestamp	action
1.	1001	Tue Nov 14 08:59:43 CST 2006	15
2.	1002	Wed Nov 15 07:36:49 CST 2006	15
3.	1003	Wed Nov 15 09:21:07 CST 2006	11
4.	1002	Wed Nov 15 14:57:36 CST 2006	16
5.	1005	Thu Nov 16 08:22:53 CST 2006	12
6.	1001	Thu Nov 16 08:36:44 CST 2006	16

Variable `timestamp` is a string which we want to convert to a %tc variable. From the table above, we know we will use function `clock()`. The *mask* in `clock()` uses the letters D, M, Y, and h, m, s, which specify the order of the day, month, year and hours, minutes, seconds. `timestamp` contains more than that and so cannot directly be converted using `clock`. First, we must create a variable that `clock()` understands:

```
. gen str ts = substr(timestamp, 5, 15) + " " + substr(timestamp, 25, 4)
. list ts
```

	ts
1.	Nov 14 08:59:43 2006
2.	Nov 15 07:36:49 2006
3.	Nov 15 09:21:07 2006
4.	Nov 15 14:57:36 2006
5.	Nov 16 08:22:53 2006
6.	Nov 16 08:36:44 2006

New variable `ts` can be translated using `clock(ts, "MD hms Y")`. `"MD hms Y"` specifies that the order of the components in `ts` is month, day, hours, minutes, seconds, and year. There is no meaning to the spaces; we could just as well have specified `clock(ts, "MDhmsY")`. You can specify spaces when they help to make what you type more readable.

Because %tc values can be so large, whenever you use the function `clock()`, you must store the results in a `double`, as we do below:

```
. gen double dt = clock(ts, "MD hms Y")
. list id dt action
```

	id	dt	action
1.	1001	1.479e+12	15
2.	1002	1.479e+12	15
3.	1003	1.479e+12	11
4.	1002	1.479e+12	16
5.	1005	1.479e+12	12
6.	1001	1.479e+12	16

Don't panic. New variable `dt` contains numeric values, and large ones, which is why it was so important that we stored the values as `doubles`. That output above just shows us what a %tc variable looks like with default formatting. If we wanted to see the numeric values better, we could change `dt` to have a `%20.0gc` format. We would then see that the first value is 1,479,113,983,000, the second 1,479,195,409,000, and so on. We will not do that. Instead, we will put a `%tc` format on our `%tc` variable:

```
. format dt %tc

. list id dt action
```

	id	dt	action
1.	1001	14nov2006 08:59:43	15
2.	1002	15nov2006 07:36:49	15
3.	1003	15nov2006 09:21:07	11
4.	1002	15nov2006 14:57:36	16
5.	1005	16nov2006 08:22:53	12
6.	1001	16nov2006 08:36:44	16

Variable `dt` is a variable we can use. Say we wanted to know how many hours it had been since the previous action:

```
. sort dt

. gen hours = hours(dt - dt[_n-1])
(1 missing value generated)

. format hours %9.2f

. list id dt action hours
```

	id	dt	action	hours
1.	1001	14nov2006 08:59:43	15	.
2.	1002	15nov2006 07:36:49	15	22.62
3.	1003	15nov2006 09:21:07	11	1.74
4.	1002	15nov2006 14:57:36	16	5.61
5.	1005	16nov2006 08:22:53	12	17.42
6.	1001	16nov2006 08:36:44	16	0.23

We subtracted `dt` from its previous value, which results in the number of milliseconds. Converting milliseconds to hours is easy enough; we just have to divide by $60 \times 60 \times 1{,}000 = 3{,}600{,}000$. It is easy to forget or mistype that constant, so we used Stata's `hours()` function, which converts milliseconds to hours. `hours()`, and other useful functions, are documented in [D] **dates and times**.

24.3   Displaying dates and times

A `%td` variable should have a `%td` format, a `%tc` variable should have a `%tc` format, and so on for all the other `%t` variable–format pairs. You do that by typing `format varname %td`, `format varname %tc`, etc.

Formats `%tc`, `%tC`, `%td`, `%tw`, `%tm`, `%tq`, `%th`, and `%ty` are called the default `%t` formats. By specifying codes following them, you can control how the variable is to be displayed.

In the previous example, we started with a string variable that contained a time stamp and looked like "Tue Nov 14 08:59:43 CST 2006". After creating a `%tc` variable from it, and putting the default `%tc` format on it, our date/times looked like "14nov2006 08:59:43". Below we specify a `%tc` format that makes our new variable look just like the original:

```
. format dt %tcDay_Mon_DD_HH:MM:SS_!C!D!T_CCYY
. list id dt action hours
```

	id	dt	action	hours
1.	1001	Tue Nov 14 08:59:43 CDT 2006	15	.
2.	1002	Wed Nov 15 07:36:49 CDT 2006	15	22.62
3.	1003	Wed Nov 15 09:21:07 CDT 2006	11	1.74
4.	1002	Wed Nov 15 14:57:36 CDT 2006	16	5.61
5.	1005	Thu Nov 16 08:22:53 CDT 2006	12	17.42
6.	1001	Thu Nov 16 08:36:44 CDT 2006	16	0.23

%t display formats are documented under *Formatting date and time values* in [D] **dates and times**.

24.4   Typing dates and times (date/time literals)

You will sometimes need to type dates and times in expressions. When we needed to calculate the age of subjects as of 01jan2000 in a previous example, for instance, we typed,

```
. gen age2000 = (td(1jan2000)-birthday)/365.25
```

although we could just as well have typed,

```
. gen age2000 = (14610-birthday)/365.25
```

because 14,610 is the %td value of 01jan2000. Typing td(1jan2000) is easier and less error prone.

Similarly, if we needed 10:55 a.m. on 01jan1960 as a %tc value, rather than typing 39,300,000, we could type tc(01jan1960 10:55). See *Typing dates and times* in [D] **dates and times** for details.

24.5   Extracting components of dates and times

Once you have a %t variable, you can use the extraction functions to obtain components of the variable. For instance, the following functions are appropriate for use with %td variables:

year(*date*)	returns four-digit year; e.g., 1980, 2002
month(*date*)	returns month; 1, 2, ..., 12
day(*date*)	returns day within month; 1, 2, ..., 31
halfyear(*date*)	returns the half of year; 1 or 2
quarter(*date*)	returns quarter of year; 1, 2, 3, or 4
week(*date*)	returns week of year; 1, 2, ..., 52
dow(*date*)	returns day of week; 0, 1, ..., 6; 0 = Sunday
doy(*date*)	returns day of year; 1, 2, ..., 366

There are other functions useful with other %t variables. See *Extracting date and time components* in [D] **dates and times**.

24.6   Converting between date and time values

You can convert between %t values. For instance, function cofd() converts a %td value to a %tc value. cofd() of 17,126 (21nov2006) returns 1,479,686,400,000 (21nov2006 00:00:00). Function dofc() of 1,479,736,920,000 (21nov2006 14:02) returns 17,126 (21nov2006).

There are other functions for converting between other %t values; see *Converting date and time values* in [D] **dates and times**.

24.7 For prior users: What's new in Stata 10 and what bites

If you used prior versions of Stata, pay particular attention to item 7.

1. The %tc and %tC values are new.

2. Many new date and time functions join the previously existing functions.

3. What used to be called elapsed dates are now called %td values.

4. The %td display format shorthand %d is undocumented but continues to work. Modern usage is %td.

5. The codes used after %t (and the old %d) display formats to control how results are displayed have changed. The old codes are undocumented but continue to work.

6. Function date(*string*, *mask* [, *baseyear*]) is greatly improved. For one, is will now translate run-together strings such as "11212006" and more. See *String-to-numeric translation functions* under *Remarks* in [D] **dates and times**.

7. In date(*string*, *mask* [, *baseyear*]), the mask used to be a permutation of the letters m, d, and y. It is now a permutation of M, D, and Y. You must capitalize or you will get missing values. (m now stands for minutes.)

8. Because of the syntax change, and because the new date() will translate many strings that the old date() would not, function date() is under version control. If you set version 9.2 or earlier, it is the old date() function that runs.

24.8 Reference

Cox, N. J. 2006. Speaking Stata: Time of day. *Stata Journal* 6: 124–137.

25 Dealing with categorical variables

25.1   Continuous, categorical, and indicator variables

Although to Stata a variable is a variable, it is helpful to distinguish among three conceptual types:

- A *continuous variable* measures something. Such a variable might measure, for instance, a person's age, height, or weight; a city's population or land area; or a company's revenues or costs.

- A *categorical variable* identifies a group to which the thing belongs. For example, you could categorize persons according to their race or ethnicity, cities according to their geographic location, or companies according to their industry. Often, but not always, categorical variables are stored as strings.

- An *indicator variable* denotes whether something is true. For example, is a person a veteran, does a city have a mass transit system, or is a company profitable?

Indicator variables are a special case of categorical variables. Consider a variable that records a person's sex. Examined one way, it is a categorical variable. A categorical variable identifies the group to which a thing belongs, and here the thing is a person and the basis for categorization is anatomy. Looked at another way, however, it is an indicator variable. It indicates whether a person is, say, female.

In fact, we can use the same logic on any categorical variable that divides the data into two groups. It is a categorical variable since it identifies whether an observation is a member of this or that group; it is an indicator variable since it denotes the truth value of the statement "the observation is in this group".

All indicator variables are categorical variables, but the opposite is not true. A categorical variable might divide the data into more than two groups. For clarity, let's reserve the term *categorical variable* for variables that divide the data into more than two groups, and let's use the term *indicator variable* for categorical variables that divide the data into exactly two groups.

Stata can convert continuous variables to categorical and indicator variables and categorical variables to indicator variables.

25.1.1   Converting continuous variables to indicator variables

Stata treats logical expressions as taking on the values *true* or *false*, which it identifies with the numbers 1 and 0; see [U] **13 Functions and expressions**. For instance, if you have a continuous variable measuring a person's age and you wish to create an indicator variable denoting persons aged 21 and over, you could type

```
. generate age21p = age>=21
```

The variable `age21p` takes on the value 1 for persons aged 21 and over and 0 for persons under 21.

Since `age21p` can take on only 0 or 1, it would be more economical to store the variable as a `byte`. Thus, it would be better to type

```
. generate byte age21p = age>=21
```

This solution has a problem. The value of `age21` is set to 1 for all persons whose `age` is missing because Stata defines missing to be larger than all other numbers. In our data, we have no such missing ages, but it still would have been safer to type

```
. generate byte age21p = age>=21 if age<.
```

That way, persons whose age is missing would also have a missing `age21p`.

❑ Technical Note

Put aside missing values and consider the following alternative to `generate age21p = age>=21` that may have occurred to you:

```
. generate age21p = 1 if age>=21
```

That does not produce the desired result. This statement makes `age21p` 1 (*true*) for all persons aged 21 and above but makes `age21p` missing for everyone else.

If you followed this second approach, you would have to combine it with

```
. replace age21p = 0 if age<21
```

to make the result identical to that produced by the single statement `gen age21p = age>=21`.

❑

25.1.2   Converting continuous variables to categorical variables

Suppose that you wish to categorize persons into four groups on the basis of their age. You want a variable to denote whether a person is 21 or under, between 22 and 38, between 39 and 64, or 65 and above. Although most people would label these categories 1, 2, 3, and 4, there is really no reason to restrict ourselves to such a meaningless numbering scheme. Let's call this new variable `agecat` and make it so that it takes on the topmost value for each group. Thus, persons in the first group will be identified with an `agecat` of 21, persons in the second with 38, persons in the third with 64, and persons in the last (drawing a number out of the air) with 75. Here is one way that will work, but it is not the best method for doing so:

```
. use http://www.stata-press.com/data/r10/agexmpl

. generate byte agecat=21 if age<=21
(176 missing values generated)

. replace agecat=38 if age>21 & age<=38
(148 real changes made)
```

```
. replace agecat=64 if age>38 & age<=64
(24 real changes made)

. replace agecat=75 if age>64 & age<.
(4 real changes made)
```

We mechanically created the categorical variable according to the definition by using the `generate` and `replace` commands. The only thing that deserves comment is the opening `generate`. We (wisely) told Stata to `generate` the new variable `agecat` as a byte, thus conserving memory.

We can create the same result with one command using the `recode()` function:

```
. use http://www.stata-press.com/data/r10/agexmpl, clear
. generate byte agecat=recode(age,21,38,64,75)
```

`recode()` takes three or more arguments. It examines the first argument (here `age`) against the remaining arguments in the list. It returns the first element in the list that is greater than or equal to the first argument or, failing that, the last argument in the list. Thus, for each observation, `recode()` asked if `age` was less than or equal to 21. If so, the value is 21. If not, is it less than or equal to 38? If so, the value is 38. If not, is it less than or equal to 64? If so, the value is 64. If not, the value is 75.

Most researchers typically make tables of categorical variables, so we will `tabulate` the result:

```
. tabulate agecat
```

agecat	Freq.	Percent	Cum.
21	28	13.73	13.73
38	148	72.55	86.27
64	24	11.76	98.04
75	4	1.96	100.00
Total	204	100.00	

There is another way to convert continuous variables into categorical variables, and it is even more automated: `autocode()` works like `recode()`, except that all you tell the function is the range and the total number of cells that you want that range broken into:

```
. use http://www.stata-press.com/data/r10/agexmpl, clear
. generate agecat=autocode(age,4,18,65)
. tabulate agecat
```

agecat	Freq.	Percent	Cum.
29.75	82	40.20	40.20
41.5	96	47.06	87.25
53.25	16	7.84	95.10
65	10	4.90	100.00
Total	204	100.00	

In one instruction, we told Stata to break `age` into four evenly spaced categories from 18 to 65. When we `tabulate agecat`, we see the result. In particular, we see that the breakpoints of the four categories are 29.75, 41.5, 53.25, and 65. The first category contains everyone aged 29.75 years or less; the second category contains persons over 29.75 who are 41.5 years old or less; the third category contains persons over 41.5 who are 53.25 years old or less; and the last category contains all persons over 53.25.

❏ Technical Note

We chose the range 18–65 arbitrarily. Although you cannot tell from the table above, there are persons in this dataset who are under 18, and there are persons over 65. Those persons are counted in the first and last cells, but we have not divided the age range in the data evenly. We could split the full age range into four categories by obtaining the overall minimum and maximum ages (by typing summarize) and substituting the overall minimum and maximum for the 18 and 65 in the autocode() function:

```
. use http://www.stata-press.com/data/r10/agexmpl, clear
. summarize age
```

Variable	Obs	Mean	Std. Dev.	Min	Max
age	204	31.57353	10.28986	2	66

```
. generate agecat2=autocode(age,4,2,66)
```

We could also sort the data into ascending order of age and tell Stata to construct four categories over the range age[1] (the minimum) to age[_N] (the maximum):

```
. use http://www.stata-press.com/data/r10/agexmpl, clear
. sort age
. generate agecat2=autocode(age,4,age[1],age[_N])
. tabulate agecat2
```

agecat2	Freq.	Percent	Cum.
18	10	4.90	4.90
34	138	67.65	72.55
50	41	20.10	92.65
66	15	7.35	100.00
Total	204	100.00	

❏

25.1.3   Converting categorical variables to indicator variables

The easiest way to convert categorical variables to indicator variables is to use the xi command, which will construct indicator variables on the fly. Here we use xi, an extremely useful command, with the logistic command; grp is a variable taking on values 1, 2, and 3:

```
. use http://www.stata-press.com/data/r10/gxmpl10, clear
. xi: logistic outcome i.grp age
i.grp           _Igrp_1-3          (naturally coded; _Igrp_1 omitted)
```

Logistic regression				Number of obs	=	189
				LR chi2(3)	=	6.54
				Prob > chi2	=	0.0879
Log likelihood = -114.06375				Pseudo R2	=	0.0279

outcome	Odds Ratio	Std. Err.	z	P>|z|	[95% Conf. Interval]	
_Igrp_2	2.106974	.9932407	1.58	0.114	.8363679	5.307878
_Igrp_3	1.767748	.6229325	1.62	0.106	.8860686	3.526738
age	.9612592	.0311206	-1.22	0.222	.9021588	1.024231

See [R] **xi**.

At other times, we will want to convert categorical variables to indicator variables permanently, so let's consider how to do that.

We should ask ourselves how this variable is stored. Is it a set of numbers, with different numbers reflecting the different categories, or is it a string? Things will be easier if it is numeric, so if it is not, we use encode to convert it; see [U] **23.2 Categorical string variables**. Making categorical variables numeric is not really necessary, but it is a good thing to do because numeric variables can be stored more compactly than string variables. More importantly, all of Stata's statistical commands know how to deal with numeric variables; some do not know what to make of a string.

Let's suppose that you have a categorical variable that divides your data into four groups. To make matters concrete, we will assume that an observation in your data is a state and that the categorical variable denotes the geographical region for each state. Each state is in one of the four census regions known as the Northeast, North Central, South, and West.

Typing one command will create four new variables, the first indicating whether the state is in the North Central, the second whether the state is in the Northeast, and so on. Such variables are sometimes called *dummy* variables, and you can use them in regressions to control for the effects of, for instance, geographic region.

Here is the dataset before we type this miraculous command:

```
. use http://www.stata-press.com/data/r10/states3
(State data)

. describe

Contains data from http://www.stata-press.com/data/r10/states3.dta
  obs:            50                          State data
  vars:            6                          17 Mar 2007 03:19
  size:         1,700 (99.8% of memory free)  (_dta has notes)

              storage   display    value
variable name   type    format     label     variable label

state          str8     %9s
reg            int      %8.0g      reg        Census Region
median_age     float    %9.0g                 Median Age
marriage_rate  long     %12.0g                Marriages per 100,000
divorce_rate   long     %12.0g                Divorces per 100,000
region         str8     %9s                   Census Region

Sorted by:  reg

. label list reg
reg:
           1 N. Centr
           2 N. East
           3 South
           4 West
```

reg is the categorical variable, and in our example, it is numeric, although that is not important for what we are about to do. The regions are numbered 1–4, and a value label, also named reg, maps those numbers into the words: N. Centr, N. East, South, and West.

(Continued on next page)

We can make the four indicator variables from this categorical variable by typing

```
. tabulate reg, generate(reg)
```

Census Region	Freq.	Percent	Cum.
N. Centr	12	24.00	24.00
N. East	9	18.00	42.00
South	16	32.00	74.00
West	13	26.00	100.00
Total	50	100.00	

```
. describe
Contains data from http://www.stata-press.com/data/r10/states3.dta
  obs:            50                          State data
  vars:           10                          17 Mar 2007 03:19
  size:        1,900 (99.8% of memory free)   (_dta has notes)
```

variable name	storage type	display format	value label	variable label
state	str8	%9s		
reg	int	%8.0g	reg	Census Region
median_age	float	%9.0g		Median Age
marriage_rate	long	%12.0g		Marriages per 100,000
divorce_rate	long	%12.0g		Divorces per 100,000
region	str8	%9s		Census Region
reg1	byte	%8.0g		reg==N. Centr
reg2	byte	%8.0g		reg==N. East
reg3	byte	%8.0g		reg==South
reg4	byte	%8.0g		reg==West

```
Sorted by:  reg
    Note:  dataset has changed since last saved
```

Typing `tabulate reg, generate(reg)` produced a table of the number of states in each region (which is, after all, what `tabulate` does), and because we specified the `generate()` option, it silently created four new variables—one for each line of the table.

Describing the data, we see that there are four new variables called `reg1`, `reg2`, `reg3`, and `reg4`. They are called this because we said `generate(reg)`. If we had said `tabulate reg, gen(junk)`, they would have been called `junk1`, `junk2`, `junk3`, and `junk4`.

Each new variable is stored as a `byte`, and each has been automatically labeled for us. The variable `reg1` takes on the value 1 if the state is in the North Central and 0 otherwise. (It also takes on the value "missing" if `reg` is missing for the observation, which never occurs in our data.)

Just to be clear about the relationship of `reg1` to `reg`, here is a tabulation:

```
. tabulate reg reg1
```

Census Region	reg==N. Centr 0	1	Total
N. Centr	0	12	12
N. East	9	0	9
South	16	0	16
West	13	0	13
Total	38	12	50

If `reg1` is 1, the region is North Central, and vice versa.

25.2   Using indicator variables in estimation

Indicator variables allow you to control for the effects of a variable in a regression. Using the indicator variables we generated in the previous example, we can control for region in the following regression:

$$y_j = \beta_0 + \beta_1 age_j + \beta_2 \delta_{2j} + \beta_3 \delta_{3j} + \beta_4 \delta_{4j} + \epsilon_j$$

where y_j represents the marriage rate in state j, age_j represents the median age of the state's population, and δ_{ij} is 1 if state j is in region i and 0 otherwise. We also eliminate the state of Nevada from our regression.

```
. regress marriage_rate median_age reg1 reg3 reg4 if state!="NEVADA"
```

Source	SS	df	MS		Number of obs =	49
					F(4, 44) =	6.16
Model	8181.65436	4	2045.41359		Prob > F =	0.0005
Residual	14615.8967	44	332.179469		R-squared =	0.3589
					Adj R-squared =	0.3006
Total	22797.551	48	474.94898		Root MSE =	18.226

marriage_r~e	Coef.	Std. Err.	t	P>|t|	[95% Conf. Interval]	
median_age	-4.256985	1.919575	-2.22	0.032	-8.125634	-.3883361
reg1	3.86654	8.680097	0.45	0.658	-13.62705	21.36013
reg3	20.03647	8.202182	2.44	0.019	3.506055	36.56688
reg4	17.15676	10.00958	1.71	0.094	-3.016221	37.32975
_cons	220.7376	60.26174	3.66	0.001	99.28805	342.1872

We see from the results above that the marriage rate, after controlling for age, is significantly higher in region 3, the South.

25.2.1   Testing the significance of indicator variables

After seeing these results, you might wonder if region, taken as a whole, significantly contributes to the explanatory power of the regression. We can find out by using the test command:

```
. test reg1=0

 ( 1)  reg1 = 0

       F(  1,    44) =    0.20
            Prob > F =    0.6582
. test reg3=0, accumulate

 ( 1)  reg1 = 0
 ( 2)  reg3 = 0

       F(  2,    44) =    4.20
            Prob > F =    0.0214
. test reg4=0, accumulate

 ( 1)  reg1 = 0
 ( 2)  reg3 = 0
 ( 3)  reg4 = 0

       F(  3,    44) =    2.97
            Prob > F =    0.0420
```

We typed three commands. The first, test reg1=0, tested the coefficient on the variable reg1 against 0. The resulting F test showed the same significance level as the corresponding t test presented in the regression output.

We next typed `test reg3=0, accumulate`, which tests whether `reg3` is zero and accumulates that test with any previous tests. Thus we are now testing the hypothesis that `reg1` and `reg3` are jointly zero. The F statistic is 4.20, and the result is significant at the 2.1% level; thus, at any significance level above 2.1%, it appears that `reg1` and `reg3` are not both zero.

We finally typed `test reg4=0, accumulate`. As before, this command tests the newly introduced constraint that `reg4` is zero and accumulates that with the previous tests. We are now testing whether `reg1`, `reg3`, and `reg4` are jointly zero. The F statistic is 2.97, and its significance level is roughly 4.2%; thus, at the 5% level, we can reject the hypothesis that, taken together, region has no effect on the marriage rate after controlling for age.

Stata's `test` command has a shorthand for tests that two or more variables are simultaneously equal to zero. Type `test` followed by the names of the variables:

```
. test reg1 reg3 reg4
 ( 1)   reg1 = 0
 ( 2)   reg3 = 0
 ( 3)   reg4 = 0
       F(  3,     44) =    2.97
             Prob > F =    0.0420
```

❏ Technical Note

Sometimes tests of this kind are embedded in an ANOVA or ANCOVA model. In the language of ANOVA, we are doing a one-way layout after controlling for the effect of age. Hence, we did not have to fit the model with `regress`: we could have used Stata's `anova` command. We would have typed `anova marriage_rate median_age reg, continuous(median_age)` to obtain the ANOVA table and the desired test directly. We could have seen the underlying regression after estimation of the model by then typing `regress` without arguments. Typing any estimation command without arguments is taken as a request to reshow the last estimation results. Since all of Stata's estimation commands are tightly coupled, after fitting an ANOVA or ANCOVA model, you can ask `regress` to show you the underlying regression coefficients.

❏

25.2.2 Importance of omitting one of the indicators

Some people prefer to fit models with dummy variables in the context of ANOVA, and others prefer the regression context. The choice is yours.

If you opt for the regression method, however, remember to leave one of the indicator variables out of the regression so that the coefficients have the interpretation of changes from a base group. If you fail to follow that advice, your regression will still be correct, but you must understand what it is you are estimating and testing.

In the example above, for instance, we omitted `reg2`, the dummy for the Northeast. Let's rerun that regression and include the `reg2` dummy:

```
. regress marriage_rate median_age reg1-reg4 if state!="NEVADA"
```

Source	SS	df	MS		
Model	8181.65436	4	2045.41359		
Residual	14615.8967	44	332.179469		
Total	22797.551	48	474.94898		

				Number of obs =	49
				F(4, 44) =	6.16
				Prob > F =	0.0005
				R-squared =	0.3589
				Adj R-squared =	0.3006
				Root MSE =	18.226

marriage_r~e	Coef.	Std. Err.	t	P>|t|	[95% Conf. Interval]	
median_age	-4.256985	1.919575	-2.22	0.032	-8.125634	-.3883361
reg1	-13.29022	7.911093	-1.68	0.100	-29.23398	2.653538
reg2	-17.15676	10.00958	-1.71	0.094	-37.32975	3.016221
reg3	2.879704	7.527591	0.38	0.704	-12.29116	18.05057
reg4	(dropped)					
_cons	237.8944	54.2438	4.39	0.000	128.5732	347.2156

If you compare the top half of the regression output with that of the previous example, you will find that they are identical. You will also find that the estimated coefficient, standard error, and related statistics for the variable median_age are identical, yet the estimates for each of the reg variables are different.

You will also notice that the models have been fitted on different variables. In the first case, dummies for regions 1, 3, and 4 were included, along with a constant term. In the latter case, dummies for regions 1, 2, and 3 were included along with a constant term, but regress mysteriously dropped the dummy for region 4.

With a constant term in the model, one of the four region dummy variables is unnecessary. The first model says that we can predict marriage_rate as

$$220.7376 - 4.256985\text{age} + c_i$$

where c_i depends on the region. For example, for region 1 $c_1 = 3.86654$. For region 2, we use $c_2 = 0$, since the dummy for region 2 was omitted from the regression. The second model says that we can predict marriage_rate as

$$237.8944 - 4.256985\text{age} + c_i'$$

where now for region 1 $c_1' = -13.29022$ and for region 4 $c_4' = 0$.

Say that we want to predict marriage_rate for region 4. Using the first model, we have

$$220.7376 - 4.2569845\text{age} + 17.1568 = 237.8944 - 4.2569845\text{age}$$

which is the same prediction we would get using the second model.

One difference between the two models is the interpretation of the coefficients on the dummy variables. In the first model, the coefficients for regions 1, 3, and 4 measure the difference between each of those regions and region 2. In the second model, the coefficients for regions 1, 2, and 3 compare those regions to region 4.

The second difference concerns the t statistics on the coefficients for the region dummy variables. For example, in the first model, the t statistic for reg1 tests whether there is a significant difference between marriage rates in regions 1 and 2 since region 2's dummy variable was omitted from the regression. In the second model, the t statistic for reg1 tests whether there is a significant difference between marriage rates in regions 1 and 4. Thus the t statistics are testing different hypotheses.

26 Overview of Stata estimation commands

Contents

26.1   Introduction

Estimation commands fit models such as linear regression and probit. Stata has many such commands, so it is easy to overlook a few. Some of these commands differ greatly from each other, others are gentle variations on a theme, and still others are equivalent to each other.

Estimation commands share features that this chapter will not discuss; see [U] **20 Estimation and postestimation commands**. Especially see [U] **20.15 Obtaining robust variance estimates**, which discusses an alternative calculation for the estimated variance matrix (and hence standard errors) that many of Stata's estimation commands provide, and [U] **20.11 Performing hypothesis tests on the coefficients**.

Here, however, this chapter will put aside all of that—and all issues of syntax—and deal solely with matching commands to their statistical concepts. Nor will it cross-reference specific commands. To find the details on a particular command, look up its name in the index.

26.2   Linear regression with simple error structures

Consider models of the form

$$y_j = \mathbf{x}_j \boldsymbol{\beta} + \epsilon_j$$

for a continuous y variable. In this category, estimation is restricted to when σ_ϵ^2 is constant across observations j. The model is called the linear regression model, and the estimator is often called the (ordinary) least-squares estimator.

regress is Stata's linear regression command. (regress produces the robust estimate of variance as well as the conventional estimate, and regress has a collection of commands that can be run after it to explore the nature of the fit.)

Also, the following commands will do linear regressions, as does regress, but offer special features:

1. ivregress fits models in which some of the regressors are endogenous, using either instrumental variables or generalized method-of-moments (GMM) estimators.

2. areg fits models $y_j = \mathbf{x}_j \boldsymbol{\beta} + \mathbf{d}_j \boldsymbol{\gamma} + \epsilon_j$, where $\mathbf{d}_j$ is a mutually exclusive and exhaustive dummy variable set. areg obtains estimates of $\boldsymbol{\beta}$ (and associated statistics) without ever forming $\mathbf{d}_j$, meaning that it also does not report the estimated γ. If your interest is in fitting fixed-effects models, Stata has a better command—xtreg—discussed in [U] **26.15.1 Linear regression with panel data** below. Most users who find areg appealing will probably want to use xtreg because it provides more useful summary and test statistics. areg duplicates the output that regress would produce if you were to generate all the dummy variables. This means, for instance, that the reported R^2 includes the effect of γ.

3. boxcox obtains maximum likelihood estimates of the coefficients and the Box–Cox transform parameters in a model of the form

$$y_i^{(\theta)} = \beta_0 + \beta_1 x_{i1}^{(\lambda)} + \beta_2 x_{i2}^{(\lambda)} + \cdots + \beta_k x_{ik}^{(\lambda)} + \gamma_1 z_{i1} + \gamma_2 z_{i2} + \cdots + \gamma_l z_{il} + \epsilon_i$$

where $\epsilon \sim N(0, \sigma^2)$. Here the *depvar* y is subject to a Box–Cox transform with parameter θ. Each of the *indepvars* $x_1, x_2, \ldots, x_k$ is transformed by a Box–Cox transform with parameter λ. The $z_1, z_2, \ldots, z_l$ are independent variables that are not transformed. In addition to the general form specified above, boxcox can fit three other versions of this model defined by the restrictions $\lambda = \theta$, $\lambda = 1$, and $\theta = 1$.

4. tobit allows estimation of linear regression models when y_i has been subject to left-censoring, right-censoring, or both. Say that y_i is not observed if $y_i < 1{,}000$, but for those observations, it is known that $y_i < 1{,}000$. tobit fits such models.

   ivtobit does the same but allows for endogenous regressors.

5. cnreg (censored-normal regression) is a generalization of tobit. The lower and upper censoring points, rather than being constants, are allowed to vary observation by observation. cnreg can fit any model tobit can fit.

6. intreg (interval regression) is a generalization of cnreg. In addition to allowing open-ended intervals, intreg allows closed intervals. Rather than observing y_j, it is assumed that y_{0j} and y_{1j} are observed, where $y_{0j} \leq y_j \leq y_{1j}$. Survey data might report that a subject's monthly income was in the range \$1,500–\$2,500. intreg allows such data to be used to fit a regression model. intreg allows $y_{0j} = y_{1j}$ and so can reproduce results reported by regress. intreg allows y_{0j} to be $-\infty$ and y_{1j} to be $+\infty$ and so can reproduce results reported by cnreg and tobit.

7. truncreg fits the regression model when the sample is drawn from a restricted part of the population and so is similar to tobit, except that here the independent variables are not observed. Under the normality assumption for the whole population, the error terms in the truncated regression model have a truncated-normal distribution.

8. cnsreg allows you to place linear constraints on the coefficients.

9. eivreg adjusts estimates for errors in variables.

10. nl provides the nonlinear least-squares estimator of $y_j = f(\mathbf{x}_j, \boldsymbol{\beta}) + \epsilon_j$.

11. rreg fits robust regression models, which are not to be confused with regression with robust standard errors. Robust standard errors are discussed in [U] **20.15 Obtaining robust variance estimates**. Robust regression concerns point estimates more than standard errors, and it implements a data-dependent method for downweighting outliers.

12. qreg produces quantile-regression estimates, a variation that is not linear regression at all but is an estimator of $y_j = \mathbf{x}_j\boldsymbol{\beta} + \epsilon_j$. In the basic form of this model, sometimes called median regression, $\mathbf{x}_j\boldsymbol{\beta}$ measures not the predicted mean of y_j conditional on $\mathbf{x}_j$, but its median. As such, qreg is of most interest when ϵ_j does not have constant variance. qreg allows you to specify the quantile, so you can produce linear estimates for the predicted 1st, 2nd, ..., 99th percentile.

Another command, bsqreg, is identical to qreg but presents bootstrapped standard errors.

The sqreg command estimates multiple quantiles simultaneously; standard errors are obtained via the bootstrap.

The iqreg command estimates the difference between two quantiles; standard errors are obtained via the bootstrap.

13. vwls (variance-weighted least squares) produces estimates of $y_j = \mathbf{x}_j\boldsymbol{\beta} + \epsilon_j$, where the variance of ϵ_j is calculated from group data or is known a priori. vwls is therefore of most interest to categorical-data analysts and physical scientists.

26.3   ANOVA, ANCOVA, MANOVA, and MANCOVA

ANOVA and ANCOVA are related to linear regression, but we classify them separately. The related Stata commands are anova, oneway, and loneway. The manova command provides MANOVA and MANCOVA (multivariate ANOVA and ANCOVA).

anova fits ANOVA and ANCOVA models, one-way and up—including two-way factorial, three-way factorial, etc.—and fits nested and mixed-design models and repeated-measures models.

oneway fits one-way ANOVA models. It is quicker at producing estimates than anova, although anova is so fast that this probably does not matter. The important difference is that oneway can report multiple-comparison tests.

loneway is an alternative to oneway. The results are numerically the same, but loneway can deal with more levels (limited only by dataset size; oneway is limited to 376 levels and anova to 798, but for anova to reach 798 requires a lot of memory), and loneway reports some additional statistics, such as the intraclass correlation.

manova fits MANOVA and MANCOVA models, one-way and up—including two-way factorial, three-way factorial, etc.—and it fits nested and mixed-design models.

26.4   Generalized linear models

The generalized linear model is

$$g\{E(y_j)\} = \mathbf{x}_j\boldsymbol{\beta}, \qquad y_j \sim F$$

where $g()$ is called the link function and F is a member of the exponential family, both of which you specify before estimation. glm fits this model.

The GLM framework encompasses a surprising array of models known by other names, including linear regression, Poisson regression, exponential regression, and others. Stata provides dedicated estimation commands for many of these. Stata has, for instance, regress for linear regression, poisson for Poisson regression, and streg for exponential regression, and that is not all of the overlap.

glm by default uses maximum likelihood estimation and alternatively estimates via iterated reweighted least squares (IRLS) when the irls option is specified. For each family F there is a corresponding link function $g()$, called the canonical link, for which IRLS estimation produces results identical to maximum likelihood estimation. You can, however, match families and link functions as you wish, and, when you match a family to a link function other than the canonical link, you obtain a different but valid estimator of the standard errors of the regression coefficients. The estimator you obtain is asymptotically equivalent to the maximum likelihood estimator, which, in small samples, produces slightly different results.

For example, the canonical link for the binomial family is logit. glm, irls with that combination produces results identical to the maximum-likelihood logit (and logistic) command. The binomial family with the probit link produces the probit model, but probit is not the canonical link here. Hence, glm, irls produces standard error estimates that differ slightly from those produced by Stata's maximum-likelihood probit command.

Many researchers feel that the maximum-likelihood standard errors are preferable to IRLS estimates (when they are not identical), but they would have a difficult time justifying that feeling. Maximum likelihood probit is an estimator with (solely) asymptotic properties; glm, irls with the binomial family and probit link is an estimator with (solely) asymptotic properties, and in finite samples, the standard errors differ a little.

Still, we recommend that you use Stata's dedicated estimators whenever possible. IRLS—the theory—and glm, irls—the command—are all encompassing in their generality, meaning that they rarely use the right jargon or provide things in the way you wish they would. The narrower commands, such as logit, probit, and poisson, focus on the issue at hand and are invariably more convenient.

glm is useful when you want to match a family to a link function that is not provided elsewhere.

glm also offers several estimators of the variance–covariance matrix that are consistent, even when the errors are heteroskedastic or autocorrelated. Another advantage of a glm version of a model over a model-specific version is that many of these VCE estimators are available only for the glm implementation. You can also obtain the ML–based estimates of the VCE from glm.

26.5   Binary-outcome qualitative dependent variable models

There are many ways to write these models, one of which is

$$\Pr(y_j \neq 0) = F(\mathbf{x}_j\boldsymbol{\beta})$$

where F is some cumulative distribution. Two popular choices for $F()$ are the normal and logistic, and the models are called the probit and logit (or logistic regression) models. A third is the complementary log–log function; maximum likelihood estimates are obtained by Stata's cloglog command.

The two parent commands for the maximum likelihood estimator of probit and logit are probit and logit, although logit has a sibling, logistic, that provides the same estimates but displays results in a slightly different way. There is also an exact logistic estimator; see [U] **26.9 Exact estimators**.

Do not read anything into the names logit and logistic. Logit and logistic have two interchanged definitions in two scientific camps. In the medical sciences, logit means the minimum χ^2 estimator, and logistic means maximum likelihood. In the social sciences, it is the other way around. From our experience, it appears that neither reads the other's literature, since both talk (and write books) asserting that logit means one thing and logistic the other. Our solution is to provide both logit and logistic, which do the same thing, so that each camp can latch on to the maximum likelihood command under the name each expects.

There are two slight differences between logit and logistic. logit reports estimates in the coefficient metric, whereas logistic reports exponentiated coefficients—odds ratios. This is in accordance with the expectations of each camp and makes no substantial difference. The other difference is that logistic has a family of postlogistic commands that you can run to explore the nature of the fit. Actually, that is not exactly true because all the commands for use after logistic can also be used after logit.

If you have not already selected logit or logistic as your favorite, we recommend that you try logistic. Logistic regression (logit) models are more easily interpreted in the odds-ratio metric.

In addition to logit and logistic, Stata provides glogit, blogit, and binreg commands.

blogit is the maximum likelihood estimator (the same as logit or logistic) but applied on data organized in a different way. Rather than having individual observations, your data are organized so that each observation records the number of observed successes and failures.

glogit is the weighted-regression, grouped-data estimator.

binreg can be used to model either individual-level or grouped data in an application of the generalized linear model. The family is assumed to be binomial, and each link provides a distinct parameter interpretation. Also, binreg offers several options for setting the link function according to the desired biostatistical interpretation. The available links and interpretation options are

Option	Implied link	Parameter
or	logit	Odds ratios = $\exp(\beta)$
rr	log	Risk ratios = $\exp(\beta)$
hr	log complement	Health ratios = $\exp(\beta)$
rd	identity	Risk differences = β

Related to logit, the skewed logit estimator scobit adds a power to the logit link function and is estimated by Stata's scobit command.

Turning to probit, you have three choices: probit, dprobit, and ivprobit. probit and dprobit are maximum likelihood estimators, and it makes no substantial difference which you use; they differ only in how they report results. probit reports coefficients and dprobit reports changes in probabilities. Many researchers find changes in probabilities easier to interpret.

ivprobit is for use with endogenous regressors.

Stata also provides `bprobit` and `gprobit`. The `bprobit` command is a maximum likelihood estimator—equivalent to `probit` or `dprobit`—but works with data organized in the different way outlined above. `gprobit` is the weighted-regression, grouped-data estimator.

Continuing with probit: `hetprob` fits heteroskedastic probit models. In these models, the variance of the error term is parameterized.

Also, Stata's `biprobit` command fits bivariate probit models, meaning two correlated outcomes. `biprobit` also fits partial-observability models in which only the outcomes $(0,0)$ and $(1,1)$ are observed.

26.6   Conditional logistic regression

`clogit` is Stata's conditional logistic regression estimator. In this model, observations are assumed to be partitioned into groups, and a predetermined number of events occur in each group. The model measures the risk of the event according to the observation's covariates, x_j. The model is used in matched case–control studies (`clogit` allows $1:1$, $1:k$, and $m:k$ matching) and is used in natural experiments whenever observations can be grouped into pools in which a fixed number of events occur.

26.7   Multiple-outcome qualitative dependent variable models

For more than two outcomes, Stata provides ordered logit, ordered probit, rank-ordered logit, multinomial logistic regression, multinomial probit regression, McFadden's choice model (conditional fixed-effects logistic regression), and nested logistic regression.

`oprobit` and `ologit` provide maximum-likelihood ordered probit and logit. These are generalizations of probit and logit models known as the proportional odds model and are used when the outcomes have a natural ordering from low to high. The idea is that there is an unmeasured $z_j = x_j\beta$, and the probability that the kth outcome is observed is $\Pr(c_{k-1} < z_j < c_k)$, where $c_0 = -\infty$, $c_k = +\infty$, and $c_1, \ldots, c_{k-1}$ along with β are estimated from the data.

`rologit` fits the rank-ordered logit model for rankings. This model is also known as the Plackett–Luce model, the exploded logit model, and choice-based conjoint analysis.

`asroprobit` fits the probit model for rankings, a more flexible estimator than `rologit` because `asroprobit` allows covariances among the rankings. `asroprobit` is also similar to `asmprobit` (below), which is used for outcomes that have no natural ordering. The `as` in the name signifies that `asroprobit` also allows alternative-specific regressors—variables that have different coefficients for each alternative.

`slogit` fits the stereotype logit model for data that are not truly ordered, as data are for `ologit`, but for which you are not sure that it is unordered, in which case `mlogit` would be appropriate.

`mlogit` fits maximum-likelihood multinomial logistic models, also known as polytomous logistic regression. It is intended for use when the outcomes have no natural ordering and you know only the characteristics of the outcome chosen (and, perhaps, the chooser).

`asclogit` fits McFadden's choice model, also known as conditional logistic regression. In the context denoted by the name *McFadden's choice model*, the model is used when the outcomes have no natural ordering, just as multinomial logistic regression, but the characteristics of the outcomes chosen and not chosen are known (along with, perhaps, the characteristics of the chooser).

In the context denoted by the name *conditional logistic regression*—mentioned above—subjects are members of pools, and one or more are chosen, typically to be infected by some disease or to have some other unfortunate event befall them. Thus the characteristics of the chosen and not chosen are known, and the issue of the characteristics of the chooser never arises. Either way, it is the same model.

In their choice-model interpretations, `mlogit` and `clogit` assume that the odds ratios are independent of other alternatives, known as the independence of irrelevant alternatives (IIA) assumption. This assumption is often rejected by the data and the nested logit model does not impose this condition. `nlogit` is also popular for fitting the random utility choice model.

`asmprobit` is for use with outcomes that have no natural ordering and with regressors that are alternative specific. It is weakly related to `mlogit`. Unlike `mlogit`, `asmprobit` does not assume the IIA.

`mprobit` is also for use with outcomes that have no natural ordering but with models that do not have alternative-specific regressors.

26.8   Count dependent variable models

These models concern dependent variables that count the number of occurrences of an event. In this category, we include Poisson and negative binomial regression. For the Poisson model,

$$E(\text{count}) = E_j \exp(\mathbf{x}_j \boldsymbol{\beta})$$

where E_j is the exposure time. `poisson` fits this model. There is also an exact Poisson estimator; see [U] **26.9 Exact estimators**.

Negative-binomial regression refers to estimating with data that are a mixture of Poisson counts. One derivation of the negative binomial model is that individual units follow a Poisson regression model but there is an omitted variable that follows a gamma distribution with variance α. Negative-binomial regression estimates β and α. `nbreg` fits such models. A variation on this, unique to Stata, allows you to model α. `gnbreg` fits those models.

Zero inflation refers to count models in which the number of 0 counts is more than would be expected in the regular model, and that is due to there being a probit or logit process that must first generate a positive outcome before the counting process can begin.

Stata's `zip` command fits zero-inflated Poisson models.

Stata's `zinb` command fits zero-inflated negative binomial models.

`ztp` and `ztnb` fit zero-truncated Poisson and negative binomial models. In zero-inflated models, you observe too many zeros, so you fit a separate model to them. In zero-truncated models, you do not observe the zeros.

26.9   Exact estimators

Exact estimators refer to models which, rather than being estimated by asymptotic formulas, are estimated by enumerating the conditional distribution of the sufficient statistics and then computing the maximum likelihood estimate using that distribution. Standard errors cannot be estimated, but confidence intervals can be and are obtained from the enumerations.

`exlogistic` fits logistic models of binary data in this way.

`expoisson` fits Poisson models of count data in this way.

In small samples, exact estimates have better coverage than the asymptotic estimates, and exact estimates are the only way to obtain estimates, tests, and confidence intervals of covariates that perfectly predict the observed outcome.

26.10   Linear regression with heteroskedastic errors

We now consider the model $y_j = \mathbf{x}_j\boldsymbol{\beta} + \epsilon_j$, where the variance of ϵ_j is nonconstant.

First, `regress` can fit such models if you specify the `vce(robust)` option. What Stata calls robust is also known as the White correction for heteroskedasticity.

For scientists who have data where the variance of ϵ_j is known a priori, `vwls` is the command. `vwls` produces estimates for the model given each observation's variance, which is recorded in a variable in the data.

Finally, `qreg` performs quantile regression, which in the presence of heteroskedasticity is most of interest. Median regression (one of `qreg`'s capabilities) is an estimator of $y_j = \mathbf{x}_j\boldsymbol{\beta} + \epsilon_j$ when ϵ_j is heteroskedastic. Even more useful, you can fit models of other quantiles and so model the heteroskedasticity. Also see the `sqreg` and `iqreg` commands; `sqreg` estimates multiple quantiles simultaneously. `iqreg` estimates differences in quantiles.

26.11   Stochastic frontier models

`frontier` fits stochastic production or cost frontier models on cross-sectional data. The model can be expressed as

$$y_i = \mathbf{x}_i\boldsymbol{\beta} + v_i - su_i$$

where

$$s = \begin{cases} 1 & \text{for production functions} \\ -1 & \text{for cost functions} \end{cases}$$

u_i is a nonnegative disturbance standing for technical inefficiency in the production function or cost inefficiency in the cost function. Although the idiosyncratic error term v_i is assumed to have a normal distribution, the inefficiency term is assumed to be one of the three distributions: half-normal, exponential, or truncated-normal. Also, when the nonnegative component of the disturbance is assumed to be either half-normal or exponential, `frontier` can fit models in which the error components are heteroskedastic conditional on a set of covariates. When the nonnegative component of the disturbance is assumed to be from a truncated-normal distribution, `frontier` can also fit a conditional mean model, where the mean of the truncated-normal distribution is modeled as a linear function of a set of covariates.

For panel-data stochastic frontier models, see [U] **26.15.1 Linear regression with panel data**.

26.12   Regression with systems of equations (correlated errors)

By correlated errors, we mean that observations are grouped, and that within group, the observations might be correlated but, across groups, they are uncorrelated. `regress` with the `vce(robust)` option can produce "correct" estimates, i.e., inefficient estimates with correct standard errors and lots of robustness; see [U] **20.15 Obtaining robust variance estimates**. Obviously, if you know the correlation structure (and are not mistaken), you can do better, so `xtreg` and `xtgls` are also of interest here; we discuss them in [U] **26.15.1 Linear regression with panel data** below.

Turning to simultaneous multiple-equation models; Stata can produce three-stage least-squares (3SLS) and two-stage least-squares (2SLS) estimates by using the reg3 and ivregress commands. Two-stage models can be estimated by either reg3 or ivregress. Three-stage models require use of reg3. The reg3 command can produce constrained and unconstrained estimates.

When we have correlated errors across equations but no endogenous right-hand-side variables,

$$y_{1j} = \mathbf{x}_{1j}\boldsymbol{\beta} + \epsilon_{1j}$$
$$y_{2j} = \mathbf{x}_{2j}\boldsymbol{\beta} + \epsilon_{2j}$$
$$\vdots$$
$$y_{mj} = \mathbf{x}_{mj}\boldsymbol{\beta} + \epsilon_{mj}$$

where $\epsilon_{k.}$ and $\epsilon_{l.}$ are correlated with correlation ρ_{kl}, a quantity to be estimated from the data. This is called Zellner's seemingly unrelated regressions, and sureg fits such models. When $\mathbf{x}_{1j} = \mathbf{x}_{2j} = \cdots = \mathbf{x}_{mj}$, the model is known as multivariate regression, and the corresponding command is mvreg.

The equations need not be linear and, then, use the command nlsur.

Estimation in the presence of autocorrelated errors is discussed in [U] **26.14 Models with time-series data**.

26.13   Models with endogenous sample selection

What has become known as the Heckman model refers to linear regression in the presence of sample selection: $y_j = \mathbf{x}_j\boldsymbol{\beta} + \epsilon_j$ is not observed unless some event occurs that itself has probability $p_j = F(\mathbf{z}_j\boldsymbol{\gamma} + \nu_j)$, where ϵ and ν might be correlated and $\mathbf{z}_j$ and $\mathbf{x}_j$ may contain variables in common.

heckman fits such models by maximum likelihood or Heckman's original two-step procedure.

This model has recently been generalized to replace the linear regression equation with another probit equation, and that model is fitted by heckprob.

Another important case of endogenous sample selection is the treatment-effects model, which considers the effect of an endogenously chosen binary treatment on another endogenous, continuous variable, conditional on two sets of independent variables. treatreg fits a treatment-effects model by using either a two-step consistent estimator or full maximum likelihood.

26.14   Models with time-series data

ARIMA refers to models with autoregressive integrated moving-average processes, and Stata's arima command fits models with ARIMA disturbances via the Kalman filter and maximum likelihood. These models may be fitted with or without confounding covariates.

Stata's prais command performs regression with AR(1) disturbances by using the Prais–Winsten or Cochrane–Orcutt transformation. Both two-step and iterative solutions are available, as well as a version of the Hildreth–Lu search procedure. The Prais–Winsten estimates for the model are an improvement over the Cochrane–Orcutt estimates in that the first observation is preserved in the estimation. This is particularly important with trended data in small samples.

prais automatically produces the Durbin–Watson d statistic, which can also be obtained after regress by using estat dwatson.

newey produces linear regression estimates with the Newey–West variance estimates that are robust to heteroskedasticity and autocorrelation of specified order.

Stata provides estimators for regression models with autoregressive conditional heteroskedastic (ARCH) disturbances

$$y_t = \mathbf{x}_t\boldsymbol{\beta} + \mu_t$$

where μ_t is distributed $N(0, \sigma_t^2)$ and σ_t^2 is given by some function of the lagged disturbances.

Stata's arch command provides different parameterizations of the conditional heteroskedasticity, and it also allows ARMA disturbances and/or multiplicative heteroskedasticity.

Stata provides var and svar for fitting vector autoregressive (VAR) and structural vector autoregressive (SVAR) models. See [TS] **var** for information on Stata's suite of commands for forecasting, specification testing, and inference on VAR and SVAR models. Stata also provides vec for fitting vector error-correction models; see [TS] **vec**. See [TS] **irf** for information on Stata's suite of commands for estimating, analyzing, and presenting impulse–response functions and forecast error variance decompositions. There is also a set of commands for performing Granger causality tests, lag-order selection, and residual analysis.

26.15   Panel-data models

26.15.1   Linear regression with panel data

This section could just as well be called "linear regression with complex error structures". Commands in this class begin with the letters xt.

xtreg fits models of the form

$$y_{it} = \mathbf{x}_{it}\boldsymbol{\beta} + \nu_i + \epsilon_{it}$$

xtreg can produce the between-regression estimator, the within-regression (fixed effects) estimator, or the GLS random-effects (matrix-weighted average of between and within results) estimator. It can also produce the maximum-likelihood random-effects estimator.

xtregar can produce the within estimator and a GLS random-effects estimator when the ϵ_{it} are assumed to follow an AR(1) process.

xtivreg contains the between-2SLS estimator, the within-2SLS estimator, the first-differenced-2SLS estimator, and two GLS random-effects-2SLS estimators to handle cases in which some of the covariates are endogenous.

xtmixed is a generalization of xtreg that allows for multiple levels of panels, random coefficients, and variance-component estimation in general. In the xtmixed framework, residuals (random effects) can occur anywhere and have any level of subscript.

xtabond is for use with dynamic panel-data models (models in which there are lagged dependent variables) and can produce the one-step, one-step robust, and two-step Arellano–Bond estimators. xtabond can handle predetermined covariates, and it reports both the Sargan and autocorrelation tests derived by Arellano and Bond.

xtdpdsys is an extension of xtabond and produces estimates with smaller bias when the coefficients of the AR process are large. xtpdsys is also more efficient than xtabond. Whereas xtabond uses moment conditions based on the differenced errors, xtpdsys uses moment conditions based on both the differenced errors and their levels.

xtdpd is an extension of xtdpdsys and can be used to estimate the parameters of a broader class of dynamic panel-data models. xtdpd can be used to fit models with serially correlated idiosyncratic errors, whereas xtdpdsys and xtabond assume no serial correlation. Or xtdpd can be used with models where the structure of the predetermined variables is more complicated than that assumed by xtdpdsys or xtabond.

xtgls produces generalized least-squares estimates for models of the form

$$y_{it} = \mathbf{x}_{it}\beta + \epsilon_{it}$$

where you may specify the variance structure of ϵ_{it}. If you specify that ϵ_{it} is independent for all i and t, xtgls produces the same results as regress up to a small-sample degrees-of-freedom correction applied by regress but not by xtgls.

You may choose among three variance structures concerning i and three concerning t, producing a total of nine different models. Assumptions concerning i deal with heteroskedasticity and cross-sectional correlation. Assumptions concerning t deal with autocorrelation and, more specifically, AR(1) serial correlation.

Alternative methods report the OLS coefficients and a version of the GLS variance–covariance estimator. xtpcse produces panel-corrected standard error (PCSE) estimates for linear cross-sectional time-series models, where the parameters are estimated by OLS or Prais–Winsten regression. When you are computing the standard errors and the variance–covariance estimates, the disturbances are, by default, assumed to be heteroskedastic and contemporaneously correlated across panels.

In the jargon of GLS, the random-effects model fitted by xtreg has exchangeable correlation within i—xtgls does not model this particular correlation structure. xtgee, however, does.

xtgee fits population-averaged models, and it optionally provides robust estimates of variance. Moreover, xtgee allows other correlation structures. One that is of particular interest to those with many data goes by the name *unstructured*. The within-panel correlations are simply estimated in an unconstrained way. [U] **26.15.3 Generalized linear models with panel data** will discuss this estimator further since it is not restricted to linear regression models.

xthtaylor uses instrumental variables estimators to estimate the parameters of panel-data random-effects models of the form

$$y_{it} = \mathbf{X}_{1it}\beta_1 + \mathbf{X}_{2it}\beta_2 + \mathbf{Z}_{1i}\delta_1 + \mathbf{Z}_{2i}\delta_2 + u_i + e_{it}$$

The individual effects u_i are correlated with the explanatory variables $\mathbf{X}_{2it}$ and $\mathbf{Z}_{2i}$ but are uncorrelated with $\mathbf{X}_{1it}$ and $\mathbf{Z}_{1i}$, where $\mathbf{Z}_1$ and $\mathbf{Z}_2$ are constant within panel.

xtfrontier fits stochastic production or cost frontier models for panel data. You may choose from a time-invariant model or a time-varying decay model. In both models, the nonnegative inefficiency term is assumed to have a truncated-normal distribution. In the time-invariant model, the inefficiency term is constant within panels. In the time-varying decay model, the inefficiency term is modeled as a truncated-normal random variable multiplied by a specific function of time. In both models, the idiosyncratic error term is assumed to have a normal distribution. The only panel-specific effect is the random inefficiency term.

26.15.2 Censored linear regression with panel data

xttobit fits random-effects tobit models and generalizes that to observation-specific censoring.

xtintreg performs random-effects interval regression and generalizes that to observation-specific censoring. Interval regression, in addition to allowing open-ended intervals, also allows closed intervals.

26.15.3 Generalized linear models with panel data

[U] **26.4 Generalized linear models** above discussed the model

$$g\{E(y_j)\} = \mathbf{x}_j\boldsymbol{\beta}, \qquad y_j \sim F \tag{1}$$

where $g()$ is the link function and F is a member of the exponential family, both of which you specify before estimation.

There are two approaches to generalizing the above to panel data: one, in which the correlation within panel is modeled in general terms, and the other, in which the intrapanel correlation is itself modeled, and in detail. The former provides estimates of β; the latter provides estimates of β and the variance components.

The first is known as the generalized estimation equation (GEE) approach.

The second is known as the generalized linear mixed model (GLMM).

The corresponding commands in Stata are xtgee for GEE and xtmixed, xtmelogit, and xtmepoisson for GLMM.

26.15.4 Qualitative dependent-variable models with panel data

xtprobit fits random-effects probit regression via maximum likelihood. It also fits population-averaged models via GEE. This last is nothing more than xtgee with the binomial family, probit link, and exchangeable error structure.

xtlogit fits random-effects logistic regression models via maximum likelihood. It also fits conditional fixed-effects models via maximum likelihood. Finally, as with xtprobit, it fits population-averaged models via GEE.

xtcloglog estimates random-effects complementary log-log regression via maximum likelihood. It also fits population-averaged models via GEE.

clogit is also of interest since it provides the conditional fixed-effects logistic estimator.

26.15.5 Count dependent-variable models with panel data

xtpoisson fits two different random-effects Poisson regression models via maximum likelihood. The two distributions for the random effect are gamma and normal. It also fits conditional fixed-effects models, and it fits population-averaged models via GEE. This last is nothing more than xtgee with the Poisson family, log link, and exchangeable error structure.

xtnbreg fits random-effects negative binomial regression models via maximum likelihood (the distribution of the random effects is assumed to be beta). It also fits conditional fixed-effects models, and it fits population-averaged models via GEE.

26.15.6　Random-coefficient models with panel data

xtrc fits Swamy's random-coefficients linear-regression model. In this model, rather than only the intercept varying across groups, all the coefficients are allowed to vary.

26.16　Survival-time (failure time) models

Commands are provided to fit Cox proportional hazards models, as well as several parametric survival models including exponential, Weibull, Gompertz, log-normal, log-logistic, and generalized gamma (see [ST] **stcox** and [ST] **streg**). The commands all allow for right-censoring, left-truncation, gaps in histories, and time-varying regressors. The commands are appropriate for use with single- or multiple-failure-per-subject data. Conventional and robust standard errors are available with and without clustering.

Both the Cox model and the parametric models (as fitted using Stata) allow for two additional generalizations. First, the models may be modified to allow for latent random effects, or *frailties*. Second, the models may be stratified in that the baseline hazard function may vary completely over a set of strata. The parametric models also allow the modeling of ancillary parameters.

stcox and streg require that the data be stset so that the proper response variables may be established. After you stset the data, the response is taken as understood, and you need only supply the regressors (and other options) to stcox and streg.

26.17　Survey data

Stata's svy command fits statistical models for complex survey data. svy is a prefix command, so to obtain linear regression, you type

. svy: regress ...

or to obtain probit regression, you type

. svy: probit ...

but first you must type a svyset command to define the survey design characteristics. Prefix svy works with many estimation commands, and everything is documented together in the *Survey Data Reference Manual*.

svy supports the following variance-estimation methods:

1. Taylor-series linearization
2. Balanced repeated replication (BRR)
3. Jackknife

See [SVY] **variance estimation** for details.

svy supports the following survey design characteristics:

1. With- and without- replacement sampling
2. Sampling weights
3. Stratification
4. Poststratification
5. Clustering
6. Multiple stages of clustering without replacement

7. BRR and jackknife replication weights

See [SVY] **svyset** for details.

Subpopulation estimation is available for all estimation commands.

Tabulations and summary statistics are also available, including means, proportions, ratios, and totals over multiple subpopulations, and direct standardization of means, proportions, and ratios.

26.18   Multivariate and cluster analysis

Most of Stata's multivariate capabilities are to be found in the *Multivariate Statistics Reference Manual*, although there are some exceptions.

1. mvreg fits multivariate regressions.

2. manova fits MANOVA and MANCOVA models, one-way and up—including two-way factorial, three-way factorial, etc.—and it fits nested and mixed-design models. Also see [U] **26.3 ANOVA, ANCOVA, MANOVA, and MANCOVA** above.

3. canon estimates canonical correlations and their corresponding loadings. Canonical correlation attempts to describe the relationship between two sets of variables.

4. pca extracts principal components and reports eigenvalues and loadings. Some people consider principal components a descriptive tool—in which case standard errors as well as coefficients are relevant—and others look at it as a dimension-reduction technique.

5. factor fits factor models and provides principal factors, principal-component factors, iterated principal-component factors, and maximum-likelihood solutions. Factor analysis is concerned with finding few common factors $\widehat{\mathbf{z}}_k$, $k = 1, \ldots, q$ that linearly reconstruct the original variables $\mathbf{y}_i$, $i = 1, \ldots, L$.

6. tetrachoric, in conjunction with pca or factor, allows you to perform PCA or factor analysis on binary data.

7. rotate provides a wide variety of orthogonal and oblique rotations after factor and pca. Rotations are often used to produce more interpretable results.

8. procrustes performs Procrustes analysis, one of the standard methods of multidimensional scaling. It can perform orthogonal or oblique rotations, as well as translation and dilation.

9. mds performs metric and nonmetric multidimensional scaling for dissimilarity between observations with respect to a set of variables. A wide variety of dissimilarity measures are available and, in fact, are the same as those for cluster.

10. ca performs correspondence analysis, an exploratory multivariate technique for analyzing cross-tabulations and the relationship between rows and columns.

11. mca performs multiple correspondence analysis (MCA) and joint correspondence analysis (JCA).

12. cluster provides cluster analysis; both hierarchical and partition clustering methods are available. Strictly speaking, cluster analysis does not fall into the category of statistical estimation. Rather, it is a set of techniques for exploratory data analysis. Stata's cluster environment has many different similarity and dissimilarity measures for continuous and binary data.

13. discrim and candisc perform discriminant analysis. candisc performs linear discriminant analysis (LDA). discrim also performs LDA, and it performs quadratic discriminant analysis (QDA), kth nearest neighbor (KNN), and logistic discriminant analysis. The two commands differ in default output. discrim shows the classification summary, candisc shows the canonical linear discriminant functions, and both will produce either.

26.19 Pharmacokinetic data

There are four estimation commands for analyzing pharmacokinetic data. See [R] **pk** for an overview of the pk system.

1. pkexamine calculates pharmacokinetic measures from time-and-concentration subject-level data. pkexamine computes and displays the maximum measured concentration, the time at the maximum measured concentration, the time of the last measurement, the elimination time, the half-life, and the area under the concentration–time curve (AUC).

2. pksumm obtains the first four moments from the empirical distribution of each pharmacokinetic measurement and tests the null hypothesis that the distribution of that measurement is normally distributed.

3. pkcross analyzes data from a crossover design experiment. When one is analyzing pharmaceutical trial data, if the treatment, carryover, and sequence variables are known, the omnibus test for separability of the treatment and carryover effects is calculated.

4. pkequiv performs bioequivalence testing for two treatments. By default, pkequiv calculates a standard confidence interval symmetric about the difference between the two treatment means. pkequiv also calculates confidence intervals symmetric about zero and intervals based on Fieller's theorem. Also, pkequiv can perform interval hypothesis tests for bioequivalence.

26.20 Specification search tools

There are three other commands that are not really estimation commands but are combined with estimation commands to assist in specification searches: stepwise, fracpoly, and mfp.

stepwise:, one of Stata's prefix commands, provides stepwise estimation. You can use the stepwise prefix with some, but not all, estimation commands. In [R] **stepwise** is a table of the estimation commands that are currently supported, but do not take it too literally. It was accurate as of the day that Stata 10 was released, but if you install the official updates, stepwise may now work with other commands, too. If you want to use stepwise with some estimation command, you should try it. Either it will work or you will get the message that the estimation command is not supported by stepwise.

fracpoly and mfp are commands to assist you in performing fractional-polynomial functional specification searches.

26.21 Obtaining new estimation commands

This chapter has discussed all the estimation commands included in Stata 10 the day that it was released; by now, there may be more. To obtain an up-to-date list, type search estimation.

And, of course, you can always write your own commands; see [R] **ml**.

26.22 Reference

Gould, W. W. 2000. sg124: Interpreting logistic regression in all its forms. *Stata Technical Bulletin* 53: 19–29. Reprinted in *Stata Technical Bulletin Reprints*, vol. 9, pp. 257–270.

27 Commands everyone should know

Contents

27.1   43 commands

Putting aside the statistical commands that might particularly interest you, here are 43 commands that everyone should know:

Getting online help
> help, hsearch,
> net search, search   [U] **4 Stata's online help and search facilities**

Keeping Stata up to date
> ado, net, update   [U] **28 Using the Internet to keep up to date**
> adoupdate   [R] **adoupdate**

Operating system interface
> pwd, cd   [D] **cd**

Using and saving data from disk
> save   [D] **save**
> use   [D] **use**
> append, merge   [U] **22 Combining datasets**
> compress   [D] **compress**

Inputting data into Stata   [U] **21 Inputting data**
> input   [D] **input**
> edit   [D] **edit**
> infile   [D] **infile (free format)**; [D] **infile (fixed format)**
> infix   [D] **infix (fixed format)**
> insheet   [D] **insheet**

Basic data reporting
> describe   [D] **describe**
> codebook   [D] **codebook**
> list   [D] **list**
> browse   [D] **edit**
> count   [D] **count**
> inspect   [D] **inspect**
> table   [R] **table**
> tabulate   [R] **tabulate oneway** and [R] **tabulate twoway**

(Continued on next page)

27.2   The by construct

If you do not understand the by *varlist*: construct, _n, and _N, and their interaction, and if you process data where observations are related, you are missing out on something. See

Say that you have a dataset with multiple observations per person, and you want the average value of each person's blood pressure (bp) for the day. You could

```
. egen avgbp = mean(bp), by(person)
```

but you could also

```
. by person, sort: gen avgbp = sum(bp)/_N
. by person: replace avgbp = avgbp[_N]
```

Yes, typing two commands is more work than typing just one, but understanding the two-command construct is the key to generating more complicated things that no one ever thought about adding to egen.

Say that your dataset also contains time recording when each observation was made. If you want to add the total time the person is under observation (last time minus first time) to each observation, type

```
. by person (time), sort: gen ttl = time[_N]-time[1]
```

Or, suppose you want to add how long it has been since the person was last observed to each observation:

```
. by person (time), sort: gen howlong = time - time[_n-1]
```

If instead you wanted how long it would be until the next observation, type

```
. by person (time), sort: gen whennext = time[_n+1] - time
```

by *varlist*:, _n, and _N are often the solution to difficult calculations.

28 Using the Internet to keep up to date

Contents

28.1 Overview

Stata can read files over the Internet. Just to prove that to yourself, type the following:

```
. use http://www.stata.com/manual/chapter28, clear
```

You have just reached out and gotten a dataset from our web site. The dataset is not in HTML format, nor does this have anything to do with your browser. We just copied the Stata data file `chapter28.dta` onto our server, and now people all over the world can use it. If you have a web page, you can do the same thing. It is a convenient way to share datasets with colleagues.

Now type the following:

```
. update query
```

We promise that nothing bad will happen. `update` will read a short file from www.stata.com that will allow Stata to report whether your copy of Stata is up to date. Is your copy up to date? Now you know. If it is not, we will show you how to update it—it is no harder than typing `update`.

Now type the following:

```
. net from http://www.stata.com
```

That will go to www.stata.com and tell you what is available from our user-download site. The material there is not official, but it is useful. More useful is to type

```
. search kernel regression, net
```

or equivalently,

```
. net search kernel regression
```

That will search the entire web for additions to Stata having to do with kernel regression, whether it be from the *Stata Journal*, *Stata Technical Bulletin*, Statalist, archive sites, or private user sites.

To summarize: Stata can read files over the Internet:

1. You can share datasets, do-files, etc., with colleagues all over the world. This requires no special expertise, but you do need to have a web page.

2. You can update Stata; it is free, easy, and fast.

3. You can find and add new features to Stata; it is also free, easy, and fast.

Finally, you can create a site to distribute new features for Stata.

28.2   Sharing datasets (and other files)

There is just nothing to it: you copy the file as-is (in binary) onto the server and then let your colleagues know the file is there. This works for .dta files, .do files, .ado files, and, in fact, all files.

On the receiving end, you can use the file (if it is a .dta dataset) or you can copy it:

    . use http://www.stata.com/manual/chapter28, clear
    . copy http://www.stata.com/manual/chapter28.dta mycopy.dta

Stata includes a copy-file command and it works over the Internet just as use does; see [D] **copy**.

❏ Technical Note

If you are concerned about transmission errors, you can create a checksum file before you copy the file onto the server. In placing chapter28.dta on our site, we started with chapter28.dta in our working directory and typed

    . checksum chapter28.dta, save

This created the new file chapter28.sum. We then placed both files on our server. We did not have to create this second file, but, since we did, when you use the data, Stata will be able to detect transmission errors and warn you if there are problems.

How would Stata know? chapter28.sum is a short file containing the result of a mathematical calculation made on the contents of chapter28.dta. When your Stata receives chapter28.dta, it repeats the calculation and then compares that result with what is recorded in chapter28.sum. If the results are different, then there must have been a transmission error.

Creating a checksum file is optional.

See [D] **checksum**.

❏

28.3   Official updates

Although we follow no formal schedule for the release of updates, the fact is that we update Stata about once every 2 weeks. You do not have to update that often, although we recommend that you do. There are two ways to check whether your copy of Stata is up to date:

    type
        . update query

    or
        Select **Help > Official Updates**
        Click on **http://www.stata.com**

After that if an update is available, you should

type   or
. update all   click on *update*

After the updates have been downloaded, you may be prompted to type update swap. Be sure to do this if Stata asks you to, or the update will be incomplete.

After you have updated your Stata, to find out what has changed

type   or
. help whatsnew   Select **Help > What's New?**

28.3.1 Example

When you type update from http://www.stata.com or when you select **Help > Official Updates**, and click on **http://www.stata.com**, Stata presents a report:

```
. update from http://www.stata.com
(contacting http://www.stata.com)

Stata executable
    folder:                C:\Program Files\Stata10\
    name of file:          wstata.exe
    currently installed:   28 Jul 2007
    latest available:      28 Jul 2007

Ado-file updates
    folder:                C:\Program Files\Stata10\ado\updates\
    names of files:        (various)
    currently installed:   28 Jul 2007
    latest available:      28 Jul 2007

Recommendation
    Do nothing; all files up-to-date.
```

There are two components of official Stata: the binary Stata executable and the ado-files that we shipped with it. Ado-files are just programs written in Stata. For instance, when you use generate, you are using a command that was compiled into the Stata executable. When you use stcox, you are using a command that was implemented as an ado-file.

Both components of our Stata are up to date.

28.3.2 Updating ado-files

When you obtain the above report, you might see

```
. update from http://www.stata.com
(contacting http://www.stata.com)

Stata executable
    folder:                C:\Program Files\Stata10\
    name of file:          wstata.exe
    currently installed:   28 Jul 2007
    latest available:      28 Jul 2007

Ado-file updates
    folder:                C:\Program Files\Stata10\ado\updates\
    names of files:        (various)
    currently installed:   28 Jul 2007
    latest available:      08 Oct 2007

Recommendation
    Type -update all-
```

Stata tells you that your executable is up to date, but your ado-files need updating. Stata recommends that you update. If you go with the point-and-click alternative, at the bottom of the screen you will see

```
Recommendation
    update ado-files
```

where *update ado-files* is clickable.

Anyway, what you are to do next is type `update ado-files` or click on *update ado-files*. Either way, you will see something like the following:

```
. update ado-files
```

```
> update ado
(contacting http://www.stata.com)
Ado-file update log
    1.  verifying C:\Program Files\Stata10\ado\updates\ is writeable
    2.  obtaining list of files to be updated
    3.  downloading relevant files to temporary area
        downloading filename.ado
        downloading filename.sthlp
        ...
        downloading filename.ado
    4.  examining files
    5.  installing files
    6.  setting last date updated
Updates successfully installed.
Recommendation
    Type -help whatsnew- to learn about the new features
```

```
> update executable
(contacting http://www.stata.com)
executable already up to date
```

That is all there is to it, but do type `help whatsnew` to learn about the new features. (If you go the point-and-click path, click on *whatsnew*.)

Here is what happens if you type `update all` and you are already up to date:

```
. update all
```

```
> update ado
(contacting http://www.stata.com)
ado-files already up to date
```

```
> update executable
(contacting http://www.stata.com)
executable already up to date
```

28.3.3   Frequently asked questions about updating the ado-files

1. Could something go wrong and make my Stata become unusable?

    No. The updates are copied to a temporary place on your computer, Stata examines them to make sure they are complete before copying them to the official place. Thus, either the updates are installed or they are not.

2. I do not believe you. Pretend that something you did not anticipate goes wrong, such as the power fails at the instant Stata is doing the local disk to local disk copy.

> If the improbable should happen, you can erase the update directory and then your Stata is back to being just as it was shipped. Updates go into a different directory from that of the originals and the originals are never erased.

> Stata tells you where it is installing your updates. You can also find out by typing sysdir. The directory you want is the one listed opposite UPDATES.

> (By the way, power failure should not cause a problem; the marker that the update is applied is set last, so you could also just type update all again and Stata would refetch the partially installed update.)

3. How much is downloaded?

> It depends on how many updates have been made since you last updated. Typical updates are only a few small files. If you wait 6 months between updates, you might have to download around 1 MB.

4. I am using Unix or a networked version of Stata. When I try to update all, I am told that the directory is not writeable. Can I copy the updates into another directory and then copy them to the official directory myself?

> Yes, assuming that you are a system administrator. Ordinarily, you would type update all and Stata would download whatever updates might be available, ado and/or executable. update ado allows you to download just the ado updates and to specify an alternate location for them to be installed. Type update ado, into(*dirname*). Stata will download the updates just as it would ordinarily but will place them in the directory you specify. We recommend that *dirname* be a new, empty directory, because later you will need to copy the entire contents of the directory to the official place. The official place is the directory listed next to UPDATES if you type sysdir. When you copy the files, copy over any existing files. Previously existing files in the official update directory are just previous updates. Also remember to make the files globally readable if necessary. See [R] **update**.

28.3.4 Updating the executable

Ado-file updates are released about every other week; updates for the executable are rarer than that. If the executable needs updating, Stata will mention it when you type update:

```
. update from http://www.stata.com
(contacting http://www.stata.com)

Stata executable
    folder:               C:\Program Files\Stata10\
    name of file:         wstata.exe
    currently installed:  28 Jul 2007
    latest available:     12 Nov 2007
Ado-file updates
    folder:               C:\Program Files\Stata10\ado\updates\
    names of files:       (various)
    currently installed:  28 Jul 2007
    latest available:     28 Jul 2007
Recommendation
    Type -update all-
```

Here is what happens when you type `update all`:

```
. update all
```

```
> update ado
(contacting http://www.stata.com)
ado-files already up to date
```

```
> update executable
(contacting http://www.stata.com)
Executable update log
    1.  verifying C:\Program Files\Stata10\ is writeable
    2.  downloading new executable
New executable successfully downloaded
Instructions
    1.  Type -update swap-
```

Just follow the instructions, which will vary depending on your computer. Here `update swap` is a command that automatically copies the newly downloaded executable over the current one. It then briefly restarts Stata to begin using the new executable. You must type `update swap` after downloading an executable update.

28.3.5   Frequently asked questions about updating the executable

1. If I understand this, `update all` does not really install the updated executable; it just copies one file onto my computer, and that one file happens to be the new executable, right?

   Probably. There can be more than one file such as a DLL. All the files are copied to the same place. When no DLLs are downloaded, this statement is true. On most systems, using the `update swap` command after downloading a new executable will perform the copy and actually install the executable.

2. How big is the downloaded file?

   Two to 35 MB, depending on the operating system.

3. What happens if I type `update all` and my executable is already up to date?

   Nothing. You are told "executable already up to date".

4. I am using Unix or a networked version of Stata. When Stata tries to update the executable, I am told that the directory is not writeable. Can I download the updated executable to another directory and then copy it to the official directory myself?

   Yes, assuming that you are a system administrator. Ordinarily, you would type `update all` and Stata would download whatever updates might be available, ado and/or executable. `update executable` allows you to download just the executable updates and to specify an alternate location for them to be installed. Type `update executable, into(`*dirname*`)`. We recommend that *dirname* be a new, empty directory, because there may be more than one file and later you will need to copy all of them to the official place. The official place is the directory listed next to `STATA` if you type `sysdir`. When you copy the files, copy over any existing files; we recommend that you make a backup of the originals first. See [R] **update**.

28.3.6   Updating both ado-files and the executable

When you type update, you may be told both the ado-files and executable need updating:

```
. update from http://www.stata.com
(contacting http://www.stata.com)

Stata executable
      folder:                   C:\Program Files\Stata10\
      name of file:             wstata.exe
      currently installed:      28 Jul 2007
      latest available:         12 Nov 2007

Ado-file updates
      folder:                   C:\Program Files\Stata10\ado\updates\
      names of files:           (various)
      currently installed:      28 Jul 2007
      latest available:         12 Nov 2007

Recommendation
      Type -update all-
```

Remember that if an executable update is downloaded, you will need to type update swap to finish installing it.

You could skip the update from step. You could just type update all and follow the instructions. If nothing needed updating, you would see

```
. update all
```

```
> update ado
(contacting http://www.stata.com)
ado-files already up to date
```

```
> update executable
(contacting http://www.stata.com)
executable already up to date

. _
```

28.4   Downloading and managing additions by users

Try the following:

type
      . net from http://www.stata.com

or
      Select **Help > SJ and User-written Programs**
      Click on one of the links

28.4.1   Downloading files

We are not the only ones developing additions to Stata. Stata is supported by a large and highly competent user community. An important part of this is the *Stata Journal* (SJ) and the *Stata Technical Bulletin* (STB). The *Stata Journal* is a refereed, quarterly journal containing articles of interest to Stata users. For more details and subscription information, visit the *Stata Journal* web site at http://www.stata-journal.com.

The *Stata Journal* is a printed and electronic journal with corresponding software. If you want the journal, you must subscribe, but the software is available for free from our web site at http://www.stata-journal.com.

The predecessor to the *Stata Journal* was the *Stata Technical Bulletin* (STB). The STB was also a printed and electronic journal with corresponding software. Individual STB issues may still be purchased. The STB software is available for free from our web site at http://www.stata.com.

Below are instructions for installing the *Stata Journal* and the *Stata Technical Bulletin* software from our web site.

Installing the Stata Journal software

1. Select **Help > SJ and User-written Programs**.

2. Click on *Stata Journal.*

3. Click on *sj2-2.*

4. Click on *st0001_1.*

5. Click on *click here to install.*

or type

1. Type: `. net from http://www.stata-journal.com/software`

2. Type: `. net cd sj2-2`

3. Type: `. net describe st0001_1`

4. Type: `. net install st0001_1`

The above could be shortened to

```
. net from http://www.stata-journal.com/software/sj2-2
. net describe st0001_1
. net install st0001_1
```

You could also type

```
. net sj 2-2
. net describe st0001_1
. net install st0001_1
```

Installing the STB software

1. Select **Help > SJ and User-written Programs**.

2. Click on *STB.*

3. Click on *stb58.*

4. Click on *sg84_3.*

5. Click on *click here to install.*

or type

1. Type: `. net from http://www.stata.com`

2. Type: `. net cd stb`

3. Type: `. net cd stb58`

4. Type: `. net describe sg84_3`

5. Type: `. net install sg84_3`

The above could be shortened to

```
. net from http://www.stata.com/stb/stb58
. net describe sg84_3
. net install sg84_3
```

28.4.2   Managing files

You now have the `concord` command, because we just downloaded and installed it. Convince yourself of this by typing

```
. help concord
```

and you might try it out, too. Let's now list the additions you have installed—that is probably just concord—and then get rid of concord.

In command mode, you can type

```
. ado dir
[1] package sg84_3 from http://www.stata.com/stb/stb58
      STB-58 sg84_3.  Concordance correlation coefficient: minor corrections
```

If you had more additions installed, they would be listed. Now knowing that you have *sg84_3* installed, you can obtain a more thorough description by typing

```
. ado describe sg84_3
```

```
[1] package sg84_3 from http://www.stata.com/stb/stb58
```

```
TITLE
      STB-58 sg84_3.  Concordance correlation coefficient:  minor corrections
DESCRIPTION/AUTHOR(S)
      STB insert by Thomas J. Steichen, RJRT
                    Nicholas J. Cox, University of Durham, UK
      Support:  steicht@rjrt.com, n.j.cox@durham.ac.uk
      After installation, see help concord
INSTALLATION FILES
      c/concord.ado
      c/concord.sthlp
INSTALLED ON
      5 Oct 2002
```

You can erase *sg84_3* by typing

```
. ado uninstall sg84_3
package sg84_3 from http://www.stata.com/stb/stb58
      STB-58 sg84_3.  Concordance correlation coefficient:  minor corrections
(package uninstalled)
```

You can do all of this from the point-and-click interface, too. Pull down **Help** and select **SJ and User-written Programs** and then click on *List*. From there, you can click on *sg84_3* to see the detailed description of the package and from there you can click on *click here to uninstall* if you want to erase it.

For more information on the `ado` command and the corresponding menu, see [R] **net**.

28.4.3 Finding files to download

There are two ways to find useful files to download. One is simply to thumb through sites. That is inefficient but entertaining. If you want to do that,

1. Select **Help > SJ and User-written Programs**.

2. Click on *Other Locations*.

3. Click on *links*.

What you are doing is starting at our download site and then working out from there. We maintain a list of other sites and those sites will have more links. You can do this from command mode, too:

```
. net from http://www.stata.com
. net cd links
```

The efficient way to find files—at least if you know what you are looking for—is to search. There are two ways to do that. If you suspect what you are looking for might already be in Stata (or published in the SJ), use Stata's `search` command:

```
. search concordance correlation
```

Equivalently, you could select **Help > Search**. Either way, you will learn about *sg84_3* and you can even click to install it.

If you want to search for additions over the net, i.e., the SJ and archive sites and user sites, type

```
. net search concordance correlation
```

or select **Help > Search**, and this time click *Search net resources*, rather than the default "*Search documentation and FAQs*".

28.4.4 Updating additions by users

After you have installed some user-written features, you should periodically check whether any updates or bug fixes area available for those commands. You can do this with the `adoupdate` command. Simply type `adoupdate` to see if any updates are available, and if they are, type `adoupdate, update` to obtain the updates. See [R] **adoupdate** for more details.

28.5 Making your own download site

There are two reasons you may wish to create your own download site:

1. You have datasets and the like, you want to share them with colleagues, and you want to make it easier for colleagues to download the files.

2. You have written Stata programs, etc., that you wish to share with the Stata user community.

Making a download site is easy; the full instructions are found in [R] **net**.

At the beginning of this chapter, we pretended that you had a dataset you wanted to share with colleagues. We said you just had to copy the dataset onto your server and then let your colleagues know the dataset is there.

Let's now pretend that you had two datasets, `ds1.dta` and `ds2.dta`, and you wanted your colleagues to be able to learn about and fetch the datasets by using the `net` command or by pulling down **Help** and selecting **SJ and User-written Programs**.

First, you would copy the datasets to your home page just as before. Then you would create three more files, one to describe your site named `stata.toc` and two more to describe each "package" you want to provide:

```
———————————————————————————————————————— top of stata.toc ————
v 3
d My name and affiliation (or whatever other title I choose)
d Datasets for the PAR study
p ds1 The base dataset
p ds2 The detail dataset
————————————————————————————————————————— end of stata.toc ————
```

```
————————————————————————————————————————— top of ds1.pkg ————
v 3
d ds1.  The base dataset
d My name or whatever else I wanted to put
d This dataset contains the baseline values for ...
d Distribution-Date: 26sep2007
p ds1.dta
————————————————————————————————————————— end of ds1.pkg ————
```

```
————————————————————————————————————————— top of ds2.pkg ————
v 3
d ds1.  The detail dataset
d My name or whatever else I wanted to put
d This dataset contains the follow-up information ...
d Distribution-Date: 26sep2007
p ds2.dta
————————————————————————————————————————— end of ds2.pkg ————
```

The `Distribution_Date` line in the description should be changed whenever you change your package. This line is used by `adoupdate` to determine if a user who has installed your package needs to update it.

Here is what users would see when they went to your site:

```
. net from http://www.myuni.edu/hande/~aparker

http://www.myuni.edu/hande/~aparker
My name and whatever else I wanted to put

Datasets for the PAR study
PACKAGES you could -net describe-:
    ds1              The base dataset
    ds2              The detail dataset
```

```
. net describe ds1

package ds1 from http://www.myuni.edu/hande/~aparker

TITLE
      ds1.  The base dataset
DESCRIPTION/AUTHOR(S)
      My name and whatever else I wanted to put
      This dataset contains the baseline values for ...
      Distribution-Date: 26sep2007
ANCILLARY FILES                                (type net get ds1)
      ds1.dta
```

```
. net get ds1
checking ds1 consistency and verifying not already installed...

copying into current directory...
     copying  ds1.dta
ancillary files successfully copied.

  . _
```

See [R] **net**.

Subject and author index

This is the subject and author index for the *Stata User's Guide*. You may also want to consult the combined subject index (and the combined author index) in the *Stata Quick Reference and Index*. The combined index indexes the *Getting Started* manuals and all the Reference manuals except the *Mata Reference Manual*.

Semicolons set off the most important entries from the rest. Sometimes no entry will be set off with semicolons, meaning that all entries are equally important.